T0214014

Calculus of One Variable

M. Thamban Nair

Calculus of One Variable

Second Edition

M. Thamban Nair
Department of Mathematics
Indian Institute of Technology Madras
Chennai, Tamil Nadu, India

ISBN 978-3-030-88639-4   ISBN 978-3-030-88637-0   (eBook)
https://doi.org/10.1007/978-3-030-88637-0

Jointly published with ANE Books Pvt. Ltd.
In addition to this printed edition, there is a local printed edition of this work available via Ane Books in
South Asia (India, Pakistan, Sri Lanka, Bangladesh, Nepal and Bhutan) and Africa (all countries in the
African subcontinent).
ISBN of the ANE Books Pvt. Ltd. edition: 9788194891840

This Springer imprint is published by the registered company Springer Nature Switzerland AG
The registered company address is: Gewerbestrasse 11, 6330 Cham, Switzerland

Dedicated to the fond memories of my

Father

P. Chandu Nair (Late),

Mother

Meloth Parvathi Amma (Late),

Father-in-law

N. Gopalan Nair (Late)

and

Mother-in-law

Santha G Nair (Late)

Preface to the Second Edition

Since its first publication, more than five years ago, the book has been used by many students all over India for their undergraduate courses in mathematics, science and engineering, either as a text or as a reference book. The present version is a re-written version of the original one with many additions here and there making it better readable and also giving better motivations and clarifications for many of the results. Some of the subsections are renamed, and a few subsections are added. I am thankful to my Ph.D. student Subhankar Mondal for critically reading the revised text and suggesting many improvements.

Unlike in the first edition, where results (theorems and corollaries), examples, exercises and remarks are numbered as p.q, where p and q denote the chapter number and occurrence, respectively; in this edition, we number them as p.q.r, where p, q, r denote the chapter number, section number and occurrence, respectively.

Chennai, India
October 2020

M. Thamban Nair

Preface to the First Edition

This book grew out of my class notes for the first semester course in mathematics for B.Tech. students at IIT Madras, which I have been teaching several times since 1996. The notes have been in circulation among the students of IIT Madras, and they have been using them as supplementary material along with the prescribed texts. Students have found the notes to be very useful, particularly the manner in which new concepts are introduced, the style of presentation, the way examples have been worked out and the interspersed remarks.

The book can be used as a first course in calculus for science and engineering students. It can also be used as a supplementary text for B.Sc. mathematics course.

Why another book?

There are many books on calculus meant for students of engineering and science. So, why another book?

To my assessment, most of the books meant for engineers are designed only for a single course or a combination of courses under a particular university with a few exceptions. They mostly consist of mathematical techniques for solving problems without much elaboration on the underlying mathematical principles. On the other hand, books meant for students pursuing bachelors of science contain elaborate description of underlying mathematical principles which may not be suitable for the bachelor of engineering courses. In this book, I have tried to strike a balance between the above two approaches.

Key features.

The book provides clear understanding of the basic concepts of differential and integral calculus starting with the concepts of sequences and series of numbers and also introduces slightly advanced topics such as sequences and series of functions, power series and Fourier series which would be of use for other courses in mathematics for science and engineering programs.

Here are some of the salient features of the book:

- Precise definitions of basic concepts are given.
- Several motivating examples are provided for understanding the concepts and also for illustrating the results.
- Proofs of theorems are given with sufficient motivation—not just for the sake of proving them alone.
- Remarks in the text supply additional information on the topics under discussion.
- Exercises are interspersed within the text for making the students attempt them while the lectures are in progress.
- A large number of problems at the end of each chapter are meant as home assignments.

The student-friendly approach of the exposition of the book would definitely be of great use not only for students, but also for the teachers.

IIT Madras   M. Thamban Nair
June 2014   tham.nair@gmail.com

Acknowledgments

While preparing the notes, I have benefited from suggestions and comments from many of my colleagues, especially professors S. H. Kulkarni, A. Singh and P. Veeramani. Also, Research Scholars Viswanathan and Ajoy Jana red the manuscript thoroughly and pointed out many typos and made some suggestions for improvements. I am grateful to all of them. Also, curtesy to *Google/Wikipedia* for all the figures in the text.

I gratefully acknowledge the support received from the Centre for Continuing Education (CCE), IIT Madras, under its book writing scheme.

Also, I thank my wife Dr. Sunita Nair and daughters Priya and Sneha for their constant support in all my academic endeavors and my mother Meloth Parvathi Amma and mother-in-law Shantha G. Nair for their blessings.

Comments and suggestions from the readers are most welcome.

Note to the Reader

Set-Theoretic Notations. Throughout the book, we shall use standard set-theoretic notations such as

$$\cup, \quad \cap, \quad \subseteq, \quad \subset, \quad \in$$

to denote "union," "intersection," "subset of," "proper subset of," "belong(s) to," respectively. For sets S_1 and S_2, the set $\{x \in S_1 : x S_2\}$ is denoted by $S_1 \setminus S_2$. If f is a function with domain S_1 and codomain S_2, then we use the notation $f : S_1 \to S_2$, and it is also called a "map" from S_1 to S_2.

Also, we use the standard notations such as

$\mathbb{N}$	:	set of all positive integers
$\mathbb{Z}$	:	set of all integers
$\mathbb{Q}$	:	set of all rotational numbers
$\mathbb{R}$	:	set of all real numbers
$\mathbb{C}$	:	set of all complex numbers
$:=$	:	is defined by
$\forall$	:	for all
$\exists$	:	there exists or there exist
$\Rightarrow$	:	implies or imply
$\Leftrightarrow$	:	if and only if
$\mapsto$	:	maps to

For "if and only if," sometimes we shall also use the symbol "iff".

To mark the end of a proof of a Theorem or Corollary, we use the symbol ∎, and to mark the end of an Example or a Definition or a Remark, the symbol ◇ is used. The symbol ◁ is used to signal the end of an exercise. **Bold face** is often used for defining a term, and *italics* is used when a new term is used in a sentence which may be defined subsequently.

About Numbering. Four sequences of numbers have been used for specifying mathematical results (Theorems and Corollaries), examples, exercises and remarks, respectively.

Contents

About the Author

M. Thamban Nair is a Professor of Mathematics at the Indian Institute of Technology Madras, India. Prof. Nair has been a research mathematician and teacher for over 30 years for postgraduate- and undergraduate-level mathematics courses. He won many awards including the C. L. Chandana Award for distinguished and outstanding contributions in mathematics research and teaching in India for the year 2003. He was also a Post Doctoral Fellow at the University of Grenoble (France) and Visiting Professor at Australian National University, Australia, University of Kaiserslautern, Germany, Sun Yat-sen University, Guangzhou, China, and University of Saint Etienne, France. He gave several invited talks at various conferences in India and abroad. Professor Nair has got over 80 journal publications and 5 books to his credit. The broad area of Prof. Nair's research is in applicable functional analysis, more specifically, spectral approximation, operator equations, and inverse and ill-posed problems.

Chapter 1
Sequence and Series of Real Numbers

The concept of convergence of sequences and series is fundamental in mathematics. They appear naturally while studying other concepts such as continuity, differentiation and integration of functions, and also while approximating certain unknown quantities using known quantities obtained via numerical or experimental methods.

We may recall that, at the school level, one comes across numbers and mathematical constants that do not have finite decimal places. For instance, $1/3$ cannot be expressed in terms of finite number of decimal places. In fact, we were taught to write $1/3$ in decimal expansion, using repeated division, as

$$\frac{1}{3} = 0.333 \cdots .$$

If we take only a finite number of decimal places, how much close that number would be to the number $1/3$? By taking more and more decimal places, is it possible to reach closer and closer to $1/3$? We expect so. How do we justify? In this chapter, we shall deal with such issues in detail.

1.1 Sequence of Real Numbers

Suppose that, for each positive integer n, we are given a real number a_n. Then, the list of numbers

$$a_1, a_2, \ldots, a_n, \ldots$$

is called a *sequence*. A more precise definition of a sequence is the following:

Definition 1.1.1 A **sequence** of real numbers is a function from the set $\mathbb{N}$ of natural numbers to the set $\mathbb{R}$ of real numbers. ◇

Notation 1.1.1 If $f : \mathbb{N} \to \mathbb{R}$ is a sequence, and if $a_n = f(n)$ for $n \in \mathbb{N}$, then we write the sequence f as

$$(a_1, a_2, \ldots) \quad \text{or} \quad (a_n) \quad \text{or} \quad \{a_n\},$$

© The Author(s), under exclusive license to Springer Nature Switzerland AG 2021
M. T. Nair, *Calculus of One Variable*,
https://doi.org/10.1007/978-3-030-88637-0_1

and the term a_n is called the nth **term** of the sequence (a_n). In specific cases, where one knows an expression for a_n, one may write

$$(a_1, a_2, \ldots, a_n, \ldots) \quad \text{instead of} \quad (a_1, a_2, \ldots).$$ ◊

Remark 1.1.1 It is to be borne in mind that a sequence $(a_1, a_2, \ldots)$ is different from the set $\{a_n : n \in \mathbb{N}\}$. For instance, a number may be repeated in a sequence (a_n), but it need not be written repeatedly in the set $\{a_n : n \in \mathbb{N}\}$. As an example, $(1, \frac{1}{2}, 1, \frac{1}{3}, \ldots, 1, \frac{1}{n}, \ldots)$ is a sequence (a_n) with $a_{2n-1} = 1$ and $a_{2n} = 1/(n+1)$ for each $n \in \mathbb{N}$, whereas the set $\{a_n : n \in \mathbb{N}\}$ is same as the set $\{1/n : n \in \mathbb{N}\}$. ◊

Remark 1.1.2 We can also talk about a sequence of elements from any non-empty set S, such as sequence of sets, sequence of functions and so on. Thus, given a non-empty set S, a sequence in S is a function $f : \mathbb{N} \to S$. In this chapter, we shall consider only sequence of real numbers. In some of the later chapters we shall consider sequences of functions as well. ◊

Example 1.1.1 Let us consider a few examples of sequences (Fig. 1.1):

(i) (a_n) with $a_n = 1$ for all $n \in \mathbb{N}$.

(ii) (a_n) with $a_n = n$ for all $n \in \mathbb{N}$.

(iii) (a_n) with $a_n = 1/n$ for all $n \in \mathbb{N}$.

(iv) (a_n) with $a_n = n/(n+1)$ for all $n \in \mathbb{N}$.

(v) (a_n) with $a_n = (-1)^n$ for all $n \in \mathbb{N}$. This sequence takes values 1 and -1 alternately. ◊

Fig. 1.1 (a_n) converges to a

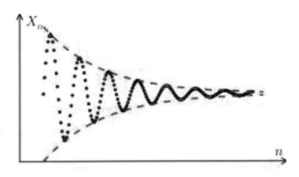

1.1.1 Convergence and Divergence

In certain sequences the nth term comes closer and closer to a particular number as n becomes larger and larger. For example, in the sequence $(\frac{1}{n})$, the nth term comes closer and closer to 0, whereas in $(\frac{n}{n+1})$, the nth term comes closer and closer to 1 as n becomes larger and larger. If you look at the sequence $((-1)^n)$, the terms oscillate between -1 and 1 as n varies, whereas in (n^2) the terms become larger and larger.

Now, we make precise the the statement "a_n comes closer and closer to a number a as n becomes larger and larger", that is, "a_n can be made arbitrarily close to a by taking n large enough", by defining the notion of *convergence* of a sequence.

Definition 1.1.2 A sequence (a_n) of real numbers is said to **converge** to a real number a if for every $\varepsilon > 0$, there exists a positive integer N, that may depend on ε, such that

$$|a_n - a| < \varepsilon \quad \forall\, n \geq N.$$

A sequence that converges is called a **convergent sequence**, and a sequence that does not converge is called a **divergent sequence**. ◊

Notation 1.1.2 (i) If (a_n) converges to a, then we write

$$a_n \to a \quad \text{as} \quad n \to \infty$$

that we may read as "a_n tends to a as n tends to infinity", that we also write in short as $a_n \to a$.

(ii) If (a_n) does not converge to a, then we write $a_n \nrightarrow a$. ◊

Remark 1.1.3 We must keep in mind that the symbol ∞ is not a number; it is only a notation used in the context of describing some properties of real numbers, such as in Definition 1.1.2. ◊

Remark 1.1.4 In Definition 1.1.2, the expression $|a_n - a| < \varepsilon$ can be replaced by $|a_n - a| \leq \varepsilon$ or by $|a_n - a| < c_0\varepsilon$ for some $c_0 > 0$. In other words, the following statements are equivalent.

(i) For every $\varepsilon > 0$, there exists $N \in \mathbb{N}$ such that $|a_n - a| < \varepsilon$ for all $n \geq N$.
(ii) For every $\varepsilon > 0$, there exists $N \in \mathbb{N}$ such that $|a_n - a| \leq \varepsilon$ for all $n \geq N$.
(iii) For every $\varepsilon > 0$, there exists $N \in \mathbb{N}$ such that $|a_n - a| \leq c_0\varepsilon$ for all $n \geq N$ for some $c_0 > 0$.

Clearly, (i) implies (ii). To see (ii) implies (i), assume (ii) and let $\varepsilon > 0$ be given. Then, by (ii), with $\varepsilon/2$ in place of ε, there exists $N \in \mathbb{N}$ such that $|a_n - a| \leq \varepsilon/2$ for all $n \geq N$. In particular, (i) holds. Now, (iii) follows from (i) by taking $c_0\varepsilon$ in place of ε, and (i) follows from (iii) by taking ε/c_0 in place of ε. ◊

Before further discussion on convergence of sequences, let us observe an important property of convergent sequences.

Theorem 1.1.1 *If (a_n) converges to a, then (a_n) cannot converge to b with $b \neq a$.*

Proof Suppose $a_n \to a$ and $a_n \to b$ as $n \to \infty$. Let $\varepsilon > 0$ be given and let N_1 and N_2 in $\mathbb{N}$ be such that

$$|a_n - a| < \varepsilon \quad \forall n \geq N_1 \quad \text{and} \quad |a_n - b| < \varepsilon \quad \forall n \geq N_2.$$

Then, we have

$$\begin{aligned}
|a - b| &= |(a - a_n) + (a_n - b)| \\
&\leq |a - a_n| + |a_n - b| \\
&< 2\varepsilon
\end{aligned}$$

for any $n \geq N := \max\{N_1, N_2\}$. Thus, we have proved that $|a - b| < 2\varepsilon$ for every $\varepsilon > 0$. This is possible only when $a = b$. ∎

Remark 1.1.5 In the proof of Theorem 1.1.1, we used an important property of real numbers:

For $a \in \mathbb{R}$, if $0 \leq a < r$ for every $r > 0$, then $a = 0$.

This follows from the fact that, if $a > 0$, then $0 \leq a < r$ is not satisfied for $r = a/2$. Similarly, it can be shown that

$$a, b \in \mathbb{R}, \quad a < b + r \quad \forall r > 0 \quad \Rightarrow \quad a \leq b. \qquad \Diamond$$

> If (a_n) converges, then there exists a *unique* $a \in \mathbb{R}$ such that $a_n \to a$

Thus, we can define the concept of the *limit* of a convergent sequence.

Definition 1.1.3 If (a_n) converges to a, then a is called the **limit** of (a_n), and we write

$$\lim_{n \to \infty} a_n = a. \qquad \Diamond$$

It can be easily seen that

$$a_n \to a \iff |a_n - a| \to 0 \Rightarrow |a_n| \to |a|.$$

Exercise 1.1.1 Verify the above. ◁

Suppose (a_n) converges to a. Then, by Definition 1.1.2, taking $\varepsilon = 10^{-k}$ for some $k \in \mathbb{N}$, there exists some $N_k \in \mathbb{N}$ such that

$$|a_n - a| < 10^{-k} \quad \forall n \geq N_k.$$

In other words:

> If $n \geq N_k$ and if a_n and a are expressed in decimal expansion, then the first k decimal places of a_n and a are the same

Note that, in the definition of convergence, different ε can result in different N, i.e., the number N may vary as ε varies. This fact will be revealed in the following examples.

Example 1.1.2 Consider the sequence (a_n) with $a_n = 1/n$ for $n \in \mathbb{N}$. We expect that $a_n \to 0$ as $n \to \infty$. Let us prove this. For this, let $\varepsilon > 0$ be given. Then

$$\frac{1}{n} < \varepsilon \iff n > \frac{1}{\varepsilon}.$$

Thus, taking a positive integer $N > 1/\varepsilon$, we obtain $|a_n - 0| < \varepsilon$ for all $n \geq N$. Similarly, for any $\varepsilon > 0$,

$$\left| \frac{(-1)^n}{n} - 0 \right| = \frac{1}{n} < \varepsilon \quad \forall n \geq N,$$

where N is a positive integer such that $N > 1/\varepsilon$. Thus we have shown that

$$\frac{1}{n} \to 0 \quad \text{and} \quad \frac{(-1)^n}{n} \to 0.$$

Note that the positive integer N in both the examples depend on the ε chosen.

Using the above arguments, it can also be verified that for any real number x, $x/n \to 0$ as $n \to \infty$. ◊

Example 1.1.3 Consider the sequence (a_n) with $a_n = n/(n+1)$ for $n \in \mathbb{N}$. We claim that $a_n \to 1$ as $n \to \infty$. Note that

$$|a_n - 1| = \frac{1}{n+1}$$

so that, for $\varepsilon > 0$,

$$|a_n - 1| < \varepsilon \iff n + 1 > \frac{1}{\varepsilon}.$$

Thus, taking a positive integer N with $N \geq 1/\varepsilon$, we obtain $|a_n - 1| < \varepsilon$ for all $n \geq N$. Note that,

$$n \geq 100 \implies |a_n - 1| < \frac{1}{100},$$

$$n \geq 1000 \implies |a_n - 1| < \frac{1}{1000}.$$

More generally, for a given $k \in \mathbb{N}$,

$$|a_n - 1| < \frac{1}{10^k} \quad \forall n \geq 10^k.$$

It is also clear that $|a_n - 1| \geq 1/10^k$ if $n < 10^k$. $\qquad\qquad \diamond$

In the following example and also at many occasions in the due course we shall use the identity

$$1 + x + \cdots + x^n = \frac{1 - x^{n+1}}{1 - x}$$

for $x \neq 1$ and $n \in \mathbb{N}$, that follows from the fact

$$(1 - x)(1 + x + \cdots + x^n) = 1 - x^{n+1}.$$

Example 1.1.4 For $n \in \mathbb{N}$, let

$$a_n = \frac{9}{10} + \frac{9}{10^2} + \cdots + \frac{9}{10^n},$$

i.e., $a_n = 0.99 \cdots 9$ with 9 repeated n times. You may guess that $a_n \to 1$ as $n \to \infty$. This is true: Note that

$$\frac{9}{10} + \frac{9}{10^2} + \cdots + \frac{9}{10^n} = \frac{9}{10}\left(1 + \frac{1}{10} + \cdots + \frac{1}{10^{n-1}}\right)$$
$$= \frac{9}{10}\left[\frac{1 - \frac{1}{10^n}}{1 - \frac{1}{10}}\right]$$
$$= 1 - \frac{1}{10^n}.$$

Hence, $|a_n - 1| = 1/10^n$ so that

$$|a_n - 1| < \varepsilon \iff 10^n > \frac{1}{\varepsilon}.$$

Let $N \in \mathbb{N}$ be such that $10^N > 1/\varepsilon$. Then, we have $|a_n - 1| < \varepsilon$ for all $n \geq N$. Therefore, $a_n \to 1$ as $n \to \infty$. $\qquad\qquad \diamond$

Example 1.1.5 For $n \in \mathbb{N}$, let

$$a_n = \frac{3}{10} + \frac{3}{10^2} + \cdots + \frac{3}{10^n},$$

i.e., $a_n = 0.33 \cdots 3$ with 3 repeated n times. Note that

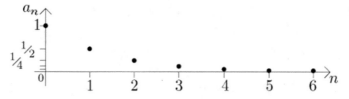

Fig. 1.2 Sequence $(1/2^n)$

$$\frac{3}{10} + \frac{3}{10^2} + \cdots + \frac{3}{10^n} = \frac{3}{10}\left[\frac{1 - \frac{1}{10^n}}{1 - \frac{1}{10}}\right] = \frac{1}{3}\left[1 - \frac{1}{10^n}\right].$$

Thus, $|a_n - 1/3| = 1/(3 \times 10^n)$ so that, for any given $\varepsilon > 0$, we can choose a positive integer N (depending on ε) such $|a_n - 1/3| < \varepsilon$ for all $n \geq N$. Therefore, $a_n \to 1/3$ as $n \to \infty$.   ◊

In the examples above we have used an important property of real numbers, known as the *Archimedean Property* :

Archimedean Property: Given any $x \in \mathbb{R}$, there exists $n \in \mathbb{N}$ such that $n > x$.

We shall use this property throughout the text without mentioning it explicitly.

Let us show the convergence of a few more sequences which we shall come across very often (Fig. 1.2).

Example 1.1.6 Consider the sequence $(1/n^2)$. Note that, for any $\varepsilon > 0$,

$$\frac{1}{n^2} < \varepsilon \iff n^2 > \frac{1}{\varepsilon} \iff n \geq \frac{1}{\sqrt{\varepsilon}}.$$

Hence, if we take $N \geq 1/\sqrt{\varepsilon}$, we obtain

$$\frac{1}{n^2} < \varepsilon \quad \forall n \geq N.$$

Thus, $1/n^2 \to 0$.

More generally, for any $k \in \mathbb{N}$, we see that

$$\frac{1}{n^k} < \varepsilon \quad \forall n \geq N_k,$$

where $N_k \in \mathbb{N}$ is such that $N \geq 1/\varepsilon^{1/k}$, so that $1/n^k \to 0$.   ◊

Example 1.1.7 Consider the sequence $(1/2^n)$. Note that, for any $\varepsilon > 0$,

$$1/2^n < \varepsilon \iff 2^n > 1/\varepsilon.$$

Since $2^n > n$, if we take $N \in \mathbb{N}$ such that $N > 1/\varepsilon$, we obtain

$$n \geq N \quad \Rightarrow \quad 2^n \geq 2^N > N > \frac{1}{\varepsilon} \quad \Rightarrow \quad \frac{1}{2^n} < \varepsilon.$$

Thus, $1/2^n \to 0$. We shall show that, $1/a^n \to 0$ for any $a > 1$. ◊

It can happen that for a convergent sequence (a_n), same N works for different ε's. To illustrate this point, look at the following examples.

Example 1.1.8 Let $a_n = 1$ for all $n \in \mathbb{N}$. We can easily guess that (a_n) converges to 1. Note that for any $\varepsilon > 0$,

$$|a_n - 1| < \varepsilon \quad \forall n \geq 1.$$

Thus, $N = 1$ works for any $\varepsilon > 0$. ◊

Example 1.1.9 For $n \in \mathbb{N}$, let

$$a_n = \begin{cases} n \text{ for } 1 \leq n \leq 99, \\ 1 \text{ for } n \geq 100. \end{cases}$$

You must have guessed that, (a_n) converges to 1. Yes; it is true: For any given $\varepsilon > 0$,

$$|a_n - 1| < \varepsilon \quad \forall n \geq 100.$$

Thus, $N = 100$ works for any $\varepsilon > 0$. ◊

In Example 1.1.8, all the terms are the same, whereas in Example 1.1.9, the terms are the same after some stage. This observation prompts us to make the following definition.

Definition 1.1.4 A sequence (a_n) is said to be a **constant sequence** if $a_n = a_1$ for all $n \in \mathbb{N}$; and it is called an **eventually constant sequence** if there exists $k \in \mathbb{N}$ such that $a_n = a_k$ for all $n \geq k$. ◊

Clearly, every constant sequence is eventually constant, and the converse does not hold. Further, every eventually constant sequence converges.

Remark 1.1.6 Throughout the book, by an interval we mean a subset I of $\mathbb{R}$ with the property that for every $x, y \in I$ with $x < y$,

$$z \in \mathbb{R} \text{ with } x < z < y \quad \Rightarrow \quad z \in I.$$

Thus, for $a, b \in \mathbb{R}$ with $a < b$, the sets

$$(a, b) := \{x \in \mathbb{R} : a < x < b\};$$
$$[a, b) := \{x \in \mathbb{R} : a \leq x < b\};$$
$$(a, b] := \{x \in \mathbb{R} : a < x \leq b\};$$
$$[a, b] := \{x \in \mathbb{R} : a \leq x \leq b\}$$

are intervals of finite lengths, and the sets

$$(a, \infty) := \{x \in \mathbb{R} : a < x < \infty\};$$
$$[a, \infty) := \{x \in \mathbb{R} : a \leq x < \infty\};$$
$$(-\infty, b) := \{x \in \mathbb{R} : -\infty < x < b\};$$
$$(-\infty, b] := \{x \in \mathbb{R} : -\infty < x \leq b\}$$

for any $a, b \in \mathbb{R}$ and $(-\infty, \infty) := \mathbb{R}$ are intervals of infinite length. Among them, (a, b), (a, ∞) and $(-\infty, b)$ are called *open intervals*, and $[a, b]$ is called a closed interval. $\diamond$

Remark 1.1.7 We may observe that, for $\varepsilon > 0$,

$$|a_n - a| < \varepsilon \iff a_n \in (a - \varepsilon, a + \varepsilon).$$

Thus, (a_n) converges to $a \in \mathbb{R}$ if and only if for every $\varepsilon > 0$, there exists a positive integer N such that
$$a_n \in (a - \varepsilon, a + \varepsilon) \quad \forall n \geq N.$$

In other words, $a_n \to a$ as $n \to \infty$ if and only if for every $\varepsilon > 0$, a_n belongs to the open interval $(a - \varepsilon, a + \varepsilon)$ for all n after some finite stage, and this finite stage may vary according as ε varies. $\diamond$

In view of Remark 1.1.7, the following theorem can be proved easily (see also Figures 1.3 and 1.4).

Theorem 1.1.2 *For a given sequence (a_n) and $a \in \mathbb{R}$, we have the following.*

(i) $a_n \to a$ *if and only if for every open interval I with $a \in I$, there exists $N \in \mathbb{N}$ (depending on a and I) such that $a_n \in I$ for all $n \geq N$.*

(ii) $a_n \nrightarrow a$ *if and only if there exists an open interval I_a with $a \in I$ such that $a_n \notin I_a$ for infinitely many n.*

Exercise 1.1.2 Supply details of the proof of the above theorem. $\triangleleft$

A convergent sequence remains convergent to the same point even if we replace or delete a finite number of terms. Equivalently,

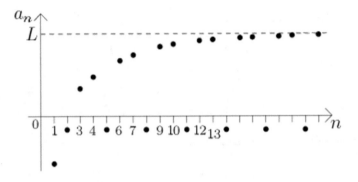

Fig. 1.3 $a_n \in (L - \varepsilon, L + \varepsilon)$ for all $n \geq N$

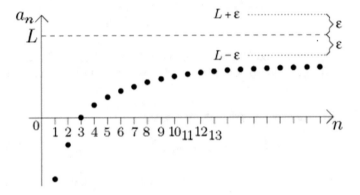

Fig. 1.4 There exists $\varepsilon > 0$ such that $a_n \notin (L - \varepsilon, L + \varepsilon)$ for infinitely many n

$a_n \to a$ if and only if $a_{k+n} \to a$ as $n \to \infty$ for any $k \in \mathbb{N}$.

Exercise 1.1.3 Verify the above statement.   ◁

Now, let us give a few examples of divergent sequences.

Example 1.1.10 Let $a_n = (-1)^{n+1}$ for $n \in \mathbb{N}$. The terms of the sequence change alternately between 1 and -1. So, one may guess that it cannot converge to any number. To see this, we observe that, for any $a \in \mathbb{R}$, if we take $0 < \varepsilon < 1/2$, either 1 or -1 will be outside the interval $(a - \varepsilon, a + \varepsilon)$; in other words, infinitely many of the terms of the sequence (a_n) lie outside $(a - \varepsilon, a + \varepsilon)$.   ◇

Example 1.1.11 Let (a_n) be defined by $a_{2n-1} = 1/n$ and $a_{2n} = 1$. If we take ε, a positive number less than 1, then, for any $a \in \mathbb{R}$, infinitely many of the terms of the sequence (a_n) lie outside $(a - \varepsilon, a + \varepsilon)$. For instance, let us take $\varepsilon = 1/2$. If $a \geq 1/2$, then infinitely many of a_{2n-1} lie outside $(a - \varepsilon, a + \varepsilon)$, and if $a < 1/2$, then every a_{2n} lie outside $(a - \varepsilon, a + \varepsilon)$ for all $n \in \mathbb{N}$. Thus, $a_n \nrightarrow a$ for any $a \in \mathbb{R}$.   ◇

Example 1.1.12 Let $a_n = n$ for $n \in \mathbb{N}$. In this case, for any given $\varepsilon > 0$ and for any $a \in \mathbb{R}$, infinitely many of the terms of the sequence (a_n) lie outside $(a - \varepsilon, a + \varepsilon)$. Similar is the case if $a_n = n^2$ or $a_n = n^2/(n+1)$ (verify). Thus, $a_n \not\to a$ for any $a \in \mathbb{R}$. ◊

Definition 1.1.5 A sequence (a_n) is said to be an **alternating sequence** if a_n changes sign alternately, that is, $a_n a_{n+1} < 0$ for every $n \in \mathbb{N}$. ◊

For example, $((-1)^n)$ and $((-1)^n/n)$ are alternating sequences. An alternating sequence may converge or diverge. We have seen that the sequence $((-1)^n)$ diverges, whereas $((-1)^n/n)$ converges to 0.

A divergent sequence may have some specific properties such as the ones defined below.

Definition 1.1.6 Consider a sequence (a_n).

(1) We say that (a_n) **diverges to infinity** if for every $M > 0$, there exists $N \in \mathbb{N}$ such that

$$a_n > M \quad \forall\, n \geq N,$$

and in that case, we write $a_n \to \infty$.

(2) We say that (a_n) **diverges to minus infinity** if for every $M > 0$, there exists $N \in \mathbb{N}$ such that

$$a_n < -M \quad \forall\, n \geq N,$$

and in that case we write $a_n \to -\infty$. ◊

It can be easily seen that

$$a_n \to \infty \iff -a_n \to -\infty.$$

Example 1.1.13 Let us look at two sequences which converge to ∞.

(i) The sequence (a_n) with $a_n = n^2/(n+1)$ diverges to infinity: Note that

$$\frac{n^2}{n+1} \geq \frac{n}{2} \quad \forall n \in \mathbb{N},$$

so that for any $M > 0$, $a_n > M$ whenever $n > 2M$. Hence, taking a positive integer N such that $N > 2M$, we have the relation $a_n > M$ for all $n \geq N$. Thus,

$$\frac{n^2}{n+1} \to \infty \quad \text{as} \quad n \to \infty.$$

(ii) The sequence (a_n) with $a_n = 2^n$ diverges to infinity: Note that $2^n \geq n$ for all $n \in \mathbb{N}$ so that for any $M > 0$, $2^n > M$ whenever $n > M$. Hence, taking a positive integer $N > M$, we have $a_n > M$ for all $n \geq N$. Thus,

$$2^n \to \infty \quad \text{as} \quad n \to \infty.$$

Similarly, for any $k \in \mathbb{N}$, $(k+1)^n \to \infty$ as $n \to \infty$.   ◊

Observe the following facts:

1. If $a_n \to \infty$ or $a_n \to -\infty$ as $n \to \infty$, then (a_n) is a divergent sequence.
2. If (a_n) and (b_n) are such that $a_n \leq b_n$ for all $n \in \mathbb{N}$, then

    (a) $a_n \to \infty$ as $n \to \infty$ implies $b_n \to \infty$ as $n \to \infty$,
    (b) $b_n \to -\infty$ as $n \to \infty$ implies $a_n \to -\infty$ as $n \to \infty$.

Exercise 1.1.4 Prove the above three statements.   ◁

The following theorem is useful for showing convergence of certain sequences using the convergence of some other sequences.

Theorem 1.1.3 *Suppose $a_n \to a$ and $b_n \to b$ as $n \to \infty$. Then the following results hold.*

(i) $a_n + b_n \to a + b$ as $n \to \infty$.
(ii) $c\,a_n \to c\,a$ as $n \to \infty$ for any real number c.
(iii) If $a_n \leq b_n$ for all $n \in \mathbb{N}$, then $a \leq b$.
(iv) **(Sandwich theorem)** *If $a_n \leq c_n \leq b_n$ for all $n \in \mathbb{N}$, and if $a = b$, then $c_n \to a$ as $n \to \infty$.*

Proof Let $\varepsilon > 0$ be given.
(i) Note that, for every $n \in \mathbb{N}$,

$$|(a_n + b_n) - (a + b)| = |(a_n - a) + (b_n - b)|$$
$$\leq |a_n - a| + |b_n - b|.$$

Since $a_n \to a$ and $b_n \to b$, the above inequality suggests that we may take $\varepsilon_1 = \varepsilon/2$, and consider $N_1, N_2 \in \mathbb{N}$ such that

$$|a_n - a| < \varepsilon_1 \quad \forall n \geq N_1 \quad \text{and} \quad |b_n - b| < \varepsilon_1 \quad \forall n \geq N_2$$

so that
$$|(a_n + b_n) - (a + b)| \leq |a_n - a| + |b_n - b| < 2\varepsilon_1 = \varepsilon$$

for all $n \geq N := \max\{N_1, N_2\}$.
    (ii) Note that
$$|c a_n - c a| = |c|\,|a_n - a| \quad \forall n \in \mathbb{N}.$$

If $c = 0$, then $(c a_n)$ is a constant sequence with every term 0, and hence $c a_n \to 0$. Next, suppose $c \neq 0$. Then we have

$$|ca_n - ca| < \varepsilon \iff |a_n - a| < \frac{\varepsilon}{|c|}.$$

Since $a_n \to a$, there exists $N \in \mathbb{N}$ such that

$$|a_n - a| < \frac{\varepsilon}{|c|} \quad \forall n \geq N.$$

Thus, $|ca_n - ca| < \varepsilon$ for all $n \geq N$.

(iii) Suppose $a_n \leq b_n$ for all $n \in \mathbb{N}$. Let $N_1, N_2 \in \mathbb{N}$ be such that

$$|a_n - a| < \varepsilon \quad \forall n \geq N_1 \quad \text{and} \quad |b_n - b| < \varepsilon \quad \forall n \geq N_2.$$

Hence,

$$a - \varepsilon < a_n < a + \varepsilon \quad \text{and} \quad b - \varepsilon < b_n < b + \varepsilon \quad \forall n \geq N = \max\{N_1, N_2\}.$$

In particular,

$$a - \varepsilon < a_n \leq b_n < b + \varepsilon \quad \forall n \geq N.$$

Thus, we obtain $a < b + 2\varepsilon$. Since this is true for every $\varepsilon > 0$, by Remark 1.1.5, we have $a \leq b$.

(iv) Suppose $a_n \leq c_n \leq b_n$ for all $n \in \mathbb{N}$ and $a = b$. We have to show that $c_n \to a$. Note that

$$c_n - a_n \leq b_n - a_n = (b_n - a) + (a - a_n) \quad \forall n \in \mathbb{N}.$$

Hence,

$$0 \leq |c_n - a_n| \leq |b_n - a| + |a - a_n| \quad \forall n \in \mathbb{N}.$$

Therefore, from the fact that $a_n \to a$ and $b_n \to a$ and the result in (iii), we obtain $|c_n - a_n| \to 0$ so that by (i), $c_n = a_n + (c_n - a_n) \to a$. ∎

Remark 1.1.8 In Theorem 1.1.3 (iii) and (iv), instead of assuming the inequalities for all $n \in \mathbb{N}$, we can assume them to hold for all $n \geq n_0$ for some $n_0 \in \mathbb{N}$ (Why?). ◊

Suppose $a_n \to a$ and $b_n \to b$ as $n \to \infty$. One may enquire whether $a_n b_n \to ab$ and whether $a_n / b_n \to a/b$ whenever $b_n \neq 0$ for all $n \in \mathbb{N}$ and $b \neq 0$. We shall prove these after a little while.

The following theorem together with Theorem 1.1.3 can be used for inferring convergence or divergence of certain sequences.

Theorem 1.1.4 *The following results hold.*

(i) *If $a > 1$, then $a^n \to \infty$.*
(ii) *If $0 < a < 1$, then $a^n \to 0$.*

Proof (i) Suppose $a > 1$. Writing $a = 1 + r$ with $r > 0$, we have

$$a^n = (1 + r)^n = 1 + nr + \frac{n(n-1)}{2!} r^2 + \cdots + r^n$$

for all $n \in \mathbb{N}$. Hence, $a^n \geq rn$ for all $n \in \mathbb{N}$. Since $rn \to \infty$, we have $a^n \to \infty$.

(ii) Let $0 < a < 1$ and $\varepsilon > 0$ be given. Then $b = 1/a$ satisfies $b > 1$ so that $b^n \to \infty$. Hence there exists $N \in \mathbb{N}$ such that $b^n > 1/\varepsilon$ for all $n \geq N$. But, $b^n > 1/\varepsilon$ if and only if $a^n < \varepsilon$. Hence,

$$0 < a^n < \varepsilon \quad \forall n \geq N.$$

Thus, $a^n \to 0$. ∎

Example 1.1.14 For $0 < r < 1$ and $n \in \mathbb{N}$, let $a_n = 1 + r + r^2 + \cdots + r^n$. Note that

$$a_n = \frac{1 - r^{n+1}}{1 - r} \quad \forall n \in \mathbb{N}.$$

By Theorem 1.1.4, $r^n \to 0$. Hence, by Theorem 1.1.3 (i)-(ii),

$$a_n \to \frac{1}{1-r} \quad \text{as} \quad n \to \infty. \qquad \Diamond$$

Exercise 1.1.5 Show that $|a| < 1$ implies $a^n \to 0$. ◁

Exercise 1.1.6 For $a \geq 0$, prove the following.
  (i) $a^n \to 0 \iff a < 1$.
  (ii) $a^n \to \infty \iff a > 1$. ◁

Exercise 1.1.7 If $a_n \to a$ and $a \neq 0$, then show that there exists $k \in \mathbb{N}$ such that $a_n \neq 0$ for all $n \geq k$. ◁

Exercise 1.1.8 Consider the sequence (a_n) with $a_n = \left(1 + \frac{1}{n}\right)^{1/n}, n \in \mathbb{N}$. Then show that $\lim_{n \to \infty} a_n = 1$.
   [Hint: Observe that $1 \leq a_n \leq (1 + 1/n)$ for all $n \in \mathbb{N}$.] ◁

Exercise 1.1.9 Consider the sequence (a_n) with $a_n = \frac{1}{n^k}, n \in \mathbb{N}$. Then show that for any given $k \in \mathbb{N}$, $\lim_{n \to \infty} a_n = 0$ (Fig. 1.5).
   [Hint: Observe that $1 \leq a_n \leq 1/n$ for all $n \in \mathbb{N}$.] ◁

Fig. 1.5 $2^n \to \infty$ as
$n \to \infty$

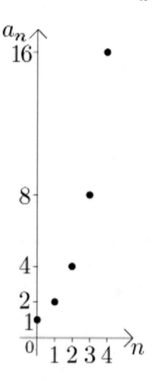

1.1.2 Some Tests for Convergence and Divergence

Theorem 1.1.5 *(Ratio test) Suppose $a_n > 0$ for all $n \in \mathbb{N}$ such that $\lim\limits_{n \to \infty} \dfrac{a_{n+1}}{a_n} = \ell$ for some $\ell \geq 0$. Then the following hold.*

(i) If $\ell < 1$, then $a_n \to 0$.
(ii) If $\ell > 1$, then $a_n \to \infty$.

Proof (i) Suppose $\ell < 1$. Let q be such that $\ell < q < 1$. Then, taking for example the open interval I containing ℓ as $I = (\ell - 1, q)$, there exists $N \in \mathbb{N}$ such that

$$\frac{a_{n+1}}{a_n} \leq q \quad \forall n \geq N.$$

Hence,

$$0 < a_n \leq q^{n-N} a_N \quad \forall n \geq N.$$

By Theorem 1.1.4, $q^{n-N} \to 0$ as $n \to \infty$. Hence, by Sandwich theorem, $a_n \to 0$ as $n \to \infty$.

(ii) Suppose $\ell > 1$. Let q be such that $1 < q < \ell$. Then, taking for example the open interval I containing ℓ as $I = (q, \ell + 1)$, there exists $N \in \mathbb{N}$ such that

$$\frac{a_{n+1}}{a_n} > q \quad \forall n \geq N.$$

Hence,

$$a_n \geq q^{n-N} a_N \quad \forall n \geq N.$$

By Theorem 1.1.4, $q^{n-N} \to \infty$ as $n \to \infty$. Hence, $a_n \to \infty$. ∎

Example 1.1.15 Let $0 < a < 1$. Then $na^n \to 0$ as $n \to \infty$. To see this, let $a_n :=$ na^n for $n \in \mathbb{N}$. Then we have

$$\frac{a_{n+1}}{a_n} = \frac{(n+1)a^{n+1}}{na^n} = \frac{(n+1)a}{n} \quad \forall n \in \mathbb{N}.$$

Hence, $\lim_{n\to\infty} \dfrac{a_{n+1}}{a_n} = a < 1$. Thus, by Theorem 1.1.5, $na^n \to 0$.

Similarly, it can be shown that, for any $k \in \mathbb{N}$, $n^k a^n \to 0$ as $n \to \infty$. ◇

Remark 1.1.9 The converse of the results (i) and (ii) in Theorem 1.1.5 does not hold. To see this, consider the following examples:

(i) Consider (a_n) with $a_n = 1/n$. Then $a_n \to 0$, but $a_{n+1}/a_n \to 1$.

(ii) Consider (a_n) with $a_n = n$. Then $a_n \to \infty$, but $a_{n+1}/a_n \to 1$.

In fact, we shall see in the next section that the condition $\ell < 1$ in Theorem 1.1.5 is too strong, in the sense that, not only we have the convergence of (a_n) to 0, but also we can show the convergence of the sequence (s_n), where $s_n = a_1 + a_2 + \cdots + a_n$.

Note that, for any sequence (a_n), if $s_n = a_1 + a_2 + \cdots + a_n$, then the convergence of (s_n) implies $a_n \to 0$, but $a_n \to 0$ does not imply the convergence of (s_n). ◇

Exercise 1.1.10 Establish the statement in the last paragraph of the above remark. ◁

The following theorem gives a sufficient condition for certain number to be a limit of a given sequence.

Theorem 1.1.6 *Let (a_n) be a sequence such that*

$$|a_{n+1} - a| \leq r|a_n - a| \quad \forall n \in \mathbb{N}$$

for some $a \in \mathbb{R}$ and for some r with $0 < r < 1$. Then $a_n \to a$.

Proof For each $n \in \mathbb{N}$, we have

$$|a_{n+1} - a| \leq r|a_n - a| \leq \cdots \leq r^n|a_1 - a|.$$

Thus,

$$0 \leq |a_{n+1} - a| \leq r^n|a_1 - a| \quad \forall n \in \mathbb{N}.$$

Since $0 < r < 1$, by Theorem 1.1.4, $r^n \to 0$. Hence, by Sandwich theorem, $a_n \to a$. ∎

Example 1.1.16 Let a sequence (a_n) be defined *iteratively* as follows :

$$a_1 = 1, \quad a_{n+1} = \frac{2 + a_n}{1 + a_n}, \quad n = 1, 2, \dots.$$

Let us assume for a moment that (a_n) converges to $a \in \mathbb{R}$. Then we obtain

$$a = \frac{2 + a}{1 + a},$$

i.e., $a(1 + a) = 2 + a$, i.e., $a^2 = 2$. Thus, if (a_n) converges, then the limit must be $a = \sqrt{2}$. Now, we prove that (a_n) actually converges to $a := \sqrt{2}$. Note that

$$
\begin{aligned}
a_{n+1} - a &= \frac{2 + a_n}{1 + a_n} - \frac{2 + a}{1 + a} \\
&= \frac{a - a_n}{(1 + a_n)(1 + a)}.
\end{aligned}
$$

Note that $a_n \geq 1$ for all $n \in \mathbb{N}$ and $a \geq 1$. Hence, we obtain

$$|a_{n+1} - a| \leq \frac{|a - a_n|}{(1 + a_n)(1 + a)} \leq \frac{|a_n - a|}{4} \quad \forall n \in \mathbb{N}.$$

Therefore, by Theorem 1.1.6, $a_n \to a = \sqrt{2}$.   ◊

Remark 1.1.10 In view of Theorem 1.1.6, one may ask the following question:

If (a_n) is such that $|a_{n+1} - a| < |a_n - a|$ for all $n \in \mathbb{N}$ for some $a \in \mathbb{R}$, then does (a_n) converge to a?

Not necessarily! To see this, consider the sequence (a_n) with

$$a_n = \frac{n + 1}{n}, \quad n \in \mathbb{N}.$$

Since

$$\frac{n + 2}{n + 1} < \frac{n + 1}{n} \quad \forall n \in \mathbb{N},$$

taking $a = 0$, we have $|a_{n+1} - a| < |a_n - a|$ for all $n \in \mathbb{N}$. But (a_n) does not converge to 0. In fact, $a_n \to 1$.   ◊

Remark 1.1.11 In Theorem 1.1.6, the sufficient condition given for inferring the convergence of a sequence (a_n) involves its limit. It would be better if we have a condition on (a_n) independent of its limit. In this connection, we shall show at the end of Sect. 1.1.4 that

$$|a_{n+2} - a_{n+1}| \leq r|a_{n+1} - a_n| \quad \forall n \in \mathbb{N}$$

is one such condition, where $0 < r < 1$.   ◊

Remark 1.1.12 In Example 1.1.16, the terms of the sequence are defined iteratively, in the sense that the $(n + 1)$-th term of the sequence is defined in terms of the nth term of the sequence. More generally, a sequence (a_n) defined iteratively can be of the form

$$a_{n+1} = \varphi(a_n), \quad n \in \mathbb{N}, \tag{1.1}$$

where φ is some function defined on $\mathbb{R}$. Suppose, φ has the property that

$$a_n \to a \quad \Rightarrow \quad \varphi(a_n) \to \varphi(a).$$

Thus, if we assume that $a_n \to a$, then (1.1) implies that

$$a = \varphi(a). \tag{1.2}$$

Thus, following the procedure as in Example 1.1.16, convergence of a sequence (a_n) which satisfies (1.1) can be established by looking for a number a satisfying (1.2), and then proving that (a_n) actually converges to a. Note that, by (1.1) and (1.2),

$$a_{n+1} - a = \varphi(a_n) - \varphi(a).$$

Thus, $a_n \to a$ if and only if $\varphi(a_n) \to \varphi(a)$. In case, there is a positive number $r < 1$ such that

$$|\varphi(a_n) - \varphi(a)| \leq r|a_n - a| \quad \text{for all} \quad n \in \mathbb{N},$$

then Theorem 1.1.6 leads to the convergence of (a_n) to a.

The above procedure also can be used to show the divergence of certain sequences, by showing that there is no a which satisfies (1.2). For example, suppose (a_n) is defined by $a_1 = 1$ and

$$a_{n+1} = \sqrt{1 + a_n^2}, \quad n \in \mathbb{N}.$$

Since there is no a satisfying $a = \sqrt{1 + a^2}$, we can assert that the given sequence does not converge.

Existence of a satisfying (1.2) alone does not imply that the sequence converges. For example, consider the sequence (a_n) defined by $a_1 = 1$ and

$$a_{n+1} = -a_n, \quad n \in \mathbb{N}.$$

We know that $a = 0$ satisfies the equation $a = -a$, but $a_n \not\to 0$. In fact, in this case, the sequence is the divergent sequence $((-1)^{n+1})$.   ◊

Let us consider another example to illustrate the procedure described in Remark 1.1.12.

Example 1.1.17 Consider the sequence (a_n) defined by

$$a_{n+1} = 1 + \frac{1}{a_n}, \quad n \in \mathbb{N},$$

with $a_1 = 1$. Here $\varphi(x) := 1 + \frac{1}{x}$, $x \neq 0$. Observe that if $a_n \to a$ for some number a, then $a = \varphi(a)$, and

$$a = \varphi(a) \iff a = \frac{1 + \sqrt{5}}{2}.$$

Now, let us try to find a number r with $0 < r < 1$ such that

$$|\varphi(a_n) - \varphi(a)| \leq r|a_n - a| \quad \text{for all} \quad n \in \mathbb{N}.$$

Note that

$$|\varphi(a_n) - \varphi(a)| \leq r|a_n - a| \iff \left| \frac{1}{a_n} - \frac{1}{a} \right| \leq r|a_n - a|$$
$$\iff \left| \frac{a - a_n}{a_n a} \right| \leq r|a_n - a|$$
$$\iff a_n a \geq \frac{1}{r}.$$

Thus, it is enough to find r such that $0 < r < 1$ and $a_n a \geq 1/r$ for all $n \in \mathbb{N}$. Since $a_n \geq 1$ and $a \geq 3/2$, we have $a_n a \geq a \geq 3/2$ so that we may take any r satisfying $2/3 \leq r < 1$. Thus,

$$a_n \to a := \frac{1 + \sqrt{5}}{2}. \qquad \Diamond$$

Remark 1.1.13 (Hemachandra–Fibonacci sequence) We observe that the terms of the sequence (a_n) in Example 1.1.17 are

$$\frac{1}{1}, \quad \frac{2}{1}, \quad \frac{3}{2}, \quad \frac{5}{3}, \quad \frac{8}{5}, \quad \frac{13}{8}, \quad \ldots.$$

These terms are ratios of consecutive terms of the sequence (b_n), given by

$$1, \quad 1, \quad 2, \quad 3, \quad 5, \quad 8, \quad 13, \quad \ldots.$$

Note that the terms of the sequence (b_n) are iteratively defined by $b_1 = 1$, $b_2 = 1$, and for $n = 3, 4, \ldots$,

$$b_n = b_{n-1} + b_{n-2}.$$

The number

$$\frac{1 + \sqrt{5}}{2} := \lim_{n \to \infty} \frac{b_{n+1}}{b_n},$$

the limit of the sequence (a_n), is called the **golden ratio**. The golden ratio is approximately equal to 1.618033, correct to $(1 + \sqrt{5})/2$ up to six decimal places, i.e.,

$$\left| \frac{1 + \sqrt{5}}{2} - 1.618033 \right| < \frac{1}{10^6}.$$

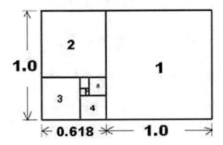

In fact, the number $(1 + \sqrt{5})/2$ is the ratio α/β of the sides α and β of a rectangle with $\beta < \alpha$ such that when the square of side β is cut out of the rectangle, the remaining rectangle has the same ratio as its sides, i.e.,

$$\frac{\alpha}{\beta} = \frac{\beta}{\alpha - \beta}.$$

Indeed, if $\gamma = \alpha/\beta$, then the above relation gives $\gamma = 1/(\gamma - 1)$, i.e.,

$$\gamma = 1 + 1/\gamma, \quad \text{equivalently,} \quad \gamma(\gamma - 1) = 1.$$

so that $\gamma = (1 + \sqrt{5})/2$. Such a rectangle is called a **golden rectangle**.

The sequence (b_n) defined above, that is the sequence,

$$1, \quad 1, \quad 2, \quad 3, \quad 5, \quad 8, \quad 13, \quad \ldots$$

is known as the **Fibonacci sequence**, after the Italian mathematician Leonardo of Pisa, also known as Fibonacci (1170 - 1250). However, this sequence of numbers appeared in Indian mathematics as early as 200 BC in the work of *Pingala* on enumerating possible patterns of Sanskrit poetry formed from syllables of two lengths, called *laghu* (short) and *guru* (long), and extensively used by *Āchārya Hemachandra* (1088–1173) about fifty years prior to Fibonacci (cf. Perkins [10]). Because of this, Fibonacci sequence is also known as **Hemachandra–Fibonacci sequence**.   ◊

Next we prove a necessary condition for the convergence of a sequence.

Theorem 1.1.7 (Boundedness test) *If (a_n) converges, then there exists $M > 0$ such that*

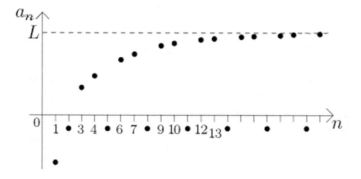

Fig. 1.6 (a_n) is a bounded sequence

$$|a_n| \leq M \quad \forall n \in \mathbb{N}.$$

(Fig. 1.6)

Proof Suppose $a_n \to a$. Note that, for all $n \in \mathbb{N}$,

$$|a_n| = |(a_n - a) + a| \leq |a_n - a| + |a|.$$

Since $|a_n - a| \to 0$, (by taking $\varepsilon = 1$) there exists $N \in \mathbb{N}$ such that $|a_n - a| \leq 1$ for all $n \geq N$. Then we have

$$|a_n| \leq 1 + |a| \quad \forall n \geq N.$$

Thus,

$$|a_n| \leq M := \max\{1 + |a|, |a_1|, |a_2|, \ldots, |a_N|\}$$

for all $n \in \mathbb{N}$. ∎

The converse of Theorem 1.1.7 is not true. Note that the sequence (a_n) with $a_n = (-1)^n$ satisfies $|a_n| = 1$ for all $n \in \mathbb{N}$, but it is not convergent.

For later use we introduce the following definition.

Definition 1.1.7 A sequence (a_n) is said to be

(1) **bounded above** if there exists a real number M such that $a_n \leq M$ for all $n \in \mathbb{N}$;
(2) **bounded below** if there exists a real number M' such that $a_n \geq M'$ for all $n \in \mathbb{N}$;
(3) **bounded** if it is bounded above and bounded below.

A sequence that is not bounded is called an **unbounded** sequence. ◇

Thus, according to Theorem 1.1.7:

> Every convergent sequence is bounded

The following statements can be verified easily:

1. A sequence (a_n) is bounded if and only if there exists $M > 0$ such that $|a_n| \le M$ for all $n \in \mathbb{N}$.
2. A sequence (a_n) is unbounded if and only if it is either not bounded above or not bounded below.
3. If (a_n) diverges to ∞, then it is not bounded above, and if it diverges to $-\infty$, then it is not bounded below.
4. A sequence (a_n) is not bounded above if and only if for every $k \in \mathbb{N}$, there exists $n_k \in \mathbb{N}$ such that $a_{n_k} > k$.
5. A sequence (a_n) is not bounded below if and only if for every $k \in \mathbb{N}$, there exists $n_k \in \mathbb{N}$ such that $a_{n_k} < -k$.

Exercise 1.1.11 Verify the above statements.   ◁

Example 1.1.18 Let us look at a few examples of bounded and unbounded sequences (verify).

(i) (a_n) with $a_n = (-1)^n$ is a bounded sequence.
(ii) (a_n) with $a_n = (-1)^n n$ is neither bounded above nor bounded below, and it neither diverges to $+\infty$ nor diverges to $-\infty$.
(iii) (a_n) with $a_n = -n$ is bounded above, but not bounded below, and it diverges to $-\infty$.
(iv) (a_n) with $a_n = n$ is bounded below, but not bounded above, and it diverges to $+\infty$.   ◇

Theorem 1.1.7 can be used to show that certain sequence is not convergent, as in the following example.

Example 1.1.19 For $n \in \mathbb{N}$, let

$$a_n = 1 + \frac{1}{2} + \frac{1}{3} + \ldots + \frac{1}{n}.$$

Then (a_n) diverges: To see this, observe that

$$\begin{aligned}
a_{2^n} &= 1 + \frac{1}{2} + \frac{1}{3} + \ldots + \frac{1}{2^n} \\
&= 1 + \frac{1}{2} + \left(\frac{1}{3} + \frac{1}{4}\right) + \left(\frac{1}{5} + \frac{1}{6} + \frac{1}{7} + \frac{1}{8}\right) + \\
&\quad \ldots + \left(\frac{1}{2^{n-1}+1} + \ldots + \frac{1}{2^n}\right) \\
&\ge 1 + \frac{n}{2}.
\end{aligned}$$

Note that

$$\frac{1}{3} + \frac{1}{4} \ge \frac{1}{2}, \quad \frac{1}{5} + \frac{1}{6} + \frac{1}{7} + \frac{1}{8} \ge \frac{1}{2}, \quad \ldots, \quad \frac{1}{2^{n-1}+1} + \ldots + \frac{1}{2^n} \ge \frac{1}{2}.$$

Hence, $a_n \ge 1 + \frac{n}{2}$. Thus, (a_n) is not bounded, so that it diverges.   ◇

Theorem 1.1.7 will be used for proving the following.

Theorem 1.1.8 *Suppose* $a_n \to a$ *and* $b_n \to b$. *Then we have the following.*

(i) $a_n b_n \to ab$ *as* $n \to \infty$.
(ii) *If* $b_n \neq 0$ *for all* $n \in \mathbb{N}$ *and* $b \neq 0$, *then*

$$\frac{1}{b_n} \to \frac{1}{b} \quad and \quad \frac{a_n}{b_n} \to \frac{a}{b}.$$

Proof (i) Note that, for every $n \in \mathbb{N}$,

$$|a_n b_n - ab| = |a_n(b_n - b) + (a_n - a)b|$$
$$\leq |a_n| |b_n - b| + |a_n - a| |b|.$$

By Theorem 1.1.7, (a_n) is bounded, say $|a_n| \leq M$ for all $n \in \mathbb{N}$. Hence, we have

$$0 \leq |a_n b_n - ab| \leq M |b_n - b| + |b||a_n - a|.$$

Since $a_n \to a$ and $b_n \to b$, by Theorem 1.1.3, we obtain $a_n b_n \to ab$ as $n \to \infty$.
   (ii) Suppose $b_n \neq 0$ for all $n \in \mathbb{N}$ and $b \neq 0$. Note that, for every $n \in \mathbb{N}$,

$$\left| \frac{1}{b_n} - \frac{1}{b} \right| = \frac{|b_n - b|}{|b_n| |b|}.$$

Since $b_n \to b$ and $b \neq 0$, we have $|b_n| \to |b|$ so that, taking $\varepsilon = |b|/2$, there exists $N \in \mathbb{N}$ such that

$$|b_n| \geq |b|/2 \quad \forall n \geq N.$$

Hence,

$$\left| \frac{1}{b_n} - \frac{1}{b} \right| \leq \frac{|b_n - b|}{(|b|/2) |b|} = \frac{2}{|b|^2} |b_n - b| \quad \forall n \geq N.$$

Hence, by Theorem 1.1.3, we obtain $1/b_n \to 1/b$ as $n \to \infty$. Now, using (i), we also obtain $a_n/b_n \to a/b$ as $n \to \infty$. ∎

Exercise 1.1.12 Prove the following.

(i) If (a_n) is not bounded above, then there exists a sequence (k_n) of natural numbers such that $k_n \leq k_{n+1}$ for all $n \in \mathbb{N}$ and $a_{k_n} \to +\infty$ as $n \to \infty$.
(ii) If (a_n) is not bounded below, then there exists sequence (k_n) of natural numbers such that $k_n \leq k_{n+1}$ for all $n \in \mathbb{N}$ and $a_{k_n} \to -\infty$ as $n \to \infty$.   ◁

If (a_n) converges to a and $a \neq 0$, then show that there

Exercise 1.1.13 Let (a_n) and (b_n) are such that $a_n \leq b_n$ for all $n \in \mathbb{N}$. Prove the following:

(i) If (b_n) is bounded above, then (a_n) bounded above.

(ii) If (a_n) is bounded below, then (b_n) bounded below.   ◁

1.1.3 Monotonic Sequences

We have seen that a bounded sequence need not converge. However, we shall show that, if the terms a_n of a bounded sequence (a_n) either increase or decrease as n increases, then the sequence does converge.

Definition 1.1.8 A sequence (a_n) is said to be a

(1) **monotonically increasing** sequence if $a_n \leq a_{n+1}$ for all $n \in \mathbb{N}$;

(2) **monotonically decreasing** sequence if $a_n \geq a_{n+1}$ for all $n \in \mathbb{N}$;

(3) **monotonic** sequence if it is either monotonically increasing or monotonically decreasing.

If strict inequality occur in (1) (resp. (2)), then we say that the sequence is **strictly increasing** (resp. **strictly decreasing**).

A monotonically increasing (respectively, a monotonically decreasing) sequence is also called an **increasing** (respectively, a **decreasing**) sequence.   ◊

We may observe that:

1. A sequence (a_n) is monotonically increasing and bounded above if and only if $(-a_n)$ is monotonically decreasing and bounded below.

2. Every monotonically increasing sequence is bounded below, and every monotonically decreasing sequence is bounded above.

Fig. 1.7 Strictly decreasing

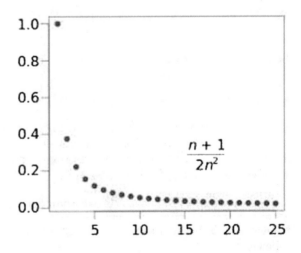

Fig. 1.8 Neither increasing
nor decreasing

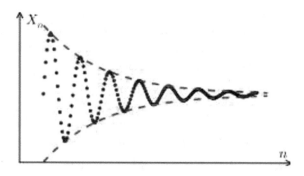

Example 1.1.20 The following statements can be easily verified (verify) (Figs. 1.7
and 1.8):

(i) (a_n) with $a_n = n/(n + 1)$ is monotonically increasing.
(ii) (a_n) with $a_n = (n + 1)/n$ is monotonically decreasing.
(iii) (a_n) with $a_n = (-1)^n n/(n + 1)$ is neither monotonically increasing nor mono-
      tonically decreasing.   ◊

Exercise 1.1.14 Let (a_n) be a monotonic sequence. Prove the following.
       (i) Suppose (a_n) converges. Then it is bounded above by the limit if it is increasing,
and bounded below by the limit if it is decreasing.
       (ii) Suppose (a_n) diverges. Then it diverges to either infinity or minus infinity
depending on whether it is increasing or decreasing.   ◁

       Note that a convergent sequence need not be monotonically increasing or mono-
tonically decreasing. For example, the sequence $((-1)^n/n)$ is convergent, but it is
neither monotonically increasing nor monotonically decreasing. However, we have
the following theorem. It helps us to show the convergence of many of the standard
sequences.

Theorem 1.1.9 (Monotone convergence theorem) *Every bounded monotonic
sequence is convergent.*

       For the proof of the above theorem we shall make use of an important property
of the set of real numbers, that is called the *least upper bound property*. For its
statement, we need to introduce the notion of the *least upper bound*.
       We have already defined the concepts of *bounded above, bounded below* and
bounded corresponding to a sequence. Now, we extend these notions to an arbitrary
subset of the set of real numbers.

Definition 1.1.9 Let S be a subset of $\mathbb{R}$. Then S is said to be

(1) **bounded above** if there exists $b \in \mathbb{R}$ such that $x \leq b$ for all $x \in S$, and in that
    case b is called an **upper bound** of S;

(2) **bounded below** if there exists $a \in \mathbb{R}$ such that $a \leq x$ for all $x \in S$, and in that case a is called a **lower bound** of S;

(3) **bounded** if it is bounded above and bounded below;

(4) **unbounded** if it is not bounded.   ◊

For a set $S \subseteq \mathbb{R}$, we observe that,

1. S is bounded iff there exists $M > 0$ such that $|x| \leq M$ for all $x \in S$;
2. S is unbounded iff either it is not bounded above or it is not bounded below;
3. b is an upper bound of S iff $-b$ is a lower bound of $\{-x : x \in S\}$.

Exercise 1.1.15 Verify the above statements.   ◁

Before going further, let us illustrate the above notions by a few simple examples.

Example 1.1.21 We may observe the following:

  (i) Intervals (a, b), $[a, b)$, $(a, b]$, $[a, b]$ with $a, b \in \mathbb{R}$ and $a < b$ are bounded sets.

 (ii) For $a, b \in \mathbb{R}$, the intervals $(-\infty, b)$, $(-\infty, b]$ are bounded above, but not bounded below, and the intervals (a, ∞), $[a, \infty)$ are bounded below but not bounded above.

(iii) The set $\mathbb{N}$ is bounded below, but not bounded above.

 (iv) The sets $\mathbb{Z}$ of integers are neither bounded above nor bounded below.

  (v) The set $\{(-1)^n n : n \in \mathbb{N}\}$ is neither bounded above nor bounded below.   ◊

We may also observe:

1. If $S \subseteq \mathbb{R}$ is bounded above (respectively, bounded below), then any subset of S is bounded above (respectively, bounded below).
2. If $S \subseteq \mathbb{R}$ is not bounded above (respectively, not bounded below), then any super set of S in $\mathbb{R}$ is not bounded above (respectively, not bounded below).

Exercise 1.1.16 Verify the above statements.   ◁

Let $S \subseteq \mathbb{R}$. If S is bounded above and if $b \in \mathbb{R}$ is an upper bound of S, then any number greater than b is also an upper bound. So, one may seek a least of the upper bounds.

Definition 1.1.10 Let S be a subset of $\mathbb{R}$. If S is bounded above, and if b_0 is an upper bound of S such that $b_0 \leq b$ for every upper bound b of S, then b_0 is called the **least upper bound** or **supremum** for S, and it is denoted by **lub**(S) or **sup**(S).   ◊

Given any set of numbers having a certain property, it is not obvious whether there is a least number having the same property. For instance, if we consider the set

$$\{x \in \mathbb{Q} : 0 < x < 1\}$$

does not have any least number. However, the set of all upper bounds for a given set $S \subseteq \mathbb{R}$, does have a least of the upper bound, provided S is non-empty. This is, in fact, one of the important properties of the set of all real numbers, called the *least upper bound property* (cf. Pugh [12]):

Least upper bound property: *Every non-empty subset of* $\mathbb{R}$ *which is bounded above has the least upper bound.*

Analogously, if S is bounded below with a lower bound $a \in \mathbb{R}$, then any number less than a is also a lower bound. In this case, we define the *greatest lower bound* as follows.

Definition 1.1.11 Let S be a subset of $\mathbb{R}$. If S is bounded below, and if a_0 is a lower bound of S such that $a_0 \geq a$ for every lower bound a of S, then a_0 is called the **greatest lower bound** or **infimum** for S, and it is denoted by $\mathbf{glb}(S)$ or $\inf(S)$. $\Diamond$

We may observe that, for a non-empty set $S \subseteq \mathbb{R}$,

b_0 is a least upper bound for S if and only if $-b_0$ is a greatest lower bound for $\tilde{S} := \{-x : x \in S\}$.

Thus, we also have the *greatest lower bound property* of $\mathbb{R}$:

Greatest lower bound property: *Every non-empty subset of* $\mathbb{R}$ *which is bounded below has the greatest lower bound.*

For a non-empty set $S \subseteq \mathbb{R}$, the following can be verified easily.

1. If S is bounded above, then S has only one least upper bound.
2. If S is bounded below, then S has only one greatest lower bound.

Exercise 1.1.17 Show that, every subset of $\mathbb{R}$ that is bounded above has the least upper bound if and only if every subset of $\mathbb{R}$ that is bounded below has the greatest lower bound. ◁

As per the Exercise 1.1.17, least upper bound property is equivalent to greatest lower bound property.

1. If S is not bounded above, then S does not have the supremum, and in that case we write $\sup(S) = \infty$.
2. If S is not bounded below, then S does not have the infimum, and in that case we write $\inf(S) = -\infty$.

Example 1.1.22 The following can be verified easily:

(i) If S is any of the intervals $(0, 1)$, $[0, 1)$, $(0, 1]$, $[0, 1]$, then 1 is the supremum for S and 0 is the infimum of S.
(ii) If $S = \{1/n : n \in \mathbb{N}\}$, then 1 is the supremum of S and 0 is the infimum for S.
(iii) For $k \in \mathbb{N}$, if $S_k = \{n \in \mathbb{N} : n \geq k\}$, then k is the infimum of S_k, and S_k has no supremum.
(iv) For $k \in \mathbb{N}$, if $S_k = \{n \in \mathbb{Z} : n \leq k\}$, then k is the supremum of S_k, and S_k has no infimum. $\Diamond$

The above examples would have convinced you the following:

> The supremum (respectively, infimum) of a set S, if exists, need not belong to S

Exercise 1.1.18 Let $S \subseteq \mathbb{R}$. Prove the following.

(i) If b_0 is the supremum of S, then there exists a sequence (x_n) in S which converges to b_0.
(ii) If a_0 is the infimum of S, then there exists a sequence (x_n) in S which converges to a_0.   ◁

Now, we prove Theorem 1.1.9.

Proof of Theorem 1.1.9 Suppose (a_n) is a bounded monotonic sequence. Then either it is monotonically increasing and bounded above or it is monotonically decreasing and bounded below.

Assume that (a_n) is monotonically increasing and bounded above. Then the set $S := \{a_n : n \in \mathbb{N}\}$ is bounded above. Hence, by the least upper bound property of $\mathbb{R}$, S has the least upper bound (the supremum). Let $b_0 := \sup(S)$, and let $\varepsilon > 0$ be given. Then, $b_0 - \varepsilon$ cannot be an upper bound of S. Hence, there exists $k \in \mathbb{N}$ such that $b_0 - \varepsilon < a_k$. Since (a_n) is monotonically increasing, we have

$$b_0 - \varepsilon < a_k \leq a_n \quad \forall n \geq k.$$

But, $a_n \leq b_0$ for all $n \in \mathbb{N}$. Hence, we obtain

$$b_0 - \varepsilon < a_k \leq a_n < b_0 + \varepsilon \quad \forall n \geq k.$$

Thus we have proved that $a_n \to b_0$ as $n \to \infty$.

Next, suppose that (a_n) is monotonically decreasing and bounded below. Then the sequence $(-a_n)$ is monotonically increasing and bounded above. Therefore, $(-a_n)$ converges to $b := \sup\{-a_n : n \in \mathbb{N}\}$. Consequently, $a_n \to a := -b = \inf\{a_n : n \in \mathbb{N}\}$.   ∎

In fact, we have proved the following:

> (i) If (a_n) is monotonically increasing and bounded above, then (a_n) converges to $\sup\{a_n : n \in \mathbb{N}\}$
> (ii) If (a_n) is monotonically decreasing and bounded below, then (a_n) converges to $\inf\{a_n : n \in \mathbb{N}\}$

A monotonic sequence that is not bounded cannot converge. Can we say something more? Yes:

> (i) A monotonically increasing sequence either converges or diverges to ∞
>
> (ii) A monotonically decreasing sequence either converges or diverges to $-\infty$

Exercise 1.1.19 Prove the statements (i) and (ii) above.   ◁

In the following examples, Theorem 1.1.9 has been used for showing convergence of certain sequences.

Example 1.1.23 We have already seen that if $0 < a < 1$, then $a^n \to 0$ as $n \to \infty$ (See Theorem 1.1.4). This can also be seen by making use of Theorem 1.1.9, as follows: Let $x_n = a^n$. Then

$$0 \le x_{n+1} = a^{n+1} = ax_n \le x_n \quad \forall n \in \mathbb{N}.$$

Thus, (x_n) is monotonically decreasing and bounded below. Hence (x_n) converges to some $x \in \mathbb{R}$. Then $x_{n+1} = ax_n \to ax$. Therefore, $x = ax$ so that $x = 0$.   ◊

Example 1.1.24 For $n \in \mathbb{N}$, let

$$a_n = 1 + \frac{1}{2^2} + \frac{1}{3^2} + \cdots + \frac{1}{n^2}.$$

Clearly, (a_n) is monotonically increasing. Now, we show that it is bounded above, so that by Theorem 1.1.9, it is convergent: We note that

$$\frac{1}{2^2} + \frac{1}{3^2} \le \frac{2}{2^2} = \frac{1}{2},$$

$$\frac{1}{4^2} + \frac{1}{5^2} + \frac{1}{6^2} + \frac{1}{7^2} \le \frac{4}{4^2} = \frac{1}{4},$$

and more generally,

$$\frac{1}{(2^{n-1})^2} + \frac{1}{(2^{n-1}+1)^2} + \cdots + \frac{1}{(2^n-1)^2} \le \frac{2^{n-1}}{(2^{n-1})^2} = \frac{1}{2^{n-1}}.$$

Hence,

$$
\begin{aligned}
a_{2^n} &= 1 + \frac{1}{2^2} + \frac{1}{3^2} + \cdots + \frac{1}{(2^n)^2} \\
&\le 1 + \frac{1}{2} + \frac{1}{2^2} + \cdots + \frac{1}{2^{n-1}} + \frac{1}{2^{2n}} \\
&= 2\left[1 - \frac{1}{2^n}\right] + \frac{1}{2^{2n}} \\
&\le 2.
\end{aligned}
$$

Since $a_n \leq a_{2^n}$ for all $n \in \mathbb{N}$, we obtain that (a_n) is monotonically increasing and also bounded above, and hence, by Theorem 1.1.9, it converges. We shall see in the final chapter of this book that the number to which this (a_n) converges is $\pi^2/6$.   ◇

Example 1.1.25 Given a sequence (a_n) with $a_n \in \{0, 1, 2, 3, 4, 5, 6, 7, 8, 9\}$, consider the sequence (b_n) with

$$b_n = \frac{a_1}{10} + \frac{a_2}{10^2} + \cdots + \frac{a_n}{10^n}$$

for $n \in \mathbb{N}$. Does (b_n) converge? By our knowledge in decimal expansion of a number, we can believe that (b_n) would converge to a number in the interval $[0, 1]$. Let us prove this:

Note that $b_n \leq b_{n+1}$ for all $n \in \mathbb{N}$. Therefore, to assert the convergence of (b_n), by Theorem 1.1.9, it is enough to show that (b_n) is bounded above. For this end, we observe that

$$
\begin{aligned}
b_n &= \frac{a_1}{10} + \frac{a_2}{10^2} + \cdots + \frac{a_n}{10^n} \\
&\leq \frac{9}{10} + \frac{9}{10^2} + \cdots + \frac{9}{10^n} \\
&= \frac{9}{10}\left[\frac{1 - \frac{1}{10^n}}{1 - \frac{1}{10}}\right] \\
&= 1 - \frac{1}{10^n}
\end{aligned}
$$

for all $n \in \mathbb{N}$. Thus, (b_n) is bounded above by 1. Therefore, (b_n) converges to a number in the interval $[0, 1]$.

More generally, for a given positive integer $k \geq 2$, let (a_n) be a sequence with $a_n \in \{0, 1, \ldots, k - 1\}$, and consider (b_n) with

$$b_n = \frac{a_1}{k} + \frac{a_2}{k^2} + \cdots + \frac{a_n}{k^n}$$

for $n \in \mathbb{N}$. Then, as in the case of $k = 10$, (b_n) is monotonically increasing and bounded above. In fact,

$$
\begin{aligned}
b_n &= \frac{a_1}{k} + \frac{a_2}{k^2} + \cdots + \frac{a_n}{k^n} \\
&\leq \frac{k - 1}{k}\left[\frac{1 - \frac{1}{k^n}}{1 - \frac{1}{k}}\right] \\
&= 1 - \frac{1}{k^n}
\end{aligned}
$$

for all $n \in \mathbb{N}$, so that (b_n) is bounded above by 1, and hence it converges to a number $b \in [0, 1]$.   ◇

Example 1.1.26 (The number e) Consider the sequences (a_n) and (b_n) defined by

$$a_n = \left(1 + \frac{1}{n}\right)^n, \quad b_n = 1 + \frac{1}{1!} + \frac{1}{2!} + \frac{1}{3!} + \cdots + \frac{1}{n!}.$$

We show that both (a_n) and (b_n) are monotonically increasing and bounded above. Hence, by Theorem 1.1.9, they converge. Further, we shall show that they have the same limit. For this we show that

(i) $b_n \leq b_{n+1} \leq 3$ for all $n \in \mathbb{N}$,
(ii) $a_n \leq b_n$ for all $n \in \mathbb{N}$,
(iii) $a_n \leq a_{n+1} \leq 3$ for all $n \in \mathbb{N}$.

Clearly, $b_n \leq b_{n+1}$ for all $n \in \mathbb{N}$. Also,

$$b_n \leq 1 + 1 + \frac{1}{2} + \frac{1}{2^2} + \cdots + \frac{1}{2^{n-1}} < 3.$$

Thus, (i) is proved. Next,

$$a_n = 1 + 1 + \frac{1}{2!} \frac{n(n-1)}{n^2} + \cdots + \frac{1}{n!} \frac{n(n-1)\ldots 2.1}{n^n}$$
$$\leq 1 + 1 + \frac{1}{2!} + \frac{1}{3!} + \cdots + \frac{1}{n!}$$
$$= b_n$$

and

$$a_n = \left(1 + \frac{1}{n}\right)^n$$
$$= 1 + n.\frac{1}{n} + \frac{n(n-1)}{2!}\frac{1}{n^2} + \cdots + \frac{n(n-1)\ldots 2.1}{n!}\frac{1}{n^n}$$
$$= 1 + 1 + \frac{1}{2!}\left(1 - \frac{1}{n}\right) + \frac{1}{3!}\left(1 - \frac{1}{n}\right)\left(1 - \frac{2}{n}\right)$$
$$+ \cdots + \frac{1}{n!}\left(1 - \frac{1}{n}\right)\cdots\left(1 - \frac{n-1}{n}\right)$$
$$\leq a_{n+1}.$$

Thus, the proofs of (i)–(iii) are over. From, (i)–(iii), we see that both (a_n) and (b_n) are monotonically increasing bounded above. Hence, by Theorem 1.1.9, both (a_n) and (b_n) converge. Let a and b be their limits. We show that $a = b$.

We have already observed that $a_n \leq b_n$. Hence, taking limits, we obtain $a \leq b$. Notice that

$$a_n = 1 + 1 + \frac{1}{2!}\left(1 - \frac{1}{n}\right) + \frac{1}{3!}\left(1 - \frac{1}{n}\right)\left(1 - \frac{2}{n}\right)$$
$$+ \cdots + \frac{1}{n!}\left(1 - \frac{1}{n}\right)\cdots\left(1 - \frac{n-1}{n}\right).$$

Hence, for m, n with $m \leq n$, we have

$$a_n \geq 1 + 1 + \frac{1}{2!}\left(1 - \frac{1}{n}\right) + \frac{1}{3!}\left(1 - \frac{1}{n}\right)\left(1 - \frac{2}{n}\right)$$
$$+ \cdots + \frac{1}{m!}\left(1 - \frac{1}{n}\right)\cdots\left(1 - \frac{m-1}{n}\right).$$

Taking limit as $n \to \infty$, we get (cf. Theorem 1.1.3 (c))

$$a \geq 1 + \frac{1}{1!} + \frac{1}{2!} + \frac{1}{3!} + \cdots + \frac{1}{m!} = b_m.$$

Now, taking limit as $m \to \infty$, we get $a \geq b$. Thus we have proved $a = b$.

The common limit of the two sequence (a_n) and (b_n) above is denoted by the letter e, after the great mathematician *Euler*.[1]   ◊

$$e := \lim_{n \to \infty} \left(1 + \frac{1}{n}\right)^n = \lim_{n \to \infty} \left(1 + \frac{1}{1!} + \frac{1}{2!} + \frac{1}{3!} + \cdots + \frac{1}{n!}\right).$$

Further examples are given in Sect. 1.1.5.

1.1.4   Subsequences

There are divergent sequences which contain terms which form convergent sequences. For example, the sequence (a_n) with $a_n = (-1)^n$ is divergent, but the sequences (a_{2n-1}) and (a_{2n}) are convergent. Those sequences extracted from a given sequence, retaining the order in which the terms occur, are called *subsequences*. More precisely, we have the following definition.

Definition 1.1.12 A **subsequence** of a sequence (a_n) is a sequence of the form (a_{k_n}), where (k_n) is a strictly increasing sequence of positive integers.   ◊

A sequence (b_n) is a subsequence of a sequence (a_n) if and only if there is a strictly increasing sequence (k_n) of positive integers such that $b_n = a_{k_n}$ for all $n \in \mathbb{N}$.

For example, given a sequence (a_n), the sequences

$$(a_{2n}), \quad (a_{2n+1}), \quad (a_{n^2}), \quad (a_{2^n})$$

are some of the subsequences of (a_n). As concrete examples,

---

[1] Leonhard Euler (15 April 1707–18 September 1783) was a Swiss mathematician and physicist. He made important discoveries in various fields in mathematics. He introduced many modern mathematical terminology and notation, including the notion of a mathematical function (Courtsey to Wikipedia).

(1) $(\frac{1}{2n})$, $(\frac{1}{2n+1})$, $(\frac{1}{n^2})$ and $(\frac{1}{2^n})$ are subsequences of $(\frac{1}{n})$;

(2) $(\frac{1}{n})$ and $(\frac{n}{n+1})$ are subsequences of $(1, \frac{1}{2}, \frac{1}{2}, \frac{2}{3}, \frac{1}{3}, \frac{3}{4}, \ldots)$.

(3) The sequence (b_n) with $b_n = 1$ for all $n \in \mathbb{N}$ and the sequence (c_n) with $c_n = -1$ for all $n \in \mathbb{N}$ are subsequences of the sequence $((-1)^n)$.

(4) $(2n)$, $(2n + 1)$, (n^2) and (2^n) are subsequences of the sequence (a_n) with $a_n = n$ for all $n \in \mathbb{N}$.

In (1), the given sequence and all the subsequences listed are convergent; in (2), the subsequences are convergent, but the original sequence is not convergent; same is the case with the case in (3); in (4), the original sequence and all the subsequences listed are divergent. In fact, in (4), every subsequence of the given sequence has to diverge to infinity (verify).

However, we have the following result.

Theorem 1.1.10 *If a sequence (a_n) converges to a, then all its subsequences converge to the same limit a.*

Proof Suppose $a_n \to a$. Consider a subsequence (a_{k_n}) of (a_n). Let $\varepsilon > 0$ be given. Since $a_n \to a$, there exists $N \in \mathbb{N}$ such that

$$|a_n - a| < \varepsilon \quad \forall n \geq N.$$

In particular, since $k_N \geq N$,

$$|a_n - a| < \varepsilon \quad \forall n \in \{k_N, k_{N+1}, \ldots\}.$$

Thus, $|a_{k_n} - a| < \varepsilon \quad \forall n \geq N$. Hence, $a_{k_n} \to a$. ∎

What about the converse of the above theorem? Obviously, if all subsequences of (a_n) converge, then (a_n) also has to converge, since (a_n) is a subsequence of itself. Thus, we have proved:

> (a_n) converges to a iff every subsequence of (a_n) converges to a

Theorem 1.1.10 can be used to assert the divergence of certain sequences. Let us consider the following example.

Example 1.1.27 Let $s_n = 1 + \frac{1}{2} + \cdots + \frac{1}{n}$ for $n \in \mathbb{N}$. We know that the sequence (s_n) is not convergent (cf. Example 1.1.19). Now, for every $n \in \mathbb{N}$,

$$s_{2n} - s_n = \frac{1}{n + 1} + \frac{1}{n + 2} + \cdots + \frac{1}{2n} \geq \frac{1}{2}.$$

Hence, by Theorem 1.1.10, (s_n) cannot converge to any point. ◇

We know that convergence of some subsequences, even if the limits are same, does not imply the convergence of the original sequence. However, in certain cases, convergence of a few of subsequences does imply the convergence of the original sequence.

Theorem 1.1.11 *Let* (a_n) *be such that* $a_{2n} \to a$ *and* $a_{2n+1} \to a$ *for some* $a \in \mathbb{R}$. *Then* $a_n \to a$.

Proof Let $\varepsilon > 0$ be given. Since $a_{2n} \to a$ and $a_{2n+1} \to a$ for some $a \in \mathbb{R}$, there exists $N \in \mathbb{N}$ such that

$$|a_{2n} - a| < \varepsilon, \quad |a_{2n+1} - a| < \varepsilon \quad \forall n \geq N.$$

Then we have

$$|a_n - a| < \varepsilon \quad \forall n \geq 2N + 1.$$

Thus, $a_n \to a$. ∎

Remark 1.1.14 In the above theorem, to obtain convergence of (a_n) it is important that both the sequences (a_{2n}) and (a_{2n+1}) converge to the same point. For instance if $a_n = (-1)^n$, then we have the convergence of (a_{2n}) and (a_{2n+1}), but (a_n) does not converge. In this case, the limits of (a_{2n}) and (a_{2n+1}) are different.   ◇

Exercise 1.1.20 Let (a_n) be such that, for some $k \in \mathbb{N}$, the sequences (a_{kn+1}), (a_{kn+2}), ..., (a_{kn+k}) converge to the same limit a. Show that $a_n \to a$.   ◁

The following result on non-convergence of a sequence to a particular $a \in \mathbb{R}$ will be of use in later chapters.

Theorem 1.1.12 *Let* (a_n) *be a sequence of real numbers and* $a \in \mathbb{R}$. *If* $a_n \not\to a$, *then there exists* $\varepsilon > 0$ *and a subsequence* (a_{k_n}) *such that* $|a_{k_n} - a| \geq \varepsilon$ *for all* $n \in \mathbb{N}$.

Proof Suppose $a_n \not\to a$. Then there exists $\varepsilon > 0$ such that $|a_n - a| \geq \varepsilon$ for infinitely many n's in $\mathbb{N}$. Thus, we can find a sequence (k_n) in $\mathbb{N}$ such that $k_n < k_{n+1}$ for all $n \in \mathbb{N}$ and $|a_{k_n} - a| \geq \varepsilon$ for all $n \in \mathbb{N}$.
In fact, if $S = \{n \in \mathbb{N} : |a_n - a| \geq \varepsilon\}$, then we may take

$$k_1 = \inf(S), \quad k_2 = \inf(S \setminus \{k_1\}), \quad k_3 = \inf(S \setminus \{k_1, k_2\}),$$

and having chosen $k_1, k_2, \ldots, k_n$, take

$$k_{n+1} = \inf(S \setminus \{k_1, \ldots, k_n\}).$$

Clearly, (k_n) is a strictly increasing sequence of positive integers such that $|a_{k_n} - a| \geq \varepsilon$ for all $n \in \mathbb{N}$. ∎

We know that a divergent sequence can have convergent subsequences. But, this cannot happen for monotonic sequences.

> If (a_n) is a monotonic sequence having at least one convergent subsequence, say with limit α, then (a_n) itself converge to a

Exercise 1.1.21 Prove the above statement.   ◁

We have seen in Theorem 1.1.7 that every convergent sequence is bounded, but a bounded sequence need not be convergent. However, we have the following theorem, which is called the *Bolzano–Weierstrass theorem*.

Theorem 1.1.13 *(Bolzano-Weierstrass theorem[2]). Every bounded sequence of real numbers has at least one convergent subsequence.*

The idea involved in the proof of the above theorem is to extract a bounded monotone subsequence and then apply Monotone convergence theorem (Theorem 1.1.9). For the details of the proof, the reader may find its proof in Binmore [3].

We use Theorem 1.1.13 to prove the following theorem, which gives another sufficient condition for the convergence of a sequence, as promised in Remark 1.1.15.

Theorem 1.1.14 *Let (a_n) be a sequence such that*

$$|a_{n+2} - a_{n+1}| \le r|a_{n+1} - a_n| \quad \forall n \in \mathbb{N}$$

for some r with $0 < r < 1$. Then (a_n) converges, and if $a = \lim_{n \to \infty} a_n$, then

$$|a_n - a| \le \frac{r^{n-1}}{1 - r}|a_2 - a_1| \quad \forall n \in \mathbb{N}.$$

Proof From the given property of (a_n) we obtain

$$|a_{n+1} - a_n| \le r^{n-1}|a_2 - a_1| \quad \forall n \in \mathbb{N}.$$

Hence, for any n, m with $n < m$, we have

$$|a_m - a_n| \le \frac{r^{n-1}}{1 - r}|a_2 - a_1|. \tag{$*$}$$

---

[2] *Bernard Bolzano* (October 5, 1781–December 18, 1848), was a Bohemian mathematician, logician, philosopher, theologian and Catholic priest of Italian Origins, and *Karl Weierstrass* (October 31, 1815–February 19, 1897) was a German mathematician who is often cited as the "father of modern analysis".

Indeed,

$$
\begin{aligned}
|a_m - a_n| &\le |a_m - a_{m-1}| + \cdots + |a_{n+1} - a_n| \\
&\le (r^{m-2} + r^{m-3} + \cdots + r^{n-1})|a_2 - a_1| \\
&\le \frac{r^{n-1}}{1-r}|a_2 - a_1|.
\end{aligned}
$$

From $(*)$, taking $n = 1$, we have $|a_m - a_1| \le |a_2 - a_1|/(1 - r)$ for all $m \in \mathbb{N}$ so that (a_n) is a bounded sequence. Hence, by Theorem 1.1.13, it has a convergent subsequence, say $a_{k_n} \to a$ for some $a \in \mathbb{R}$. Taking $m = k_n$ in $(*)$, we have

$$
|a_{k_n} - a_n| \le \frac{r^{n-1}}{1-r}|a_2 - a_1| \quad \forall n \in \mathbb{N}.
$$

In particular, $|a_{k_n} - a_n| \to 0$ as $n \to \infty$. Now, since

$$
|a - a_n| \le |a - a_{k_n}| + |a_{k_n} - a_n|,
$$

it follows that $|a - a_n| \to 0$.   ∎

Remark 1.1.15 A convergent sequence (a_n) need not satisfy the condition in Theorem 1.1.14 for any r with $0 < r < 1$. To see this one may consider the sequence (a_n) with $a_n = 1/n$ for $n \in \mathbb{N}$. In this case, we see that

$$
|a_{n+2} - a_{n+1}| \le r|a_{n+1} - a_n| \iff \frac{n}{n+2} \le r
$$

which is satisfied for all $n \in \mathbb{N}$ if and only if $r \ge 1$.   ◇

1.1.5 Further Examples

Example 1.1.28 Let a sequence (a_n) be defined iteratively as follows: $a_1 = 1$ and

$$
a_{n+1} = \frac{2a_n + 3}{4}
$$

$n \in \mathbb{N}$. We show that (a_n) is monotonically increasing and bounded above.
     Note that

$$
a_{n+1} = \frac{2a_n + 3}{4} = \frac{a_n}{2} + \frac{3}{4} \ge a_n \iff a_n \le \frac{3}{2}.
$$

Thus it is enough to show that $a_n \le 3/2$ for all $n \in \mathbb{N}$.

Clearly, $a_1 \leq 3/2$. If $a_n \leq 3/2$, then $a_{n+1} = a_n/2 + 3/4 < 3/4 + 3/4 = 3/2$. Thus, we have proved that $a_n \leq 3/2$ for all $n \in \mathbb{N}$. Hence, by Theorem 1.1.9, (a_n) converges. Let its limit be a. Then taking limit on both sides of $a_{n+1} = \frac{2a_n+3}{4}$ we have

$$a = \frac{2a+3}{4} \quad \text{i.e., } 4a = 2a+3 \quad \text{so that } a = \frac{3}{2}.$$

Another solution: Since $a := 3/2$ satisfies

$$a = \frac{2a+3}{4},$$

we obtain

$$a_{n+1} - a = \frac{2a_n+3}{4} - \frac{2a+3}{4} = \frac{1}{2}(a_n - a).$$

Thus, by Theorem 1.1.6, $a_n \to a = 3/2$. ◇

Example 1.1.29 Let a sequence (a_n) be defined as follows : $a_1 = 2$ and

$$a_{n+1} = \frac{1}{2}\left(a_n + \frac{2}{a_n}\right)$$

for $n \in \mathbb{N}$. Note that, if the sequence converges, then its limit $a \geq 0$, and then

$$a = \frac{1}{2}\left(a + \frac{2}{a}\right)$$

so that $a = \sqrt{2}$.

Since $a_1 = 2$, one may try to show that (a_n) is monotonically decreasing and bounded below.

Note that

$$a_{n+1} := \frac{1}{2}\left(a_n + \frac{2}{a_n}\right) \leq a_n \iff a_n^2 \geq 2, \quad \text{i.e., } a_n \geq \sqrt{2},$$

and

$$a_{n+1} := \frac{1}{2}\left(a_n + \frac{2}{a_n}\right) \geq \sqrt{2} \iff a_n^2 - 2\sqrt{2}a_n + 2 \geq 0$$

$$\iff (a_n - \sqrt{2})^2 \geq 0.$$

Hence, $a_{n+1} \geq \sqrt{2}$ for all $n \in \mathbb{N}$ so that $a_{n+1} \leq a_n$ for all $n \geq 2$. Hence, (a_n) is monotonically decreasing and bounded below, so that by Theorem 1.1.9, (a_n) converges and hence its limit is $\sqrt{2}$. ◇

Example 1.1.30 The sequence $(n^{1/n})$ converges and the limit is 1:

For each $n \in \mathbb{N}$, since $n^{1/n} \geq 1$, there exists $r_n \geq 0$ such that

$$n^{1/n} = 1 + r_n.$$

Then we have

$$n = (1 + r_n)^n \geq \frac{n(n-1)}{2} r_n^2,$$

so that

$$r_n^2 \leq \frac{2}{n-1} \quad \forall n \geq 2.$$

Since $2/(n-1) \to 0$, by Theorem 1.1.3(c), $r_n \to 0$. Hence $n^{1/n} \to 1$.   ◇

Example 1.1.31 For any $a > 0$, $(a^{1/n})$ converges to 1:

If $a > 1$, then we can write $a^{1/n} = 1 + r_n$ for some sequence (r_n) of positive reals. Then we have

$$a = (1 + r_n)^n \geq nr_n \quad \text{so that} \quad r_n \leq a/n.$$

Since $a/n \to 0$, by Theorem 1.1.3(c), $r_n \to 0$ and $a^{1/n} = 1 + r_n \to 1$.

In case $0 < a < 1$, then $1/a > 1$. Hence, by the first part,

$$1/a^{1/n} = (1/a)^{1/n} \to 1,$$

so that by Theorem 1.1.8 (ii), $a^n \to 1$.   ◇

Example 1.1.32 Let (a_n) be a bounded sequence of non-negative real numbers. Then $(1 + a_n)^{1/n} \to 1$ as $n \to \infty$:

Let $M > 0$ be such that $0 \leq a_n \leq M$ for all $n \in \mathbb{N}$. Then,

$$1 \leq (1 + a_n)^{1/n} \leq (1 + M)^{1/n} \quad \forall n \in \mathbb{N}.$$

Using the result in Example 1.1.31, $(1 + M)^{1/n} \to 1$. Hence, by the Sandwich theorem (Theorem 1.1.3 (iv)), $(1 + a_n)^{1/n} \to 1$.   ◇

Example 1.1.33 Let (a_n) be a sequence of non-negative terms such that $1 \leq a_n \leq n$ for all $n \in \mathbb{N}$. Then $(1 + a_n)^{1/n} \to 1$: Note that

$$1 \leq (1 + a_n)^{1/n} \leq (1 + n)^{1/n} \leq (2n)^{1/n} = 2^{1/n} n^{1/n} \quad \forall n \in \mathbb{N}.$$

By the results in Examples 1.1.30 and 1.1.31, $2^{1/n} \to 1$ and $n^{1/n} \to 1$. Hence, by the Sandwich theorem (Theorem 1.1.3 (iv)), we have the convergence $(1 + a_n)^{1/n} \to 1$.   ◇

Example 1.1.34 Let $a_n = (n!)^{1/n^2}$, $n \in \mathbb{N}$. Then $a_n \to 1$ as $n \to \infty$. We give two proofs for this.

(i) Note that, for every $n \in \mathbb{N}$,

$$1 \le (n!)^{1/n^2} \le (n^n)^{1/n^2} = n^{1/n}.$$

Since $n^{1/n} \to 1$, by the Sandwich theorem (Theorem 1.1.3 (iv)), we have $(n!)^{1/n^2} \to 1$.

(ii) By GM-AM inequality, for $n \in \mathbb{N}$,

$$(n!)^{1/n} = (1.2.\ldots.n)^{1/n} \le \frac{1+2+\ldots+n}{n} = \frac{n+1}{2} \le n.$$

Thus, $1 \le (n!)^{1/n^2} \le n^{1/n}$. Since $n^{1/n} \to 1$, by the Sandwich theorem (Theorem 1.1.3 (iv)), we have $(n!)^{1/n^2} \to 1$. $\diamond$

What about the convergence of the sequence (a_n) with $a_n = (n!)^{1/n}$?

Example 1.1.35 Let $a_n = (n!)^{1/n}$, $n \in \mathbb{N}$. We show that $a_n \not\to 1$. In fact, we show that (a_n) is unbounded.

Note that, for any $k, n \in \mathbb{N}$ with $n \ge k$,

$$(n!)^{1/n} \ge (k!)^{1/n}(k^{n-k})^{1/n} = (k!)^{1/n}k^{1-k/n} = k\left(\frac{k!}{k^k}\right)^{1/n}.$$

Since for any fixed k, $k\left(\frac{k!}{k^k}\right)^{1/n} \to k$ as $n \to \infty$, we can conclude that $(n!)^{1/n} \not\to 1$. Now, for $k \in \mathbb{N}$, let $n_k \in \mathbb{N}$ be such that $n_k \ge k$ and

$$\left(\frac{k!}{k^k}\right)^{1/n} \ge \frac{1}{2} \quad \forall n \ge n_k.$$

Thus,

$$(n!)^{1/n} \ge \frac{k}{2} \quad \forall n \ge n_k.$$

Therefore, the sequence $\left((n!)^{1/n}\right)$ is unbounded. In fact $(n!)^{1/n} \to \infty$ as $n \to \infty$. $\diamond$

Remark 1.1.16 Suppose for each $k \in \mathbb{N}$, $a_n^{(k)} \to 0$, $b_n^{(k)} \to 1$ as $n \to \infty$, and also $a_n^{(n)} \to 0$, $b_n^{(n)} \to 1$ as $n \to \infty$. In view of Theorems 1.1.3 and 1.1.8, one may think that

$$a_n^{(1)} + a_n^{(2)} + \cdots + a_n^{(n)} \to 0 \quad \text{as } n \to \infty$$

and

$$b_n^{(1)} b_n^{(2)} \cdots b_n^{(n)} \to 1 \quad \text{as } n \to \infty.$$

Unfortunately, that is not the case. To see this consider

$$a_n^{(k)} = \frac{k}{n^2}, \qquad b_n^{(k)} = k^{1/n} \quad \text{for } k, n \in \mathbb{N}.$$

Then, for each $k \in \mathbb{N}$, we have

$$a_n^{(k)} \to 0, \quad b_n^{(k)} \to 1 \quad \text{as} \quad n \to \infty.$$

Also

$$a_n^{(n)} \to 0, \quad b_n^{(n)} \to 1 \quad \text{as} \quad n \to \infty.$$

But,

$$a_n^{(1)} + a_n^{(2)} + \cdots + a_n^{(n)} = \frac{1}{n^2} + \frac{2}{n^2} + \cdots + \frac{n}{n^2} = \frac{n+1}{2n} \to \frac{1}{2} \quad \text{as} \quad n \to \infty$$

and from Example 1.1.35,

$$b_n^{(1)} b_n^{(2)} \cdots b_n^{(n)} = 1^{1/n} 2^{1/n} \cdots n^{1/n} = (n!)^{1/n} \nrightarrow 1 \quad \text{as} \quad n \to \infty. \qquad \Diamond$$

1.1.6   Cauchy Criterion

In Theorem 1.1.14, we have given a sufficient condition for the convergence of a sequence (a_n), namely, that if (a_n) satisfies

$$|a_{n+2} - a_{n+1}| \leq r|a_{n+1} - a_n| \quad \forall\, n \in \mathbb{N}$$

for some r with $0 < r < 1$, then it converges. In the proof we have observed that, (a_n) satisfies

$$|a_n - a_m| \leq \frac{r^{m-1}}{1 - r}|a_2 - a_1|$$

for all $n, m \in \mathbb{N}$ with $n > m$. Thus, $|a_n - a_m|$ can be made arbitrarily small for all large enough $n, m \in \mathbb{N}$. Now, we show that any sequence (a_n) such that $|a_n - a_m|$ can be made arbitrarily small for all large enough $n, m \in \mathbb{N}$ actually converges. First, let us formally define the requirement on the sequence.

Definition 1.1.13 A a sequence (a_n) is said to be a **Cauchy sequence**[3] if for every $\varepsilon > 0$, there exists $N \in \mathbb{N}$ such that

$$|a_n - a_m| < \varepsilon \quad \forall\, n, m \geq N. \qquad \Diamond$$

We have already observed in Remark 1.1.15 that if (a_n) converges, then it need not satisfy the assumption in Theorem 1.1.14. However, we have the following theorem.

Theorem 1.1.15 *Every convergent sequence is a Cauchy sequence.*

---

[3] Augustin-Louis Cauchy (21 August 1789 – 23 May 1857) was a French mathematician who made many contributions to calculus, specifically in terms of its rigorous foundation.

Proof Suppose (a_n) converges to a. Let $\varepsilon > 0$ be given. Then we know that there exists $N \in \mathbb{N}$ such that $|a_n - a| < \varepsilon/2$ for all $n \geq N$. Hence, we have

$$|a_n - a_m| \leq |a_n - a| + |a - a_m| < \varepsilon \quad \forall n, m \geq N.$$

This completes the proof. ∎

Now, we show that the converse of Theorem 1.1.15 is also true. The idea of the proof is akin to the idea used in the proof of Theorem 1.1.14, namely, we first show that (a_n) is a bounded sequence, so that by Bolzano–Weierstrass theorem (Theorem 1.1.13), (a_n) has a subsequence which converges to some a, and then show that (a_n) itself converges to a.

Theorem 1.1.16 (Cauchy criterion) *Every Cauchy sequence of real numbers converges.*

Proof Let (a_n) be a Cauchy sequence. Taking $\varepsilon = 1$, there exists $N \in \mathbb{N}$ such that

$$|a_n - a_m| < 1 \quad \forall n, m \geq N.$$

In particular, for all $n \geq N$,

$$|a_n| \leq |(a_n - a_N) + a_N| \leq |a_n - a_N| + |a_N| \leq 1 + |a_N|.$$

Therefore,

$$|a_n| \leq \max\{|a_1|, |a_2|, \ldots, |a_N|, 1 + |a_N|\}.$$

Thus, (a_n) is a bounded sequence. As already mentioned, by Bolzano–Weierstrass theorem (Theorem 1.1.13), (a_n) has a subsequence (a_{k_n}) which converges to some a. Now, let $\varepsilon > 0$ be given. Then there exist positive integers N_1, N_2 such that

$$|a_{k_n} - a| < \varepsilon/2 \quad \forall n \geq N_1, \quad |a_n - a_{k_n}| < \varepsilon/2 \quad \forall n \geq N_2.$$

Therefore,

$$|a_n - a| \leq |a_n - a_{k_n}| + |a_{k_n} - a| < \varepsilon/2 + \varepsilon/2 = \varepsilon$$

for all $n \geq N_3 := \max\{N_1, N_2\}$. Thus, $a_n \to a$. This completes the proof. ∎

An alternative proof of Theorem 1.1.16. Let (a_n) be a Cauchy sequence and $\varepsilon > 0$ be given. Let $n_\varepsilon \in \mathbb{N}$ be such that $|a_n - a_m| < \varepsilon$ for all $n, m \geq n_\varepsilon$. In particular, $|a_n - a_{n_\varepsilon}| < \varepsilon$ for all $n \geq n_\varepsilon$. Hence, for any $\varepsilon_1, \varepsilon_2 > 0$,

$$a_{n_{\varepsilon_1}} - \varepsilon_1 < a_n < a_{n_{\varepsilon_2}} + \varepsilon_2 \quad \forall n \geq \max\{n_{\varepsilon_1}, n_{\varepsilon_2}\}. \tag{1.1}$$

From this, we see that the set $\{a_{n_\varepsilon} - \varepsilon : \varepsilon > 0\}$ is bounded above. Let $a := \sup\{a_{n_\varepsilon} - \varepsilon : \varepsilon > 0\}$. Then, by (1.1), we have

$$a_{n_\varepsilon} - \varepsilon \le a \le a_{n_\varepsilon} + \varepsilon \qquad \forall\, \varepsilon > 0. \tag{1.2}$$

From (1.1) and (1.2), we obtain $|a_n - a| \le \varepsilon$ for all $n \ge n_\varepsilon$. Thus, in view of Remark 1.1.4, (a_n) converges to a. ∎

Remark 1.1.17 The above alternative proof of Theorem 1.1.16, which does not directly make use of Bolzano-Weierstrass theorem, was suggested by my colleague Professor P. Veeramani. ◇

Example 1.1.36 Let $s_n = 1 + \frac{1}{2} + \ldots + \frac{1}{n}$ for $n \in \mathbb{N}$. We have seen in Example 1.1.27 that

$$s_{2n} - s_n = \frac{1}{n+1} + \frac{1}{n+2} + \cdots + \frac{1}{2n} \ge \frac{1}{2}$$

for every $n \in \mathbb{N}$. This also shows that (s_n) is not a Cauchy sequence, and hence it diverges. ◇

Exercise 1.1.22 Suppose f is a function defined on an interval J such that

$$|f(x) - f(y)| \le r|x - y| \quad \forall\, x, y \in J,$$

for some r with $0 < r < 1$. For a given $a_0 \in J$, let (a_n) be defined by

$$a_1 = f(a_0), \quad a_{n+1} := f(a_n) \quad \forall\, n \in \mathbb{N}.$$

Show that (a_n) is a Cauchy sequence. Show also that the limit of the sequence (a_n) is independent of the choice of a_0. ◁

Remark 1.1.18 We know that every Cauchy sequence of real numbers converges to a real number. However, we cannot expect a Cauchy sequence of rational numbers to have a rational limit. We have seen many such examples. If S is a subset of $\mathbb{R}$ such that limit of every convergent sequence in S belongs to S, then we say that S is a *closed subset* of $\mathbb{R}$. Thus, the set of rational numbers is not a closed set. Similarly, intervals $(a, b], [a, b), (a, b)$ for $a, b \in \mathbb{R}$ with $a < b$ and intervals $(a, \infty), (-\infty, b)$ for any $a, b \in \mathbb{R}$ are not closed subsets of $\mathbb{R}$, whereas for any $a, b \in \mathbb{R}$ with $a < b$, the interval $[a, b]$ is a closed set.

To see that $[a, b]$ is a closed set let (u_n) be a sequence in $[a, b]$ that converges to some $u \in \mathbb{R}$. If $u \notin [a, b]$, then we can find an $\varepsilon > 0$ such that $(u - \varepsilon, u + \varepsilon) \cap [a, b] = \varnothing$. As $u_n \to u$, there exists $N \in \mathbb{N}$ such that $u_n \in (u - \varepsilon, u + \varepsilon)$ for all $n \ge N$. This is not possible as (u_n) is a sequence in $[a, b]$.

It can be shown that if $S \subseteq \mathbb{R}$, then

1. S is a closed set if and only if S contains all those points $x \in \mathbb{R}$ such that $(x - \varepsilon, x + \varepsilon) \cap S \ne \varnothing$ for every $\varepsilon > 0$;
2. S is not a closed set if and only if there exists $x \in \mathbb{R} \setminus S$ such that $(x - \varepsilon, x + \varepsilon) \cap S \ne \varnothing$ for every $\varepsilon > 0$.

Exercise 1.1.23 Justify the above two statements. ◁

A subset S of $\mathbb{R}$ is called an *open set* if for every $x \in S$, there exists $\varepsilon > 0$ such that $(x - \varepsilon, x + \varepsilon) \subseteq S$. It can be shown that if $S \subseteq \mathbb{R}$, then S is open if and only if its complement $\mathbb{R} \setminus S$ is closed.

Exercise 1.1.24 Justify the above statement. ◁

In mathematics and its applications, the notion of convergence, closed sets and open sets are defined not only for numbers but also for more general objects such as vectors and functions (see, e.g., Nair [8]). We shall also encounter some such situations in the later chapters of this book. ◊

1.2 Series of Real Numbers

In the last section we have come across sequences whose terms involve some other sequences. For example, we had sequences such as (a_n) with

(i) $a_n = \frac{3}{10} + \frac{3}{10^2} + \cdots + \frac{3}{10^n}$,

(ii) $a_n = 1 + \frac{1}{2} + \cdots + \frac{1}{n}$,

(iii) $a_n = 1 + \frac{1}{2^2} + \cdots + \frac{1}{n^2}$.

Recall that, the sequence in (i) and (iii) converge whereas the sequence in (ii) diverge. In (i), we have also seen that the sequence converges to $\frac{1}{3}$. Because of this we may represent the number $\frac{1}{3}$ as

$$\frac{1}{3} = \frac{3}{10} + \frac{3}{10^2} + \cdots .$$

More generally, we may have a sequence (a_n) of real numbers, and we may form a new sequence (s_n) by defining its nth term as sum of the first n terms of (a_n), that is,

$$s_n = a_1 + \cdots + a_n.$$

Then we may enquire whether this new sequence converges or not. In case this sequence (s_n) converge, then we may write its limit as

$$a_1 + a_2 + \cdots .$$

This expression has a special name!

Definition 1.2.1 A **series** of real numbers is an expression of the form

$$a_1 + a_2 + a_3 + \ldots,$$

or more compactly,

$$\sum_{n=1}^{\infty} a_n,$$

where (a_n) is a sequence of real numbers. The number a_n is called the nth **term** of the series and the sum of the first n terms of (a_n), that is,

$$s_n := a_1 + \cdots + a_n,$$

is called the nth **partial sum** of the series.   ◇

Remark 1.2.1 Some authors denote a series as the sequence (a_n, s_n), where s_n is n^{th} partial sum of the sequence (a_n).   ◇

Example 1.2.1 Following are some examples of series:

(i) $\frac{3}{10} + \frac{3}{10^2} + \cdots$,

(ii) $1 + \frac{1}{2} + \frac{1}{3} + \cdots$,

(iii) $1 + \frac{1}{2^2} + \frac{1}{3^2} + \cdots$

Note that in (i), (ii), (iii) above the partial sums are

$$\sum_{k=1}^{n} \frac{3}{10^k}, \quad \sum_{k=1}^{n} \frac{1}{k}, \quad \sum_{k=1}^{n} \frac{1}{k^2},$$

respectively.   ◇

1.2.1 Convergence and Divergence of Series

The following definition is on expected lines (Fig. 1.9):

Definition 1.2.2 A series $\sum_{n=1}^{\infty} a_n$ is said to be a **convergent series** if the corresponding sequence (s_n) of partial sums converges. If $s_n \to s$, then we say that the series $\sum_{n=1}^{\infty} a_n$ **converges** to s, and s is called the **sum** of the series, and we write this fact as

$$s = \sum_{n=1}^{\infty} a_n.$$

Fig. 1.9 $s_n = 1 + \frac{2}{3} + \left(\frac{2}{3}\right)^2 + \cdots + \left(\frac{2}{3}\right)^n$

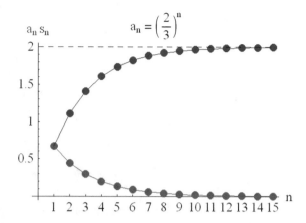

A series which does not converge is called a **divergent series**.   ◊

Observe the following: Suppose $a_n \geq 0$ for all $n \in \mathbb{N}$. Then the sequence (s_n) of partial sums of the series $\sum_{n=1}^{\infty} a_n$ is monotonically increasing. Hence, in this case, either (s_n) converges or $s_n \to \infty$.

Example 1.2.2 Consider the three series given in Example 1.2.1, i.e.,

$$\sum_{n=1}^{\infty} \frac{3}{10^n}, \quad \sum_{n=1}^{\infty} \frac{1}{n}, \quad \sum_{n=1}^{\infty} \frac{1}{n^2}.$$

Note that the partial sums of these series, say $(s_n^{(1)})$, $(s_n^{(2)})$, $(s_n^{(3)})$ with

$$s_n^{(1)} := \sum_{k=1}^{n} \frac{3}{10^k}, \quad s_n^{(2)} := \sum_{k=1}^{n} \frac{1}{k}, \quad s_n^{(3)} := \sum_{k=1}^{n} \frac{1}{k^2},$$

are the sequences considered in Examples 1.1.5, 1.1.19, 1.1.24, respectively, and we have seen that $(s_n^{(1)})$ and $(s_n^{(3)})$ converge, whereas $(s_n^{(2)})$ diverges. Thus, $\sum_{n=1}^{\infty} \frac{3}{10^n}$ and $\sum_{n=1}^{\infty} \frac{1}{n^2}$ are convergent series, whereas $\sum_{n=1}^{\infty} \frac{1}{n}$ is a divergent series.   ◊

Example 1.2.3 Consider the *geometric series*

$$1 + q + q^2 + \cdots$$

for $q \in \mathbb{R}$. We show that this series converges if and only if $|q| < 1$:

Note that

$$s_n = 1 + q + \cdots + q^{n-1} = \begin{cases} n & \text{if } q = 1, \\ (1 - q^n)/(1 - q) & \text{if } q \neq 1. \end{cases}$$

Fig. 1.10
$s_n = 1 + \frac{1}{2} + \frac{1}{2^2} + \cdots + \frac{1}{2^n}$

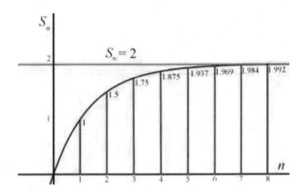

Thus, if $q = 1$, then (s_n) is not bounded; hence not convergent (Fig. 1.10). If $q = -1$, then we have

$$s_n = \begin{cases} 1 & \text{if } n \text{ odd}, \\ 0 & \text{if } n \text{ even}. \end{cases}$$

Thus, (s_n) diverges for $q = -1$ as well. Now, let $|q| \neq 1$. Then

$$s_n = \frac{1 - q^n}{1 - q} = \frac{1}{1 - q} - \frac{q^n}{1 - q}.$$

Recall that (q^n) converges if and only if $|q| < 1$, and in that case $q^n \to 0$. Hence, (s_n) converges if and only if $|q| < 1$, and in that case

$$\lim_{n \to \infty} s_n = \frac{1}{1 - q}.$$

In particular, if $|q| < 1$, then $\sum_{n=1}^{\infty} q^{n-1}$ converges with its sum $\frac{1}{1-q}$, and if $|q| \geq 1$, then $\sum_{n=1}^{\infty} q^{n-1}$ diverges. Further, we have the *error estimate* corresponding to the partial sums as

$$\left| s_n - \frac{1}{1 - q} \right| = \frac{|q|^n}{|1 - q|}. \qquad \Diamond$$

Here is a necessary condition for the convergence of a series, which can help to infer the divergence of certain series.

Theorem 1.2.1 (A necessary condition) *If $\sum_{n=1}^{\infty} a_n$ converges, then $a_n \to 0$ as $n \to \infty$. The converse is not true.*

Proof Suppose the series $\sum_{n=1}^{\infty} a_n$ converges to s, that is, $s_n \to s$, where s_n is the nth partial sum of the series. Since $a_n = s_n - s_{n-1}$ for $n = 2, 3, \ldots$, we have

$$a_n = s_n - s_{n-1} \to 0 \quad \text{as} \quad n \to \infty.$$

Thus, $a_n \to 0$.

To see that the converse does not hold, consider the series $\sum_{n=1}^{\infty} 1/n$. In this case, $a_n = 1/n \to 0$, but the series diverges.   ∎

Exercise 1.2.1 Prove the following.

(i) The series $\sum_{n=1}^{\infty} \frac{n}{n+1}$ diverges.

(ii) If $a_{n+1} > a_n > 0$ for all $n \in \mathbb{N}$, then $\sum_{n=1}^{\infty} a_n$ diverges.   ◁

The following theorem shows that if we remove the first few terms from a convergent (respectively, divergent) series, then the resulting series remain convergent (respectively, divergent).

Theorem 1.2.2 *For each $k \in \mathbb{N}$, the series $\sum_{n=1}^{\infty} a_n$ converges if and only if the series $\sum_{n=1}^{\infty} a_{k+n}$ converges, and if they converge, then*

$$\sum_{n=1}^{\infty} a_n = (a_1 + \cdots + a_k) + \sum_{n=1}^{\infty} a_{k+n}.$$

Proof Let $k \in \mathbb{N}$. Let $s_n = \sum_{i=1}^{n} a_i$ and $s_n' = \sum_{i=1}^{n} a_{k+i}$. Then we have

$$s_n' = s_{k+n} - (a_1 + \cdots + a_k) \quad \forall n \in \mathbb{N}.$$

Recall that a sequence (b_n) converges to b if and only if, for any $k \in \mathbb{N}$, the sequence (b_{k+n}) converges to b. Hence, from the above relation between (s_n) and (s_n') we obtain

$$s_n \to s \iff s_n' \to s - (a_1 + \cdots + a_k).$$

Thus, we have proved the theorem.   ∎

Modifying the arguments in the proof of the above theorem, it follows that, if $\sum_{n=1}^{\infty} b_n$ is obtained from $\sum_{n=1}^{\infty} a_n$ by omitting or adding a finite number of terms, then

$$\sum_{n=1}^{\infty} a_n \text{ converges} \iff \sum_{n=1}^{\infty} b_n \text{ converges}.$$

The following theorem is an immediate corollary to Theorem 1.2.2.

Theorem 1.2.3 *Suppose (a_n) and (b_n) are sequences such that for some $k \in \mathbb{N}$, $a_n = b_n$ for all $n \geq k$. Then $\sum_{n=1}^{\infty} a_n$ converges if and only if $\sum_{n=1}^{\infty} b_n$ converges.*

Theorem 1.2.4 *(Sum-rule) Suppose $\sum_{n=1}^{\infty} a_n$ converges to s and $\sum_{n=1}^{\infty} b_n$ converges to s'. Then for any $\alpha, \beta \in \mathbb{R}$,*

$$\sum_{n=1}^{\infty} (\alpha \, a_n + \beta \, b_n) \text{ converges to } \alpha s + \beta s'.$$

In particular, $\sum_{n=1}^{\infty}(a_n + b_n)$ converges to $s + s'$ and $\sum_{n=1}^{\infty} c\, a_n$ converges to cs for any $c \in \mathbb{R}$.

Proof Let $s_n^{(1)}$, $s_n^{(2)}$ and s_n be the nth partial sums of the series $\sum_{n=1}^{\infty} a_n$, $\sum_{n=1}^{\infty} b_n$ and $\sum_{n=1}^{\infty}(\alpha\, a_n + \beta\, b_n)$ respectively. Then we obtain

$$s_n = \sum_{k=1}^{n}(\alpha\, a_k + \beta\, b_k) = \alpha \sum_{k=1}^{n} a_k + \beta \sum_{k=1}^{n} \beta\, b_k = \alpha s_n^{(1)} + \beta s_n^{(2)} \quad \forall n \in \mathbb{N}.$$

Since $s_n^{(1)} \to s$ and $s_n^{(2)} \to s'$, we obtain $s_n = \alpha s_n^{(1)} + \beta s_n^{(2)} \to \alpha s + \beta s'$. ∎

Exercise 1.2.2 As proof of Theorem 1.2.4, suppose we write

$$\sum_{n=1}^{\infty}(\alpha\, a_n + \beta\, b_n) = \alpha \sum_{n=1}^{\infty} a_n + \beta \sum_{n=1}^{\infty} b_n = \alpha s + \beta s'.$$

What is wrong with it?   ◁

Theorem 1.2.5 *Let (a_n) be such that $a_n \to 0$ and let $s_n := \sum_{k=1}^{n} a_k$ for $n \in \mathbb{N}$. Then for $x \in \mathbb{R}$,*

$$s_n \to x \iff s_{2n} \to x \iff s_{2n-1} \to x.$$

Proof Clearly, if $s_n \to x$ then $s_{2n} \to x$ and $s_{2n-1} \to x$, as (s_{2n}) and (s_{2n-1}) are subsequences of (s_n). Next, we note that

$$s_{2n} = s_{2n-1} + a_{2n} \quad \text{and} \quad s_{2n+1} = s_{2n} + a_{2n+1} \quad \forall n \in \mathbb{N}.$$

Hence, $s_{2n} \to x$ if and only if $s_{2n-1} \to x$, and in that case, by Theorem 1.1.11 , $s_n \to x$. ∎

1.2.2 Some Tests for Convergence

Theorem 1.2.6 *(Comparison test) Let $0 \leq a_n \leq b_n$ for all $n \in \mathbb{N}$. Then*

$$\sum_{n=1}^{\infty} b_n \text{ converges } \Rightarrow \sum_{n=1}^{\infty} a_n \text{ converges} .$$

Proof Suppose s_n and s_n' are the nth partial sums of the series $\sum_{n=1}^{\infty} a_n$ and $\sum_{n=1}^{\infty} b_n$, respectively. By the assumption, we have $0 \leq s_n \leq s_n'$ for all $n \in \mathbb{N}$, and both (s_n) and (s_n') are monotonically increasing.

Suppose $\sum_{n=1}^{\infty} b_n$ converges, that is, (s_n') converges. Then, (s_n') is bounded. Hence, by the relation $0 \leq s_n \leq s_n'$ for all $n \in \mathbb{N}$, (s_n) is bounded as well as monotonically increasing. Therefore, by Theorem 1.1.9, (s_n) converges. ∎

The following corollary is immediate from the above theorem.

Corollary 1.2.7 (Comparison test) *Let $0 \leq a_n \leq b_n$ for all $n \in \mathbb{N}$. Then*

$$\sum_{n=1}^{\infty} a_n \text{ diverges} \quad \Rightarrow \quad \sum_{n=1}^{\infty} b_n \text{ diverges.}$$

Corollary 1.2.8 *Suppose (a_n) and (b_n) are sequences of positive terms.*

(i) Suppose $\ell := \lim\limits_{n \to \infty} \dfrac{a_n}{b_n}$ exists.

    (a) If $\ell > 0$, then $\sum_{n=1}^{\infty} b_n$ converges $\iff \sum_{n=1}^{\infty} a_n$ converges.
    (b) If $\ell = 0$, then $\sum_{n=1}^{\infty} b_n$ converges $\Rightarrow \sum_{n=1}^{\infty} a_n$ converges.

(ii) Suppose $\lim\limits_{n \to \infty} \dfrac{a_n}{b_n} = \infty$. Then $\sum_{n=1}^{\infty} a_n$ converges $\Rightarrow \sum_{n=1}^{\infty} b_n$ converges.

Proof (i) Assume that $\ell := \lim\limits_{n \to \infty} \dfrac{a_n}{b_n}$ exists, i.e., $0 \leq \ell < \infty$.

    (a) Suppose $\ell > 0$. Then for any $0 < \varepsilon < \ell$ there exists $n \in \mathbb{N}$ such that

$$0 \leq \ell - \varepsilon < \frac{a_n}{b_n} < \ell + \varepsilon \quad \forall n \geq N.$$

Thus, $(\ell - \varepsilon)b_n < a_n < (\ell + \varepsilon)b_n$ for all $n \geq N$. Hence, by comparison test (Theorem 1.2.6), $\sum_{n=1}^{\infty} b_n$ converges if and only if $\sum_{n=1}^{\infty} a_n$ converges.

    (b) Suppose $\ell = 0$ and let $\varepsilon > 0$ be given. Then, there exists $n \in \mathbb{N}$ such that $-\varepsilon < \frac{a_n}{b_n} < \varepsilon$ for all $n \geq N$. In particular,

$$a_n < \varepsilon b_n \quad \forall n \geq N.$$

Again, by comparison test (Theorem 1.2.6), convergence of $\sum_{n=1}^{\infty} b_n$ implies the convergence of $\sum_{n=1}^{\infty} a_n$.

    (ii) Assume that $\frac{a_n}{b_n} \to \infty$. Then there exists $N \in \mathbb{N}$ such that $\frac{a_n}{b_n} \geq 1$ for all $n \geq N$, i.e.,

$$b_n \leq a_n \quad \forall n \geq N.$$

Hence, comparison test can be applied in this case as well to obtain the required result. ∎

Example 1.2.4 We have seen that the sequence (s_n) with $s_n = \sum_{k=1}^{n} \frac{1}{k!}$ converges. This also follows from comparison test, since

$$\frac{1}{n!} \leq \frac{1}{2^{n-1}} \quad \forall n \in \mathbb{N}$$

and $\sum_{n=1}^{\infty} \frac{1}{2^{n-1}}$ converges. ◇

Example 1.2.5 Let (a_n) be with $a_n \in \{0, 1, 2, 3, 4, 5, 6, 7, 8, 9\}$, $n \in \mathbb{N}$. The, using the sequence (s_n) with $s_n = \sum_{k=1}^{n} a_k/10^k$, we see that the series $\sum_{n=1}^{\infty} \frac{a_n}{10^n}$ converges. This also follows by comparison test, since

$$\frac{a_n}{10^n} \le \frac{9}{10^n} \quad \forall n \in \mathbb{N}$$

and $\sum_{n=1}^{\infty} \frac{1}{10^n}$ converges. Note that if $s = \sum_{n=1}^{\infty} \frac{a_n}{10^n}$, then the series represents the *decimal expansion* of s.

By similar arguments, we can assert the following (Verify): For any $k \in \mathbb{N}$,

$$a_n \in \{0, 1, \ldots, k-1\},\ n \in \mathbb{N} \quad \Rightarrow \quad \sum_{n=1}^{\infty} \frac{a_n}{k^n} \quad \text{converges.}$$

Suppose $s = \sum_{n=1}^{\infty} \frac{a_n}{k^n}$. If $k = 2$, then the series represents the *binary expansion* of s and if $k = 3$, then the series represents the *ternary expansion* of s.   ◊

Example 1.2.6 (i) Since $\frac{1}{\sqrt{n}} \ge \frac{1}{n}$ for all $n \in \mathbb{N}$, and since the series $\sum_{n=1}^{\infty} \frac{1}{n}$ diverges, by comparison test, the series $\sum_{n=1}^{\infty} \frac{1}{\sqrt{n}}$ also diverges. More generally, let $p \le 1$. Since

$$\frac{1}{n^p} \ge \frac{1}{n} \quad \forall n \in \mathbb{N},$$

by comparison test,

$$\sum_{n=1}^{\infty} \frac{1}{n^p} \quad \text{diverges for} \quad p \le 1.$$

(ii) Let $p \ge 2$. We know that $\sum_{k=1}^{\infty} \frac{1}{n^2}$ converges (Example 1.1.24). Since $\frac{1}{n^p} \le \frac{1}{n^2}$ for all $n \in \mathbb{N}$, by comparison test, the series

$$\sum_{n=1}^{\infty} \frac{1}{n^p} \quad \text{converges for} \quad p \ge 2.$$ ◊

Now, we use comparison test to show the convergence of certain sequences with positive and negative terms.

Example 1.2.7 We show that the series

$$1 - \frac{1}{2} + \frac{1}{3} - \frac{1}{4} + \cdots + \frac{1}{2n-1} - \frac{1}{2n} + \cdots$$

converges. For this let s_n be the n-th partial sum of this series. Then we see that

$$s_{2n+1} = s_{2n} + \frac{1}{2n-1} \quad \forall n \in \mathbb{N},$$

where

$$s_{2n} = \left(1 - \frac{1}{2}\right) + \left(\frac{1}{3} - \frac{1}{4}\right) + \cdots + \left(\frac{1}{2n-1} - \frac{1}{2n}\right)$$
$$= \frac{1}{1 \times 2} + \frac{1}{3 \times 4} + \cdots + \frac{1}{(2n-1) \times 2n}$$
$$\leq \frac{1}{1^2} + \frac{1}{3^2} + \cdots + \frac{1}{(2n-1)^2}$$
$$\leq \sigma_{2n-1},$$

where σ_n is the n-th partial sum of the convergent series $\sum_{n=1}^{\infty}(1/n^2)$ (cf. Example 1.1.24). Hence, by the comparison test (Theorem 1.2.6) (s_{2n}) converges, and hence by Theorem 1.2.5, (s_n) also converges. Thus we have proved that the given series converges.   ◇

Example 1.2.8 The series

$$1 - \frac{1}{3} + \frac{1}{5} - \frac{1}{7} + \cdots + \frac{(-1)^{n+1}}{2n-1} + \cdots, \tag{1.1}$$

and

$$\frac{1}{2} - \frac{1}{4} + \frac{1}{6} - \frac{1}{8} + \cdots + \frac{(-1)^{n+1}}{2n} + \cdots, \tag{1.2}$$

converge: We observe that, if s_n is the n-th partial sum of the series in (1.1), then

$$s_{2n} = 1 - \frac{1}{3} + \frac{1}{5} - \frac{1}{7} + \cdots + \frac{(-1)^{2n+1}}{4n-1}$$
$$= \left(1 - \frac{1}{3}\right) + \left(\frac{1}{5} - \frac{1}{7}\right) + \cdots + \left(\frac{1}{4n-3} - \frac{1}{4n-1}\right)$$
$$= \frac{2}{1 \times 3} + \frac{2}{5 \times 7} + \cdots + \frac{2}{(4n-3) \times (4n-1)}$$
$$\leq \frac{2}{1^2} + \frac{2}{5^2} + \cdots + \frac{2}{(4n-3)^2} \leq \sigma_{4n-3},$$

where σ_n is the n-th partial sum of the convergence series $\sum_{n=1}^{\infty}(1/n^2)$ (cf. Example 1.1.24). Hence, by the comparison test (Theorem 1.2.6) (s_{2n}) converges, and hence by Theorem 1.2.5, (s_n) also converges. Thus we have proved that the given series in (1.1) converges.

To see the convergence of the series in (2), we first observe that this series is same as

$$\frac{1}{2} \sum_{n=1}^{\infty} \frac{(-1)^{n+1}}{n}.$$

We already know that the series $\sum_{n=1}^{\infty} \frac{(-1)^{n+1}}{n}$ converges (cf. Example 1.2.7). Hence $\frac{1}{2}\sum_{n=1}^{\infty} \frac{(-1)^{n+1}}{n}$ also converges.   ◇

In the following example we show the convergence of the series $\sum_{n=1}^{\infty} \frac{1}{n^p}$ for any $p > 1$. We shall use the familiarity of the integral to the extent that the reader knows the result:

$$\int_a^b x^k dx = \left[\frac{x^{k+1}}{k+1}\right]_a^b = \frac{b^{k+1} - a^{k+1}}{k+1} \quad \text{for} \quad k \neq -1.$$

Example 1.2.9 Consider the series $\sum_{n=1}^{\infty} \frac{1}{n^p}$ for $p > 1$. To discuss this general case, consider the function

$$f(x) := 1/x^p, \quad x \geq 1.$$

Note that, for each $k \in \mathbb{N}$, we have

$$k - 1 \leq x \leq k \quad \Rightarrow \quad \frac{1}{k^p} \leq \frac{1}{x^p} \quad \Rightarrow \quad \frac{1}{k^p} \leq \int_{k-1}^k \frac{dx}{x^p}.$$

Hence,

$$\sum_{k=2}^n \frac{1}{k^p} \leq \sum_{k=2}^n \int_{k-1}^k \frac{dx}{x^p} = \int_1^n \frac{dx}{x^p} = \frac{n^{1-p} - 1}{1 - p} \leq \frac{1}{p - 1}.$$

Thus,

$$s_n := \sum_{k=1}^n \frac{1}{k^p} \leq \frac{1}{p - 1} + 1 = \frac{p}{p - 1}.$$

Hence, (s_n) is monotonically increasing and bounded above, and hence (s_n) converges. Thus,

$$\sum_{n=1}^{\infty} \frac{1}{n^p} \quad \text{converges for} \quad p > 1.$$

A more general result on convergence of series in terms of integrals, known as *integral test*, will be considered in Chap. 4.   ◊

Example 1.2.10 In this example, we make use of some simple properties of the *logarithm* function

$$\log x, \quad x > 0,$$

namely, the relation $\log x^k = k \log x$ for $x \geq 1$, for showing the divergence of the series

$$\sum_{n=1}^{\infty} \frac{1}{(n+1) \log(n+1)}.$$

Let $a_n = \frac{1}{n \log n}$ for $n = 2, 3, \ldots$, and $s_n = \sum_{k=2}^{n} a_k$. Then, using the fact that $a_n \geq a_{n+1}$ for all $n \in \mathbb{N}$, we have

$$s_{2^n} = a_2 + a_3 + \cdots + a_{2^n}$$
$$= a_2 + (a_3 + a_4) + (a_5 + a_6 + a_7 + a_8) + \cdots + (a_{2^{n-1}+1} + \cdots a_{2^n})$$
$$\geq a_2 + 2a_4 + 2^2 a_8 + \cdots + 2^{n-1} a_{2^n}.$$

Thus, using the relation

$$a_{2^k} = \frac{1}{2^k \log 2^k} = \frac{1}{2^k k \log 2},$$

we have

$$s_{2^n} \geq \sum_{k=2}^{n} 2^{k-1} a_{2^k} \geq \frac{1}{2 \log 2} \sum_{k=2}^{n} \frac{1}{k}.$$

Since $\sum_{k=2}^{n} \frac{1}{k} \to \infty$, by comparison test, $s_{2^n} \to \infty$. Hence, the sequence (s_n) also diverges. $\diamond$

Another consequence of the comparison test is the following.

Theorem 1.2.9 *Let (a_n) be a sequence of positive numbers. If there exists r with $0 < r < 1$ and a positive integer N such that at least one of the following conditions is satisfied:*

(i) $\frac{a_{n+1}}{a_n} \leq r \quad \forall n \geq N$,

(ii) $a_n^{1/n} \leq r \quad \forall n \geq N$.

Then the series $\sum_{n=1}^{\infty} a_n$ converges.

Proof Let $0 < r < 1$. First, let $N \in \mathbb{N}$ be such that

$$\frac{a_{n+1}}{a_n} \leq r \quad \forall n \geq N.$$

Then $a_{n+1} \leq r a_n$ for all $n \geq N$ so that

$$a_{n+1} \leq r^{n-N+1} a_N = \left(\frac{a_N}{r^{N-1}} \right) r^n \quad \forall n \geq N.$$

Since $\sum_{n=1}^{\infty} r^n$ converges, the comparison test shows that, the series $\sum_{n=1}^{\infty} a_n$ converges.

Next, let $N \in \mathbb{N}$ be such that

$$a_n^{1/n} \leq r \quad \forall n \geq N.$$

Then $a_n \leq r^n$ for all $n \geq N$. Again, since $\sum_{n=1}^{\infty} r^n$ converges, by comparison test, $\sum_{n=1}^{\infty} a_n$ converges. ∎

Sometimes it may be easier to find the limit of either $\left(\dfrac{a_{n+1}}{a_n}\right)$ or $(a_n^{1/n})$ rather than finding an upper bound. In such cases, the following two tests due to d'Alembert[4] and Cauchy, respectively, are useful to decide the convergence of a series.

Theorem 1.2.10 *(D'Alembert's ratio test) Suppose (a_n) is a sequence of positive terms such that* $\lim\limits_{n\to\infty} \dfrac{a_{n+1}}{a_n} = \ell$ *exists.*

(i) If $\ell < 1$, then the series $\sum_{n=1}^{\infty} a_n$ converges.
(ii) If $\ell > 1$, then $a_n \nrightarrow 0$; in particular, $\sum_{n=1}^{\infty} a_n$ diverges.

Proof (i) Suppose $\ell < 1$. Let r be such that $\ell < r < 1$. Then there exists $N \in \mathbb{N}$ such that
$$\frac{a_{n+1}}{a_n} \le r \quad \forall\, n \ge N.$$

Hence, by Theorem 1.2.9, the series $\sum_{n=1}^{\infty} a_n$ converges.
   (ii) Let $\ell > 1$. Then there exists $N \in \mathbb{N}$ such that

$$\frac{a_{n+1}}{a_n} > 1 \quad \forall\, n \ge N.$$

Hence, $a_{n+1} > a_n$ for all $n \ge N$. Therefore, $a_n \nrightarrow 0$.   ∎

Theorem 1.2.11 *(Cauchy's root test) Suppose (a_n) is a sequence of positive terms such that* $\lim\limits_{n\to\infty} a_n^{1/n} = \ell$ *exists.*

(i) If $\ell < 1$, then the series $\sum_{n=1}^{\infty} a_n$ converges.
(ii) If $\ell > 1$, then $a_n \nrightarrow 0$; in particular, $\sum_{n=1}^{\infty} a_n$ diverges.

Proof (i) Suppose $\ell < 1$. Let r be such that $\ell < r < 1$. Then there exists $N \in \mathbb{N}$ such that
$$a_n^{1/n} \le r \quad \forall\, n \ge N.$$

Hence, by Theorem 1.2.9, the series $\sum_{n=1}^{\infty} a_n$ converges.
   (ii) Let $\ell > 1$. Then there exists $N \in \mathbb{N}$ such that

$$a_n^{1/n} > 1 \quad \forall\, n \ge N.$$

Hence, $a_n \ge 1$ for all $n \ge N$. Therefore, $a_n \nrightarrow 0$.   ∎

Remark 1.2.2 We remark that both d'Alembert's ratio test and Cauchy's root test are silent for the case $\ell = 1$. In this case, we cannot assert either way. For example,

---

[4] Jean-Baptiste le Rond d'Alembert (16 November 1717 – 29 October 1783) was a French mathematician, physicist and philosopher. A particular method of solution of wave equation is named after him - courtsey Wikipedia.

we know that the series $\sum_{n=1}^{\infty} \frac{1}{n}$ diverges whereas $\sum_{n=1}^{\infty} \frac{1}{n^2}$ converges, and in both the cases, we have

$$\frac{a_{n+1}}{a_n} \to 1, \quad a_n^{1/n} \to 1.$$

However, for such cases, we may be able to infer the convergence or divergence by some other means. ◇

Example 1.2.11 Let us test the convergence of the series

$$\sum_{n=1}^{\infty} \frac{n^2}{2^n} \quad \text{and} \quad \sum_{n=1}^{\infty} \frac{n!}{2^n}.$$

(i) $\sum_{n=1}^{\infty} \frac{n^2}{2^n}$: In this case $a_n = \frac{n^2}{2^n}$ so that

$$\frac{a_{n+1}}{a_n} = \frac{(n+1)^2/2^{n+1}}{n^2/2^n} = \frac{1}{2}\left(\frac{n+1}{n}\right)^2 \to \frac{1}{2}.$$

Hence, by d'Alembert's ratio test, the series converges. (ii) $\sum_{n=1}^{\infty} \frac{n!}{2^n}$: In this case $a_n = \frac{n!}{2^n}$ so that

$$\frac{a_{n+1}}{a_n} = \frac{(n+1)!/2^{n+1}}{n!/2^n} = \frac{n+1}{2} \to \infty.$$

Hence, by d'Alembert's ratio test, the series diverges. ◇

Example 1.2.12 For $x \in \mathbb{R}$, the series $\sum_{n=1}^{\infty} \frac{|x|^n}{n!}$ converges:

Clearly the series converges if $x = 0$. For $x \neq 0$, let $a_n = \frac{|x|^n}{n!}$. Then we have

$$\frac{a_{n+1}}{a_n} = \frac{|x|}{n+1} \to 0.$$

Hence, by d'Alembert's ratio test, the series converges. We shall also show that the series $\sum_{n=1}^{\infty} \frac{x^n}{n!}$ converges for any $x \in \mathbb{R}$. ◇

Example 1.2.13 The series $\sum_{n=1}^{\infty} \left(\frac{n}{2n+1}\right)^n$ converges: In this case, we have

$$a_n^{1/n} = \frac{n}{2n+1} \to \frac{1}{2} < 1.$$

Hence, by Cauchy's root test, the series converges.

The convergence of the above series can also be proved using comparison test, since

$$a_n = \left(\frac{n}{2n+1}\right)^n = \left(\frac{1}{2+1/n}\right)^n \le \frac{1}{2^n}$$

for all $n \in \mathbb{N}$ and $\sum\limits_{n=1}^{\infty} \dfrac{1}{2^n}$ converges.   ◊

Example 1.2.14 Consider the series $\sum\limits_{n=1}^{\infty} \dfrac{1}{n(n+1)}$. In this case, we have

$$\lim_{n\to\infty} \frac{a_{n+1}}{a_n} = 1 = \lim_{n\to\infty} a_n^{1/n}.$$

Hence, we are not in a position to apply ratio test and root test. However,

$$s_n := \sum_{k=1}^{n} \frac{1}{k(k+1)} = \sum_{k=1}^{n} \left(\frac{1}{k} - \frac{1}{k+1}\right) = 1 - \frac{1}{n+1} \to 1.$$

Thus, the series converges to 1.   ◊

Exercise 1.2.3 Assert the convergence of the series $\sum_{n=1}^{\infty} a_n$ in Example 1.2.11(i) and Example 1.2.13 by showing that

$$\frac{a_{n+1}}{a_n} \le \frac{8}{9} \quad \forall n \ge 3 \quad \text{and} \quad a_n^{1/n} \le \frac{1}{2} \quad \forall n \in \mathbb{N},$$

respectively.   ◁

Exercise 1.2.4 Assert the divergence of the series $\sum_{n=1}^{\infty} a_n$ in Example 1.2.11(ii) by showing either

$$\frac{a_{n+1}}{a_n} \ge 1 \quad \forall n \in \mathbb{N} \quad \text{or} \quad a_n \to \infty.$$ ◁

Exercise 1.2.5 Show the convergence of the series $\sum_{n=1}^{\infty} \frac{1}{n^2}$ by comparing it with $\sum_{n=1}^{\infty} \frac{1}{n(n+1)}$.   ◁

Example 1.2.15 Consider the series $\sum_{n=1}^{\infty} a_n$ with

$$a_{2n-1} = \frac{1}{6^{n-1}}, \quad a_{2n} = \frac{1}{3 \times 6^{n-1}}$$

for $n \in \mathbb{N}$. Note that

$$\frac{a_{2n}}{a_{2n-1}} = \frac{1}{3}, \quad \frac{a_{2n+1}}{a_{2n}} = \frac{1}{2}$$

for all $n \in \mathbb{N}$ so that $\lim\limits_{n \to \infty} \dfrac{a_{n+1}}{a_n}$ does not exist. However, $\frac{a_{n+1}}{a_n} \leq 1/2$ for all $n \in \mathbb{N}$ so that by Theorem 1.2.9, the series $\sum_{n=1}^{\infty} a_n$ converges. $\qquad\qquad\qquad\Diamond$

Example 1.2.16 Consider the series $\sum_{n=1}^{\infty} a_n$ with

$$
a_n = \begin{cases} (\frac{n}{2n+1})^n, & n \text{ odd}, \\[2mm] \frac{1}{3^n}, & n \text{ even}. \end{cases}
$$

Note that

$$
\lim_{n \to \infty} a_{2n+1}^{1/(2n+1)} = \lim_{n \to \infty} \frac{2n+1}{4n+3} = \frac{1}{2}, \quad \lim_{n \to \infty} a_{2n}^{1/2n} = \frac{1}{3}
$$

so that $\lim\limits_{n \to \infty} a_n^{1/n}$ does not exist. However, $a_n^{1/n} \leq 1/2$ for all $n \in \mathbb{N}$. Hence, by Theorem 1.2.9, the series $\sum_{n=1}^{\infty} a_n$ converges. $\qquad\qquad\qquad\Diamond$

We close this subsection with another application of the comparison test (Theorem 1.2.6).

Theorem 1.2.12 *Let (a_n) be a sequence of non-negative numbers, and let (b_n) be a sequence obtained from (a_n) by a rearrangement of the terms of (a_n). Then*

$$
\sum_{n=1}^{\infty} a_n \text{ converges} \iff \sum_{n=1}^{\infty} b_n \text{ converges},
$$

and in that case, both the series have the same limit.

Proof Suppose s_n and t_n are the n-th partial sums of $\sum_{n=1}^{\infty} a_n$ and $\sum_{n=1}^{\infty} b_n$, respectively, that is,

$$
s_n = \sum_{i=1}^{n} a_i, \quad t_n = \sum_{i=1}^{n} b_i
$$

for all $n \in \mathbb{N}$. Suppose $s_n \to s$. Since (b_n) is a rearrangement of the terms of (a_n), for each $n \in \mathbb{N}$, there exists $m \geq n$ such that $t_n \leq s_m$. Then we have $t_n \leq s$. This shows that (t_n) converges, say to t, and we obtain $t \leq s$.

Similarly, changing the roles of (s_n) and (t_n), we see that convergence of (t_n) to some t implies (s_n) converges to some s and $s \leq t$. Thus, we obtain $s = t$, and this completes the proof. $\qquad\qquad\qquad\blacksquare$

1.2.3 Alternating Series

In the last subsection we have described some tests for asserting the convergence or divergence of series of non-negative terms. In this subsection we provide a sufficient condition for convergence of series with alternatively positive and negative terms.

Definition 1.2.3   A series of the form $\sum_{n=1}^{\infty}(-1)^{n+1}u_n$, where (u_n) is a sequence of positive terms, is called an **alternating series**.   ◊

We have seen in Example 1.2.7 that the alternating series

$$1 - \frac{1}{2} + \frac{1}{3} - \frac{1}{4} + \frac{1}{5} + \cdots + \frac{(-1)^{n+1}}{n} + \cdots \tag{1.1}$$

is convergent. Note that the series

$$1 + \frac{1}{2} - \frac{1}{3} - \frac{1}{4} + \cdots + \frac{1}{4n-3} + \frac{1}{4n-2} - \frac{1}{4n-1} - \frac{1}{4n} + \cdots \tag{1.2}$$

is not an alternating series. As you can see, in the latter case, though the series in not an alternating series, it is of the form

$$\sum_{n=1}^{\infty}[(-1)^{n+1}u_n + (-1)^{n+1}v_n]$$

with

$$u_n = \frac{1}{2n-1}, \quad v_n = \frac{1}{2n}.$$

We know from the results in Example 1.2.8 that the alternating series $\sum_{n=1}^{\infty}(-1)^{n+1}u_n$ and $\sum_{n=1}^{\infty}(-1)^{n+1}v_n$ are convergent. Hence, we can assert the convergence of the original series $\sum_{n=1}^{\infty}[(-1)^{n+1}u_n + (-1)^{n+1}v_n]$.

The next theorem, due to Leibnitz,[5] provides such a sufficient condition for the convergence of alternating series.

Theorem 1.2.13   (Leibnitz's theorem) *Suppose (u_n) is a sequence of positive terms such that $u_n \geq u_{n+1}$ for all $n \in \mathbb{N}$ and $u_n \to 0$ as $n \to \infty$. Then the alternating series $\sum_{n=1}^{\infty}(-1)^{n+1}u_n$ converges, and in that case,*

$$|s - s_n| \leq u_{n+1} \quad \forall\, n \in \mathbb{N},$$

where

$$s_n = \sum_{j=1}^{n}(-1)^{j+1}u_j \quad \text{and} \quad s = \sum_{n=1}^{\infty}(-1)^{n+1}u_n,$$

Proof   We observe that

$$s_{2n+1} = s_{2n} + u_{2n+1} \quad \forall\, n \in \mathbb{N}.$$

---

[5] Gottfried Wilhelm von Leibnitz (July 1, 1646 – November 14, 1716) was a German mathematician and philosopher. He developed the infinitesimal calculus independently of Isaac Newton - courtsey Wikipedia.

Since $u_n \to 0$ as $n \to \infty$, it is enough to show that (s_{2n}) converges, and in that case, both (s_{2n}) and (s_{2n+1}) converge to the same limit. Note that, for every $n \in \mathbb{N}$,

$$s_{2n+2} = s_{2n} + (u_{2n+1} - u_{2n+2}) \geq s_{2n}$$

and

$$s_{2n} = u_1 - (u_2 - u_3) - \ldots (u_{2n-2} - u_{2n-1}) - u_{2n} \leq u_1.$$

Hence, (s_{2n}) is monotonically increasing and bounded above. Therefore (s_{2n}) converges. Thus, the series $\sum_{n=1}^{\infty} (-1)^{n+1} u_n$ converges.

To obtain the remaining part, note also that, for $n \in \mathbb{N}$,

$$s_{2n+1} = s_{2n-1} - (u_{2n} - u_{2n+1}) \leq s_{2n-1},$$

so that $\{s_{2n-1}\}$ is a monotonically decreasing sequence. Thus,

$$s_{2n-1} = s_{2n} + u_{2n} \leq s + u_{2n}, \qquad s \leq s_{2n+1} = s_{2n} + u_{2n+1},$$

$$s \leq s_{2n-1}, \quad s_{2n} \leq s.$$

Hence,

$$s_{2n-1} - s \leq u_{2n}, \qquad s - s_{2n} \leq u_{2n+1},$$

Consequently,

$$|s - s_n| \leq u_{n+1} \quad \forall \, n \in \mathbb{N}.$$

This completes the proof. ∎

Corollary 1.2.14 *Suppose (a_n) is a sequence of positive terms such that $a_n \geq a_{n+1}$ for all $n \in \mathbb{N}$ and $a_n \to 0$ as $n \to \infty$. Then the series*

$$a_1 + a_2 - a_3 - a_4 + \cdots + a_{4n-3} + a_{4n-2} - a_{4n-1} - a_{4n} + \cdots$$

converges.

Proof The given series can be written as

$$\sum_{n=1}^{\infty} [(-1)^{n+1} u_n + (-1)^{n+1} v_n],$$

where (u_n) and (v_n) are decreasing sequences of positive terms such that $u_n \to 0$ and $v_n \to 0$. Hence, by Theorem 1.2.13 can be applied to obtain the conclusion. ∎

Remark 1.2.3 The relation $|s - s_n| \leq u_{n+1}$ in Theorem 1.2.13 shows the rate of convergence of the partial sums to the sum of the series. In particular, for a given

$k \in \mathbb{N}$, if n is large enough such that $u_n < 1/10^k$, then first k decimal places of s_n and s are the same. $\Diamond$

The following example shows that the condition $u_n \geq u_{n+1}$ for all $n \in \mathbb{N}$ in Theorem 1.2.13 is not redundant, that is, the conclusion in the theorem need not hold if the above condition is dropped.

Example 1.2.17 Consider the series

$$\frac{1}{\sqrt{3}-1} - \frac{1}{\sqrt{3}+1} + \frac{1}{\sqrt{4}-1} - \frac{1}{\sqrt{4}+1} + \cdots + \frac{1}{\sqrt{n}-1} - \frac{1}{\sqrt{n}+1} + \cdots$$

which is of the form $\sum_{n=1}^{\infty} (-1)^{n+1} u_n$ with

$$u_{2n-1} = \frac{1}{\sqrt{n+2}-1}, \quad u_{2n} = \frac{1}{\sqrt{n+2}+1}.$$

Note that, in this case, we have $u_n \geq 0$ for all $n \in \mathbb{N}$ and $u_n \to 0$. However, the series diverges. To see this, we observe that

$$s_{2n} = \sum_{k=1}^{n} \left(\frac{1}{\sqrt{n+2}-1} - \frac{1}{\sqrt{n+2}+1} \right) = \sum_{k=1}^{n} \frac{2}{n+1} \to \infty.$$

Thus, (s_n) has a subsequence that diverges. Therefore, (s_n) diverges. Note that (u_n) is not a decreasing sequence. $\Diamond$

1.2.4   Madhava-Nilakantha Series

The convergence of the series

$$1 - \frac{1}{3} + \frac{1}{5} - \frac{1}{7} + \cdots + \frac{(-1)^n}{2n+1} + \cdots,$$

which is generally known as *Leibnitz-Gregory series*, was known to Indian mathematicians as early as in 15-th century, and the value of the above series was proved to be $\frac{\pi}{4}$. The above series appeared in the work of a Kerala mathematician *Madhava* around 1425 that was presented later in the year around 1550 by another Kerala mathematician *Nilakantha* (cf. [7]). The discovery of the above series is normally attributed to Leibnitz and James Gregory after nearly 300 years of its discovery. Respecting the chronology of its discovery, we shall refer this series as *Madhava-Nilakantha series*.

Let us give a simple proof for the equality

$$\frac{\pi}{4} = 1 - \frac{1}{3} + \frac{1}{5} - \frac{1}{7} + \cdots + \frac{(-1)^{n+1}}{2n-1} + \cdots$$

using some elementary rules of integration that one studies in school, which we shall study in detail in Chap. 6.

We know that $(1-r)(1+r++\cdots+r^n) = (1-r^{n+1})$ so that for $r \neq 1$,

$$\frac{1}{1-r} = 1 + r + \cdots + r^n + \frac{r^{n+1}}{1-r}.$$

Now, taking $r = -x^2$ we have

$$\frac{1}{1+x^2} = 1 - x^2 + x^4 - x^6 + \cdots + (-1)^n x^{2n} + (-1)^{n+1} \frac{x^{2n+2}}{1+x^2}.$$

On integration

$$\int \frac{dx}{1+x^2} = x - \frac{x^3}{3} + \frac{x^5}{5} - \frac{x^7}{7} + \cdots + (-1)^n \frac{x^{2n+1}}{2n+1} + \int \frac{(-1)^{n+1} x^{2n+2}}{1+x^2} dx.$$

Now, recalling

$$\int_0^1 \frac{dx}{1+x^2} = \tan^{-1}(1) = \frac{\pi}{4},$$

we have

$$\frac{\pi}{4} = 1 - \frac{1}{3} + \frac{1}{5} - \frac{1}{7} + \cdots + (-1)^n \frac{1}{2n+1} + \int_0^1 \frac{(-1)^{n+1} x^{2n+1}}{1+x^2} dx.$$

Now, observe that

$$\left| \int_0^1 (-1)^{n+1} \frac{x^{2n+2}}{1+x^2} dx \right| \leq \int_0^1 \frac{x^{2n+2}}{1+x^2} dx \leq \int_0^1 x^{2n+2} dx = \frac{1}{2n+3}.$$

Thus,

$$\left| \frac{\pi}{4} - \left(1 - \frac{1}{3} + \frac{1}{5} - \frac{1}{7} + \cdots + (-1)^n \frac{1}{2n+1} \right) \right| \leq \frac{1}{2n+3} \to 0$$

Thus, we have proved that

$$\frac{\pi}{4} = 1 + \sum_{n=1}^{\infty} \frac{(-1)^n}{2n+1}.$$

Using the procedure used above and the fact that $\int_0^1 \frac{dx}{1+x} = \log 2$, we see that

$$\log 2 = \sum_{n=1}^{\infty} \frac{(-1)^{n+1}}{n}.$$

Exercise 1.2.6 Derive the above series representation for log 2.   ◁

1.2.5   Absolute Convergence

We know that for a sequence (a_n), the series $\sum_{n=1}^{\infty} a_n$ may converge, but $\sum_{n=1}^{\infty} |a_n|$ can diverge. For example we have seen that $\sum_{n=1}^{\infty} \frac{(-1)^{n+1}}{n}$ converges whereas $\sum_{n=1}^{\infty} \frac{1}{n}$ diverges.

Definition 1.2.4 Let (a_n) be a sequence of real numbers. Then the series $\sum_{n=1}^{\infty} a_n$ is said to be

(1) **absolutely convergent**, if $\sum_{n=1}^{\infty} |a_n|$ is convergent,
(2) **conditionally convergent**, if it converges, but not absolutely.   ◊

Example 1.2.18 (i) the series

$$\sum_{n=1}^{\infty} \frac{(-1)^n}{n^2}, \quad \sum_{n=1}^{\infty} \frac{(-1)^{n+1}}{n!}, \quad \sum_{n=1}^{\infty} \frac{\sin n}{n^2}$$

are absolutely convergent, so also the series

$$\sum_{n=1}^{\infty} \frac{a^n}{n!}$$

for any $a \in \mathbb{R}$. **(ii)** We already observed in Sect. 1.2.4 that the series

$$\sum_{n=1}^{\infty} \frac{(-1)^{n+1}}{n} \quad \text{and} \quad \sum_{n=1}^{\infty} \frac{(-1)^n}{2n-1}$$

are convergent series, which also from Leibnitz theorem, but they are not absolutely convergent. Thus, these series are conditionally convergent.   ◊

From the above observations we conclude:

A convergent series need not converge absolutely.

However, we have the following theorem.

Theorem 1.2.15 *Every absolutely convergent series is convergent.*

Proof Suppose $\sum_{n=1}^{\infty} a_n$ is an absolutely convergent series. Let s_n and s_n' be the nth partial sums of the series $\sum_{n=1}^{\infty} a_n$ and $\sum_{n=1}^{\infty} |a_n|$ respectively. Then, for $n > m$, we have

$$|s_n - s_m| = \left| \sum_{j=m+1}^{n} a_n \right| \leq \sum_{j=m+1}^{n} |a_n| = |s_n' - s_m'|.$$

Since, (s_n') converges, it is a Cauchy sequence. Hence, from the above relation it follows that (s_n) is also a Cauchy sequence. Therefore, by the *Cauchy criterion*, it converges. ∎

Another proof without using Cauchy criterion. Suppose $\sum_{n=1}^{\infty} a_n$ is an absolutely convergent series. For each $n \in \mathbb{N}$, let us write a_n as difference of two positive numbers, namely,

$$a_n = (a_n + |a_n|) - |a_n|.$$

Since $\sum_{n=1}^{\infty} |a_n|$ converges, by the sum rule (Theorem 1.2.4), it is enough to show that $\sum_{n=1}^{\infty} (a_n + |a_n|)$ converges. This is true, since

$$0 \leq a_n + |a_n| \leq 2|a_n|$$

for all $n \in \mathbb{N}$ and $\sum_{n=1}^{\infty} 2|a_n|$ converges, by applying the comparison test for series of non-negative terms (Theorem 1.2.6). ∎

In view of Theorem 1.2.15, the series

$$\sum_{n=1}^{\infty} \frac{x^n}{n!} \text{ converges for each } x \in \mathbb{R}.$$

Remark 1.2.4 We shall prove in the next chapter that the value of the above sum is same as

$$\lim_{n \to \infty} \left(1 + \frac{x}{n} \right)^n,$$

and it is generally denoted by $\exp(x)$ or e^x. ◊

In view of Theorem 1.2.15 we can assert the following.

Let (a_n) be a sequence of nonzero numbers

(i) If $\lim_{n \to \infty} \left| \dfrac{a_{n+1}}{a_n} \right| = \ell$ exists and $\ell < 1$, then $\sum_{n=1}^{\infty} a_n x^n$ converges absolutely

(ii) If $\lim_{n \to \infty} |a_n|^{1/n} = \ell$ exists and $\ell < 1$, then $\sum_{n=1}^{\infty} a_n x^n$ converges absolutely.

Definition 1.2.5 By a **rearrangement** of a series $\sum_{n=1}^{\infty} a_n$ we mean a series $\sum_{n=1}^{\infty} b_n$ obtained by *rearranging the terms* of $\sum_{n=1}^{\infty} a_n$, in the sense that $b_n = a_{\varphi(n)}$, $n \in \mathbb{N}$, where φ is a bijection on $\mathbb{N}$. ◇

Recall from Theorem 1.2.12 that if $\sum_{n=1}^{\infty} a_n$ is a series of non-negative terms, then this series converges if and only if any rearrangement of it converges and all rearranged series have the same sum. However, rearrangements of a conditionally convergent series need not converge to the same limit, as the following example shows.

Example 1.2.19 Consider the series $\sum_{n=1}^{\infty} a_n$ with

$$a_n := \frac{(-1)^{n+1}}{n}, \quad n \in \mathbb{N}.$$

We know that this series converges to $s := \log 2$ (Sect. 1.2.4). Let us see if a rearrangement of it converges to a different number. For this, let us consider the following rearrangement:

$$1 - \frac{1}{2} - \frac{1}{4} + \frac{1}{3} - \frac{1}{6} - \frac{1}{8} + \cdots + \frac{1}{2n-1} - \frac{1}{4n-2} - \frac{1}{4n} + \cdots.$$

Thus, the rearranged series is written as $\sum_{n=1}^{\infty} b_n$, with

$$b_{3n-2} = \frac{1}{2n-1}, \quad b_{3n-1} = -\frac{1}{4n-2}, \quad b_{3n} = -\frac{1}{4n}$$

for $n \in \mathbb{N}$. Let s_n and s_n' be the nth partial sums of the series $\sum_{n=1}^{\infty} a_n$ and $\sum_{n=1}^{\infty} b_n$ respectively. Then we have

$$
\begin{aligned}
s_{3n}' &= 1 - \frac{1}{2} - \frac{1}{4} + \frac{1}{3} - \frac{1}{6} - \frac{1}{8} + \cdots + \frac{1}{2n-1} - \frac{1}{4n-2} - \frac{1}{4n} \\
&= \left(1 - \frac{1}{2} - \frac{1}{4}\right) + \left(\frac{1}{3} - \frac{1}{6} - \frac{1}{8}\right) + \cdots + \left(\frac{1}{2n-1} - \frac{1}{4n-2} - \frac{1}{4n}\right) \\
&= \left(\frac{1}{2} - \frac{1}{4}\right) + \left(\frac{1}{6} - \frac{1}{8}\right) + \cdots + \left(\frac{1}{4n-2} - \frac{1}{4n}\right) \\
&= \frac{1}{2}\left[\left(1 - \frac{1}{2}\right) + \left(\frac{1}{3} - \frac{1}{4}\right) + \cdots + \left(\frac{1}{2n-1} - \frac{1}{2n}\right)\right] \\
&= \frac{1}{2}\left[1 - \frac{1}{2} + \frac{1}{3} - \frac{1}{4} + \cdots + \frac{1}{2n-1} - \frac{1}{2n}\right] \\
&= \frac{1}{2} s_{2n}.
\end{aligned}
$$

Also, we have

$$s'_{3n+1} = s'_{3n} + \frac{1}{2n+1}, \qquad s'_{3n+2} = s'_{3n} + \frac{1}{2n+1} - \frac{1}{4n+2}.$$

Since, $a_n \to 0$ as $n \to \infty$, we obtain

$$\lim_{n\to\infty} s'_{3n} = \frac{s}{2}, \qquad \lim_{n\to\infty} s'_{3n+1} = \frac{s}{2}, \qquad \lim_{n\to\infty} s'_{3n+2} = \frac{s}{2}.$$

Hence, we can infer that $s'_n \to s/2$ as $n \to \infty$. ◊

Not only that. Look at the following astonishing result!

Theorem 1.2.16 *Suppose $\sum_{n=1}^{\infty} a_n$ is a conditionally convergent series. Then, for each $s \in \mathbb{R}$, there exists a rearrangement of $\sum_{n=1}^{\infty} a_n$ such that the rearranged series converges to s.*

The proof of the above two theorem (Theorems 1.2.16 is quite involved, and hence we omit its proof. Interested readers may refer Delninger [4] for the proof.

1.3 Additional Exercises

1.3.1 Sequences

1. Let (a_n) be a sequence of real numbers which converges to a, i.e., $a_n \to a$ as $n \to \infty$. Prove:

   (a) $|a_n - a| \to 0$ and $|a_n| \to |a|$ as $n \to \infty$.
   (b) There exists $k \in \mathbb{N}$ such that $|a_n| > |a|/2$ for all $n \geq k$.
   (c) If $a \neq 0$, then $a_n \neq 0$ for all large enough n and $1/a_n \to 1/a$.

2. Prove that

   (a) $a_n \to a$ if and only if for every open interval I containing a, there exists a positive integer N (depending on I) such that $a_n \in I$ for all $n \geq N$.
   (b) $a_n \not\to a$ if and only if there exists an open interval I containing a such that infinitely may a_n's are not in I.

3. In each of the following, establish the convergence or divergence of the sequence (a_n), where a_n is:

$$\text{(i)} \quad \frac{(-1)^n}{n+1} \qquad \text{(ii)} \quad \frac{2n}{3n^2+1}, \qquad \text{(iii)} \quad \frac{2n^2+3}{3n^2+1}.$$

4. Suppose $a_n \to a$ and $a_n \geq 0$ for all $n \in \mathbb{N}$. Show that $a \geq 0$ and $\sqrt{a_n} \to \sqrt{a}$.

5. Let $0 < a < 1$. If $b_n > 0$ for all $n \in \mathbb{N}$ and $\frac{b_{n+1}}{b_n} \to \ell$ with $0 \le \ell < 1/a$, then show that $b_n a^n \to 0$.

6. Let (a_n) be a sequence defined recursively by $a_{n+2} = a_{n+1} + a_n$ for $n \in \mathbb{N}$ with $a_1 = a_2 = 1$. Show that $a_n \to \infty$.

7. For $n \in \mathbb{N}$, let $a_n = \sqrt{n+1} - \sqrt{n}$. Show that (a_n) and $(\sqrt{n}a_n)$ are convergent sequences. Find their limits.

8. For $n \in \mathbb{N}$, let $x_n = \sum_{k=1}^{n} \frac{1}{n+k}$. Show that (x_n) is convergent.

9. Prove the following:

   (a) If (a_n) is increasing and unbounded, then $a_n \to +\infty$.
   (b) If (a_n) is decreasing and unbounded, then $a_n \to -\infty$.

10. If every subsequence of (a_n) has at least one subsequence which converges to x, then (a_n) also converges to x.

11. Suppose (a_n) is an increasing sequence. Prove the following.

    (a) If (a_n) has a bounded subsequence, then (a_n) is convergent.
    (b) If (a_n) does not diverge to $+\infty$, then (a_n) has a subsequence which is bounded above.
    (c) If (a_n) is divergent, then $a_n \to +\infty$.

12. If (a_n) is a sequence with positive terms, prove that

$$a_n \to 0 \iff \frac{a_n}{1 + a_n} \to 0.$$

13. Let $a_1 = 1$ and $a_{n+1} = \sqrt{2 + a_n}$ for all $n \in \mathbb{N}$. Show that (a_n) converges. Also, find its limit.

14. Let $a_1 = 1$ and $a_{n+1} = \frac{1}{4}(2a_n + 3)$ for all $n \in \mathbb{N}$. Show that (a_n) is monotonically increasing and bounded above. Find its limit.

15. Let $a_1 = 1$ and $a_{n+1} = \dfrac{a_n}{1 + a_n}$ for all $n \in \mathbb{N}$. Show that (a_n) converges. Find its limit.

16. Suppose (a_n) is a sequence such that the subsequences (a_{2n-1}) and (a_{2n}) converge to the same limit, say a. Show that (a_n) also converges to a.

17. Let (a_n) be a monotonically increasing sequence such that (a_{3n}) is bounded. Is (a_n) convergent? Why?

18. Let (a_n) be defined by

$$a_1 = 0, \quad a_{n+1} = a_n + \frac{1}{4}(1 - 2a_n^2), \quad n = 1, 2, \ldots.$$

Show that (a_n) converges to $1/\sqrt{2}$.

19. Give an example in support of the statement: If (a_n) is a sequence such that $a_{n+1} - a_n \to 0$ as $n \to \infty$, then (a_n) need not converge.

20. For $a > 0$, let

$$b_n = \frac{a^n - 1}{a^n + 1}, \quad n \in \mathbb{N}.$$

Find values of a for which (b_n) converges, and also find the limits for such values.

21. For $0 < a < b$, let

$$a_n = (a^n + b^n)^{1/n}, \quad n \in \mathbb{N}.$$

Show that (a_n) converges to b.

[*Hint:* Note that $(a^n + b^n)^{1/n} = b(1 + \left(\frac{a}{b}\right)^n)^{1/n}.$]

22. For $0 < a < b$, let

$$a_{n+1} = (a_n b_n)^{1/2} \quad \text{and} \quad b_{n+1} = \frac{a_n + b_n}{2}, \quad n \in \mathbb{N}$$

with $a_1 = a$, $b_1 = b$. Show that (a_n) and (b_n) converge to the same limit.

[*Hint:* First observe that $a_n \le a_{n+1} \le b_{n+1} \le b_n$ for all $n \in \mathbb{N}$.]

23. Let $a_1 = 1/2$ and $b = 1$, and for $n \in \mathbb{N}$, let

$$a_{n+1} = (a_n b_n)^{1/2}, \quad b_{n+1} = \frac{2a_n b_n}{a_n + b_n}.$$

Show that (a_n) and (b_n) converge to the same limit.

[*Hint:* First observe that $b_n \le b_{n+1} \le a_{n+1} \le a_n$ for all $n \in \mathbb{N}$.]

24. Let $D \subseteq S \subseteq \mathbb{R}$, and let $a \in S$. Show that the following statements are equivalent:

   (a) Every open interval containing a contains at least one point from D.
   (b) There exists a sequence (a_n) in D such that $a_n \to a$.
   [If every $a \in D$ has the above listed property, then D is said to be *dense* in S.]

25. Show that given any $a \in \mathbb{R}$, there exists a sequence (a_n) of rational numbers such that $a_n \to a$; in other words, the set of rational numbers is dense in $\mathbb{R}$.

   [*Hint:* Use Archimedean Property.]

1.3.2 Series

1. Suppose (a_n) and (b_n) are sequences of positive numbers. Prove that, if (a_n) is bounded and $\sum_{n=1}^{\infty} b_n$ is convergent, then $\sum_{n=1}^{\infty} a_n b_n$ convergent.

2. Let (a_n) and (b_n) are sequences such that $\sum_{n=1}^{\infty} a_n^2$ and $\sum_{n=1}^{\infty} b_n^2$ are convergent series. Show that $\sum_{n=1}^{\infty} |a_n b_n|$ converges.

   [*Hint:* Observe $2ab \le a^2 + b^2$.]

3. In each of the following check whether the statement or its converse is true. Justify the answers.

   (a) Convergence of $\sum_{n=1}^{\infty} a_n$ imply the convergence of $\sum_{n=1}^{\infty} a_n^2$.
   (b) Convergence of $\sum_{n=1}^{\infty} a_n$ and $\sum_{n=1}^{\infty} b_n$ imply the convergence of $\sum_{n=1}^{\infty} a_n b_n$.

(c) Divergence of $\sum_{n=1}^{\infty} a_n$ and $\sum_{n=1}^{\infty} b_n$ imply the divergence of $\sum_{n=1}^{\infty} a_n b_n$.

4. Test the following series for convergence

(a) $\dfrac{1}{2} + \dfrac{2}{2^2} + \dfrac{3}{2^3} + \cdots + \dfrac{n}{2^n} + \cdots$

(b) $\dfrac{1}{\sqrt{10}} + \dfrac{1}{\sqrt{20}} + \dfrac{1}{\sqrt{30}} + \cdots + \dfrac{1}{\sqrt{10n}} + \cdots$

(c) $\dfrac{1}{\sqrt[3]{7}} + \dfrac{1}{\sqrt[3]{8}} + \dfrac{1}{\sqrt[3]{9}} + \cdots + \dfrac{1}{\sqrt[3]{n+6}} + \cdots$

(d) $\dfrac{1}{2} + \dfrac{2}{5} + \dfrac{3}{10} + \cdots + \dfrac{n}{n^2+1} + \cdots$

(e) $\dfrac{1}{2} + \dfrac{2}{2^2} + \dfrac{3}{2^3} + \cdots + \dfrac{n}{2^n} + \cdots$

(f) $\dfrac{1}{\sqrt{10}} + \dfrac{1}{\sqrt{20}} + \dfrac{1}{\sqrt{30}} + \cdots + \dfrac{1}{\sqrt{10n}} + \cdots$

(g) $\dfrac{1}{\sqrt[3]{7}} + \dfrac{1}{\sqrt[3]{8}} + \dfrac{1}{\sqrt[3]{9}} + \cdots + \dfrac{1}{\sqrt[3]{n+6}} + \cdots$

5. Test the following series for convergence:

(a) $\displaystyle\sum_{n=1}^{\infty} \dfrac{(n!)^2}{(2n)!}$

(b) $\displaystyle\sum_{n=1}^{\infty} \dfrac{(n!)^2}{(2n)!} 5^n$

(c) $\displaystyle\sum_{n=1}^{\infty} \dfrac{(-2)^n}{n^2}$

(d) $\displaystyle\sum_{n=1}^{\infty} \dfrac{\sqrt{2n!}}{n!}$

(e) $\displaystyle\sum_{n=1}^{\infty} \dfrac{(-1)^{(n-1)}}{\sqrt{n}}$

(f) $\displaystyle\sum_{n=1}^{\infty} \dfrac{1}{(n+1)\sqrt{n}}$

(g) $\displaystyle\sum_{n=1}^{\infty} \dfrac{(-1)^{n-1}}{\sqrt{n(n+1)(n+2)}}$

(h) $\displaystyle\sum_{n=1}^{\infty} \dfrac{1}{n}\left(\sqrt{n+1} - \sqrt{n}\right)$

(i) $\displaystyle\sum_{n=1}^{\infty} \left(\sqrt{n^4+1} - \sqrt{n^4-1}\right)$

(j) $\displaystyle\sum_{n=1}^{\infty} \left(1 + \dfrac{1}{n}\right)^{n^2}$

(k) $\displaystyle\sum_{n=1}^{\infty} \left(\sqrt[3]{n^3+1} - n\right)$

(l) $\displaystyle\sum_{n=1}^{\infty} (-1)^{n+1} \left(\sqrt{n+1} - \sqrt{n}\right)$

6. Find the sum of the series:

$$\dfrac{1}{1.2.3} + \dfrac{1}{2.3.4} + \cdots + \dfrac{1}{n.(n+1).(n+2)} + \cdots$$

7. Find out whether the following series converge absolutely or conditionally:

   (a) $\displaystyle\sum_{n=1}^{\infty}(-1)^{n+1}\frac{1}{\log(n+1)}$

   (b) $\displaystyle\sum_{n=1}^{\infty}\frac{(-1)^{n+1}}{(2n-1)^2}$,

   (c) $\displaystyle\sum_{n=1}^{\infty}\frac{(-1)^{n+1}n}{3^{n-1}}$,

   (d) $\displaystyle\sum_{n=1}^{\infty}\frac{(-1)^n}{n(\log n)^2}$,

   (e) $\displaystyle\sum_{n=1}^{\infty}\frac{(-1)^n\log n}{n\log\log n}$,

   (f) $\displaystyle\sum_{n=1}^{\infty}(-1)^{n+1}\frac{1}{n2^n}$

8. Let (a_n) be a sequence of non-negative numbers and (a_{k_n}) be a subsequence (a_n). Show that, if $\sum_{n=1}^{\infty} a_n$ converges, then $\sum_{n=1}^{\infty} a_{k_n}$ also converges. Is the converse true? Why?

9. Let (a_n) be a sequence of nonzero numbers. If $\displaystyle\lim_{n\to\infty}\left|\frac{a_{n+1}}{a_n}\right| = \ell$ exists, then show that

   (a) $\sum_{n=1}^{\infty} a_n x^n$ converges absolutely if $|x| < 1/\ell$,
   (b) $\sum_{n=1}^{\infty} a_n x^n$ diverges if $|x| > 1/\ell$.

10. Find the set of all points x such that the following series convergent/absolutely convergent:

   (i) $\displaystyle\sum_{n=1}^{\infty}\frac{x^n}{n}$,

   (ii) $\displaystyle\sum_{n=1}^{\infty}\frac{x^n}{n^2}$,

   (iii) $\displaystyle\sum_{n=1}^{\infty}\frac{x^n}{n!}$

Chapter 2
Limit, Continuity and Differentiability of Functions

In this chapter, we deal with the concepts of limit, continuity and differentiability of functions. Apart from establishing some of the basic properties related to these concepts, we shall also consider many of its applications to the study of qualitative nature of functions such as increasing or decreasing, and determination of points at which they attain maxima and minima. Also, the issue of approximating a function by a polynomial is considered.

2.1 Limit of a Function

Suppose f is a real valued function defined on a subset D of $\mathbb{R}$. We are going to define the *limit* of $f(x)$ as $x \in D$ *approaches* a point a which is not necessarily in D.

In the last chapter we have considered the notion of convergence of a sequence of numbers to a particular number so that the terms of the sequence *approaches* that number. Now, the question:

What do we mean by saying that points of a set $D \subseteq \mathbb{R}$ approaches a particular point $a \in \mathbb{R}$?

2.1.1 Limit Point of a Set

By saying the points of a set $D \subseteq \mathbb{R}$ approaches a particular point $a \in \mathbb{R}$, we shall mean that a is a *limit point* of the set D, in the following sense.

Definition 2.1.1 Let $D \subseteq \mathbb{R}$ and $a \in \mathbb{R}$. A point $a \in \mathbb{R}$ is said to be a **limit point** of D if every open interval containing the point a contains at least one point from D other than a. ◊

Thus, $a \in \mathbb{R}$ is a limit point of D if and only if for any $\delta > 0$,

$$D \cap \{x \in \mathbb{R} : 0 < |x - a| < \delta\} \neq \emptyset.$$

© The Author(s), under exclusive license to Springer Nature Switzerland AG 2021
M. T. Nair, *Calculus of One Variable*,
https://doi.org/10.1007/978-3-030-88637-0_2

Remark 2.1.1 Very often, instead of saying "x is an element of D", we may say that "x is a point in D", with the geometrical connotation associated with it, as we identify $\mathbb{R}$ by, the so called, *real line*.   ◊

Example 2.1.1 The reader is urged to verify the following:

(i) For $a, b \in \mathbb{R}$ with $a < b$, the interval $[a, b]$ is the set of all limit points of each of the intervals (a, b), $(a, b]$, $[a, b)$ and $[a, b]$.

(ii) For $a \in \mathbb{R}$, the interval $[a, \infty)$ is the set of all limit points of each of the intervals (a, ∞) and $[a, \infty)$.

(iii) For $b \in \mathbb{R}$, the interval $(-\infty, b]$ is the set of all limit points of each of the intervals $(-\infty, b)$ and $(-\infty, b]$.

(iv) The set of limit points of $\mathbb{R}$ is itself.

(v) The set of all limit points of the set $D = (0, 1) \cup \{2\}$ is the closed interval $[0, 1]$.

(vi) If $D = \{\frac{1}{n} : n \in \mathbb{N}\}$, then 0 is the only limit point of D.

(vii) If $D = \{\frac{n}{n+1} : n \in \mathbb{N}\}$, then 1 is the only limit point of D.

(viii) A finite subset of $\mathbb{R}$ does not have any limit points.

(ix) The set $\mathbb{N}$ and $\mathbb{Z}$ have no limit points.

(x) The set $\mathbb{R}$ is the set of all limit points of the set $\mathbb{Q}$ and $\mathbb{R} \setminus \mathbb{Q}$.   ◊

Definition 2.1.2 Let $a \in \mathbb{R}$.

(1) Any open interval containing a point a is called a **neighbourhood** of a or an **open** neighbourhood of a.

(2) If the point a is deleted from a neighbourhood of a, then the remaining part of that neighbourhood is called a **deleted neighbourhood** of a.   ◊

Thus, for $a \in \mathbb{R}$ and $\delta > 0$, the open interval $(a - \delta, a + \delta)$ is a neighbourhood of a, which is also called a δ-**neighbourhood** of a, and the set $(a - \delta, a + \delta) \setminus \{a\}$ is a deleted neighbourhood of a, which is also called a **deleted** δ-**neighbourhood** of a.

With the above terminologies, we can state the following:

> A point $a \in \mathbb{R}$ is a limit point of $D \subseteq \mathbb{R}$ if and only if every deleted neighbourhood of a contains at least one point of D

In particular, if D contains either a deleted neighbourhood of a or if D contains an open interval with a as one of its end points, then a is a limit point of D.

Notation 2.1.1 In the sequel, for $a \in \mathbb{R}$, we shall use the notation I_a for an open neighbourhood of a, and $\hat{I}_a$ for a deleted neighbourhood of a.   ◊

Now we give a characterization of limit points in terms of convergence of sequences.

Theorem 2.1.1 *Let $D \subseteq \mathbb{R}$. A point $a \in \mathbb{R}$ is a limit point of D if and only if there exists a sequence (a_n) in $D \setminus \{a\}$ such that $a_n \to a$ as $n \to \infty$.*

Proof Suppose $a \in \mathbb{R}$ is a limit point of D. Then for each $n \in \mathbb{N}$, there exists $a_n \in D \setminus \{a\}$ such that $a_n \in (a - 1/n, a + 1/n)$, i.e., $|a_n - a| < 1/n$. Hence, $a_n \to a$.

Conversely, suppose that there exists a sequence (a_n) in $D \setminus \{a\}$ such that $a_n \to a$. Hence, for every $\delta > 0$, there exists $N \in \mathbb{N}$ such that $a_n \in (a - \delta, a + \delta)$ for all $n \geq N$. In particular, every deleted δ-neighbourhood of a contains a_n for every $n \geq N$. Hence, a is a limit point of D. ∎

Exercise 2.1.1 Let $D \subseteq \mathbb{R}$. Prove that a point $a \in \mathbb{R}$ is a limit point of D if and only if there exists a sequence (a_n) in D such that (a_n) is not eventually constant and $a_n \to a$ as $n \to \infty$. ◁

2.1.2 Limit of a Function at a Point

Definition 2.1.3 Let f be a real valued function defined on a set $D \subseteq \mathbb{R}$, and let $a \in \mathbb{R}$ be a limit point of D. We say that $b \in \mathbb{R}$ is a **limit of $f(x)$ as x approaches** a or **limit of f at** a, if for every $\varepsilon > 0$, there exists $\delta > 0$ such that

$$|f(x) - b| < \varepsilon \quad \text{whenever} \quad x \in D, \, 0 < |x - a| < \delta. \qquad ◊$$

Let us observe the following important property.

Theorem 2.1.2 *A function cannot have more than one limit at a given point.*

Proof Suppose b_1 and b_2 are limits of a function $f : D \to \mathbb{R}$ at a given point a, where a is a limit point of D. Let $\varepsilon > 0$ be given. By the definition of the limit, there exist $\delta_1 > 0$ and $\delta_2 > 0$ such that

$$x \in D, \, 0 < |x - a| < \delta_1 \quad \Rightarrow \quad |f(x) - b_1| < \varepsilon,$$

$$x \in D, \, 0 < |x - a| < \delta_2 \quad \Rightarrow \quad |f(x) - b_2| < \varepsilon.$$

Thus,

$$\begin{aligned}
|b_1 - b_2| &= |(b_1 - f(x)) + (f(x) - b_2)| \\
&\leq |b_1 - f(x)| + |f(x) - b_2| \\
&< 2\varepsilon
\end{aligned}$$

whenever, $x \in D$, and $0 < |x - a| < \delta := \min\{\delta_1, \delta_2\}$. ∎

Notation 2.1.2 If b is the limit of $f(x)$ as x approaches a, the we denote this fact as

$$\lim_{x \to a} f(x) = b$$

Fig. 2.1 $\lim_{x \to 0} |x| = 0$

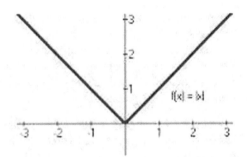

or (Fig. 2.1)

$$f(x) \to b \quad \text{as} \quad x \to a. \qquad \qquad \Diamond$$

Observe that (verify):

$\lim_{x \to a} f(x) = b$ if and only if for every open interval I_b containing b there exists an open interval I_a containing a such that $x \in \hat{I}_a \cap D \implies f(x) \in I_b$.

Convention: In the following, whenever we talk about limit of a function f as x approaches $a \in \mathbb{R}$, we assume that f is defined on a set $D_f \subseteq \mathbb{R}$ and a is a limit point of D_f.

Theorem 2.1.3 *Suppose* $\lim_{x \to a} f(x) = b$. *If* $b \neq 0$, *then there exists a deleted neighbourhood* $\hat{I}_a$ *of* a *such that* $f(x) \neq 0$ *for every* $x \in \hat{I}_a \cap D_f$. *In fact, for every* r *with* $0 < r < |b|$, *there exists a deleted neighbourhood* $\hat{I}_a$ *of* a *such that*

$$|f(x)| \geq r \quad \forall x \in \hat{I}_a \cap D_f.$$

Proof Suppose $\lim_{x \to a} f(x) = b$ with $b \neq 0$. Let $0 < r < |b|$. Now, for every $x \in D_f$,

$$|f(x)| = |b - (b - (f(x))| \geq |b| - |f(x) - b|.$$

Hence,

$$|f(x)| \geq r \quad \text{whenever} \quad |b| - |f(x) - b| \geq r.$$

But,

$$|b| - |f(x) - b| \geq r \iff |f(x) - b| \leq |b| - r.$$

Now, let $0 < \varepsilon \leq |b| - r$. Then there exists $\delta > 0$ such that $|f(x) - b| < \varepsilon$ whenever $x \in D_f$, $0 < |x - a| < \delta$. Thus, for $x \in D$ with $0 < |x - a| < \delta$, we have $|f(x) - b| < \varepsilon \leq |b| - r$ so that $|f(x)| \geq r$. ∎

> If $\lim_{x \to a} f(x) = b \neq 0$, then there exists a deleted neighbourhood $\hat{I}_a$ of a such that $f(x) \neq 0$ for all $x \in \hat{I}_a \cap D_f$

Example 2.1.2 In (i) and (ii) below, let I be an interval and a is either in I or a is an end point of I.

(i) Let $f(x) = x$. We show that $\lim_{x \to a} f(x) = a$: Since

$$|f(x) - a| = |x - a| \quad \forall x \in I,$$

it is clear that, for any $\varepsilon > 0$, taking $\delta = \varepsilon$,

$$x \in I, \ 0 < |x - a| < \delta \quad \Rightarrow \quad |f(x) - a| < \varepsilon.$$

Hence, $\lim_{x \to a} f(x) = a$.

(ii) Let $f(x) = x^2$. We show that $\lim_{x \to a} f(x) = a^2$: We have

$$|f(x) - a^2| \leq |x + a| \, |x - a| \leq (|x - a| + 2|a|)|x - a|.$$

Now, let $\varepsilon > 0$ be given. Then, $(|x - a| + 2|a|)|x - a| < \varepsilon$ implies $|f(x) - a^2| < \varepsilon$. Note that

$$|x - a| \leq 1 \quad \Rightarrow \quad (|x - a| + 2|a|)|x - a| < (1 + 2|a|)|x - a|$$

Thus, $|x - a| \leq 1$ and $(1 + 2|a|)|x - a| < \varepsilon$ implies $|f(x) - a^2| < \varepsilon$. Hence, taking $\delta = \min\{1, \varepsilon/(1 + 2|a|)\}$, we have

$$|f(x) - a^2| < \varepsilon \quad \text{whenever} \quad x \in I, \ |x - a| < \delta.$$

Hence, $\lim_{x \to a} f(x) = a^2$.

(iii) Let $f(x) = 1/x$ for $x \in \mathbb{R} \setminus \{0\}$. Let $a \in \mathbb{R} \setminus \{0\}$. We show that $\lim_{x \to a} \dfrac{1}{x} = \dfrac{1}{a}$:
Let $x \in \mathbb{R} \setminus \{0\}$. W observe that

$$|f(x) - f(a)| = \left| \frac{1}{x} - \frac{1}{a} \right| = \frac{|x - a|}{|ax|}.$$

Note that

$$|ax| = |a| |a + (x - a)| \geq |a|(|a| - |x - a|).$$

Hence,

$$|x - a| \leq |a|/2 \Rightarrow |ax| \geq |a|(|a| - |x - a|) \geq \frac{|a|^2}{2}$$

$$\Rightarrow |f(x) - f(a)| = \frac{|x - a|}{|ax|} \leq \frac{2|x - a|}{|a|^2}.$$

Thus, for any $\varepsilon > 0$,

$$|x - a| \leq |a|/2 \quad \text{and} \quad \frac{2|x - a|}{|a|^2} < \varepsilon \quad \Rightarrow \quad |f(x) - f(a)| < \varepsilon.$$

Therefore, for any $\varepsilon > 0$, taking $\delta = \min\{|a|/2, \ \varepsilon|a|^2/2\}$, we have

$$0 < |x - a| < \delta \quad \Rightarrow \quad |f(x) - f(a)| = \frac{|x - a|}{|ax|} < \varepsilon,$$

showing that $\lim_{x \to a} \frac{1}{x} = \frac{1}{a}$ ◇

More examples will be considered in Sect. 2.1.4 after proving some properties of the limit. Before that let us look at what the statement "$\lim_{x \to a} f(x)$ does not exist" means. From the definition of the limit, we can state the the following.

Theorem 2.1.4 *Suppose f is a real valued function defined on a set D and $a \in \mathbb{R}$ is a limit point of D. Then $\lim_{x \to a} f(x)$ does not exist if and only if for any $b \in \mathbb{R}$, there exists an $\varepsilon_0 > 0$ with the property that for any $\delta > 0$, there is at least one $x_\delta \in (a - \delta, a + \delta) \cap D$ with $x_\delta \neq a$, but $f(x_\delta) \notin (b - \varepsilon_0, b + \varepsilon_0)$.*

Exercise 2.1.2 Write the details of the proof of the above theorem.   ◁

We illustrate Theorem 2.1.4 by two simple examples.

Example 2.1.3 Let $f : [-1, 1] \to \mathbb{R}$ be defined by

$$f(x) = \begin{cases} 0, & -1 \leq x \leq 0, \\ 1, & 0 < x \leq 1. \end{cases}$$

We show that $\lim_{x \to 0} f(x)$ does not exist. For this let $b \in \mathbb{R}$. Let us consider the following cases:

Case (i) : $b = 0$. In this case, if $0 < \varepsilon < 1$, then $(b - \varepsilon, b + \varepsilon)$ does not contain 1 so that $f(x) \notin (b - \varepsilon, b + \varepsilon)$ for any $x > 0$.

Case (ii): $b = 1$. In this case, if $0 < \varepsilon < 1$, then $(b - \varepsilon, b + \varepsilon)$ does not contain 0 so that $f(x) \notin (b - \varepsilon, b + \varepsilon)$ for any $x < 0$.

Case (iii) : $b \neq 0$, $b \neq 1$. In this case, if $0 < \varepsilon < \min\{|b|, |b - 1|\}$, then $(b - \varepsilon, b + \varepsilon)$ does not contain 0 and 1 so that $f(x) \notin (b - \varepsilon, b + \varepsilon)$ for any $x \neq 0$. Thus, b is not a limit of $f(x)$ as x approaches 0.   ◇

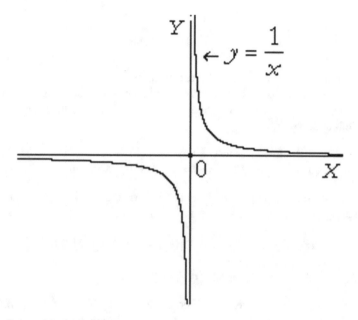

Fig. 2.2 $\lim_{x \to 0} 1/x$ does not exist

Example 2.1.4 Let $f : \mathbb{R} \setminus \{0\} \to \mathbb{R}$ is defined by

$$f(x) = \frac{1}{x}, \quad x \in \mathbb{R} \setminus \{0\}.$$

We show that $\lim_{x \to 0} f(x)$ does not exist: Let $b \in \mathbb{R}$. Note that

$$|f(x) - b| \geq |f(x)| - |b| > 1 \quad \text{whenever} \quad |f(x)| > 1 + |b|.$$

But, (Fig. 2.2)

$$|f(x)| > 1 + |b| \iff |x| < \frac{1}{1 + |b|}.$$

Thus, for any $b \in \mathbb{R}$,

$$|f(x) - b| > 1 \quad \text{whenever} \quad |x| < \frac{1}{1 + |b|}.$$

Thus, we have proved that it is not possible to find a $\delta > 0$ such that $|f(x) - b| < 1$ for all x with $|x| < \delta$. This is true for any $b \in \mathbb{R}$. Hence, $\lim_{x \to 0} f(x)$ does not exist. By similar argument, we see that, for any $k \in \mathbb{N}$,

$$\lim_{x \to 0} \frac{1}{x^k} \quad \text{does not exist.} \qquad \qquad \Diamond$$

In Example 2.1.4, we see that the function takes arbitrary large values in any deleted neighbourhood of 0. In the following theorem, we show that, if a function has such property in a deleted neighbourhood of a point, then the function cannot have a limit at that point.

Theorem 2.1.5 (Boundedness test) *If* $\lim_{x \to a} f(x) = b$, *then there exists a deleted neighbourhood* $\hat{I}_a$ *of a and* $M > 0$ *such that* $|f(x)| \leq M$ *for all* $x \in \hat{I}_a \cap D$.

Proof Suppose $\lim_{x \to a} f(x) = b$. Then there exists a deleted neighbourhood $\hat{I}_a$ of a such that $|f(x) - b| < 1$ for all $x \in \hat{I}_a \cap D$. Hence,

$$|f(x)| \leq |f(x) - b| + |b| < 1 + |b| \quad \forall x \in \hat{I}_a \cap D.$$

Thus, $|f(x)| \leq M = 1 + |b|$ for all $x \in \hat{I}_a \cap D$.   ∎

The necessary condition given in Theorem 2.1.5 may be called as *boundedness* of f in a deleted neighbourhood of a point. More generally, we have the following definition.

Definition 2.1.4 Let $f : D \to \mathbb{R}$ and $S \subseteq D$.

(1) The function f is said to be **bounded on a subset** $S \subseteq D$ if the set $\{f(x) : x \in S\}$ is a bounded subset of $\mathbb{R}$. If f is bounded on D, then we say that f is a **bounded function**.

(2) The function f is said to be **unbounded on a subset** $S \subseteq D$ if it is not bounded on S. $\Diamond$

Thus, $f : D \to \mathbb{R}$ is bounded on $S \subseteq D$ if and only if there exists $M > 0$ (in general, depends on S) such that

$$|f(x)| \leq M \quad \forall x \in S,$$

and $f : D \to \mathbb{R}$ is unbounded on S if and only if there exists a sequence (x_n) in S such that $|f(x_n)| \to \infty$ as $n \to \infty$.

In view of the Definition 2.1.4, Theorem 2.1.5 can be restated as follows:

If $\lim_{x \to a} f(x)$ exists, then f is bounded on some deleted neighbourhood of a

---

2.1.3   Limit of a Function in Terms of Sequences

Let a be a limit point of $D \subseteq \mathbb{R}$ and $f : D \rightarrow \mathbb{R}$. Suppose $\lim_{x \rightarrow a} f(x) = b$. Since a is a limit point of D, we know by Theorem 2.1.1 that there exists a sequence (x_n) in $D \setminus \{a\}$ such that $x_n \rightarrow a$. Does $f(x_n) \rightarrow b$? The answer is in the affirmative.

Theorem 2.1.6 *If* $\lim_{x \rightarrow a} f(x) = b$, *then for every sequence* (x_n) *in* D *with* $x_n \rightarrow a$, *we have* $f(x_n) \rightarrow b$.

Proof Suppose $\lim_{x \rightarrow a} f(x) = b$. Let (x_n) be a sequence in D such that $x_n \rightarrow a$. Let $\varepsilon > 0$ be given. We have to show that there exists $N \in \mathbb{N}$ such that $|f(x_n) - b| < \varepsilon$ for all $n \geq N$.

Since $\lim_{x \rightarrow a} f(x) = b$, we know that there exists $\delta > 0$ such that

$$x \in D, \ 0 < |x - a| < \delta \quad \Rightarrow \quad |f(x) - b| < \varepsilon.$$

Also, since $x_n \rightarrow a$, corresponding to the above δ, there exists $N \in \mathbb{N}$ such that $|x_n - a| < \delta$ for all $n \geq N$. Hence, we have $|f(x_n) - b| < \varepsilon$ for all $n \geq N$. ∎

The converse of the above theorem is also true.

Theorem 2.1.7 *Suppose that, for every sequence* (x_n) *in* D *for which* $x_n \rightarrow a$, *we have* $f(x_n) \rightarrow b$. *Then* $\lim_{x \rightarrow a} f(x) = b$.

Proof Assume for a moment that f does not have the limit b as x approaches a. Then, by Theorem 2.1.4, there exists $\varepsilon_0 > 0$ such that for every $\delta > 0$, there exists at least one $x_\delta \in D$ such that

$$0 < |x_\delta - a| < \delta \quad \text{and} \quad |f(x_\delta) - b| \geq \varepsilon_0.$$

In particular, for every $n \in \mathbb{N}$, there exists $x_n \in D$ such that

$$0 < |x_n - a| < \frac{1}{n} \quad \text{and} \quad |f(x_n) - b| \geq \varepsilon_0.$$

Thus, $x_n \rightarrow a$ but $f(x_n) \nrightarrow b$. This is a contradiction to our hypothesis in the theorem. ∎

Remark 2.1.2 Here are some implications of Theorem 2.1.6. Suppose (x_n) is a sequence in $D \setminus \{a\}$ such that $x_n \rightarrow a$.

1. If $(f(x_n))$ does not converge, then $\lim_{x \rightarrow a} f(x)$ does not exist.
2. If $(f(x_n))$ does not converge to a given $b \in \mathbb{R}$, then either $\lim_{x \rightarrow a} f(x)$ does not exist or $\lim_{x \rightarrow a} f(x)$ exists but $\lim_{x \rightarrow a} f(x) \neq b$.
3. If (y_n) is another sequence in $D \setminus \{a\}$ which converges to a and the sequences $(f(x_n))$ and $(f(y_n))$ converge to different points, then $\lim_{x \rightarrow a} f(x)$ does not exist.

Fig. 2.3 Limit does not exist
at 0

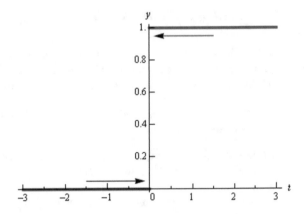

Suppose we are able to show the convergence of $(f(x_n))$ to some b for any arbitrary (*not for a specific*) sequence (x_n) in $D \setminus \{a\}$ which converges to a, then by Theorem 2.1.7, we can assert that $\lim_{x \to a} f(x) = b$ (Fig. 2.3).   ◇

Example 2.1.5 We have seen in Example 2.1.3 that $f : [-1, 1] \to \mathbb{R}$ is defined by

$$f(x) = \begin{cases} 0, & -1 \le x \le 0, \\ 1, & 0 < x \le 1 \end{cases}$$

does not have a limit at the point 0. Now, we show this fact by making use of Theorem 2.1.6.

Suppose (x_n) is a sequence of negative numbers and (y_n) is a sequence of positive numbers such that both of them converge to 0. Then we have $f(x_n) = 0$ and $f(y_n) = 1$ for all $n \in \mathbb{N}$. Hence, $\lim_{n \to \infty} f(x_n)$ and $\lim_{n \to \infty} f(y_n)$ exist, but they are different. Hence $\lim_{x \to 0} f(x)$ does not exist.   ◇

Example 2.1.6 Consider the function f in Example 2.1.4, i.e.,

$$f(x) = \frac{1}{x}, \quad x \in \mathbb{R} \setminus \{0\}.$$

Let $x_n = 1/n$, $n \in \mathbb{N}$. Then we have $x_n \to 0$, but $f(x_n) = n \to \infty$. Hence, we can assert the fact that $\lim_{x \to 0} f(x)$ does not exist, by Theorem 2.1.6 also.   ◇

Example 2.1.7 Consider the function f defined by $f(x) = \sin(1/x)$ for $x > 0$. Let $x_n = 2/n\pi$, $n \in \mathbb{N}$. Then we have $x_n \to 0$, and

$$f(x_n) = \sin(n\pi/2) \quad \forall n \in \mathbb{N}.$$

Note that $f(x_{2n-1}) = (-1)^{n-1}$ and $f(x_{2n}) = 0$ for all $n \in \mathbb{N}$. Hence, the sequence $(f(x_n))$ does not converge. Therefore, by Theorem 2.1.6, $\lim_{x \to 0} f(x)$ does not exist. $\Diamond$

2.1.4 Some Properties

For considering certain properties of the limit, and also for later use, we recall some definitions from set theory:

Suppose f and g are (real valued) functions with domains D_f and D_g, respectively, and let $\alpha \in \mathbb{R}$. Suppose $D_f \cap D_g \neq \varnothing$. Then, we define functions $f + g$, fg and αf as

$$(f + g)(x) = f(x) + g(x), \quad x \in D_f \cap D_g,$$
$$(fg)(x) = f(x)g(x), \quad x \in D_f \cap D_g,$$
$$(\alpha f)(x) = \alpha f(x), \quad x \in D_f.$$

The function $f + g$ is called the **sum** of f and g, and fg is called the **product** of f and g. If f is a nonzero function, that is, $f(x) \neq 0$ for some $x \in D_f$, then we define the function $1/f$ by

$$\left(\frac{1}{f}\right)(x) = \frac{1}{f(x)}, \quad x \in D'_f,$$

where $D'_f := \{x \in D_f : f(x) \neq 0\}$. Thus, we can also define the function f/g by

$$\frac{f}{g} = f\frac{1}{g}$$

on the set $D_{f/g} := \{x \in D_f \cap D_g : g(x) \neq 0\}$.

If $D := \{x \in D_f : f(x) \in D_g\} \neq \varnothing$, then we define the **composition** of g and f, denoted by $g \circ f$, by

$$(g \circ f)(x) = g(f(x)), \quad x \in D.$$

Thus, domain of $g \circ f$ is the set $D_{g \circ f} := \{x \in D_f : f(x) \in D_g\}$.

> In the following, when we talk about the functions $f + g$, fg, f/g and $g \circ f$, we mean their definitions as above with appropriate domains of definitions; we may not write the domains explicitly. Functions are assumed to be defined on subsets of the set $\mathbb{R}$ of real numbers, and are real valued.

The following three theorems can be proved using Theorems 2.1.6 and 2.1.7, and the results on convergence of sequences of real numbers (supply details). However, to get accustomed with the ε-δ arguments, we provide the detailed proof using the definition itself.

Theorem 2.1.8 *Suppose $\lim_{x \to a} f(x) = b$ and $\lim_{x \to a} g(x) = c$. Then we have the following.*

(i) $\lim_{x \to a}[f(x) + g(x)] = b + c$,

(ii) $\lim_{x \to a} f(x)g(x) = bc$.

(iii) *If $c \neq 0$, then $g(x) \neq 0$ in a deleted neighbourhood of a and*

$$\lim_{x \to a} \frac{f(x)}{g(x)} = \frac{b}{c}.$$

Proof (i) Note that for $x \in D_f \cap D_g$,

$$|(f(x) + g(x)) - (b + c)| = |[f(x) - b] + [g(x) - c]| \leq |f(x) - b| + |g(x) - c|. \tag{2.1}$$

Since $\lim_{x \to a} f(x) = b$ and $\lim_{x \to a} g(x) = c$, there exist $\delta_1 > 0, \delta_2 > 0$ such that

$$|f(x) - b| < \varepsilon/2 \quad \text{whenever} \quad x \in D_f, \, 0 < |x - a| < \delta_1, \tag{2.2}$$

$$|g(x) - c| < \varepsilon/2 \quad \text{whenever} \quad x \in D_g, \, 0 < |x - a| < \delta_2. \tag{2.3}$$

From (2.1), (2.2), (2.3), we obtain

$$|(f(x) + g(x)) - (b + c)| < \varepsilon$$

whenever $x \in D_f \cap D_g$ with $0 < |x - a| < \delta := \min\{\delta_1, \delta_2\}$. Thus, $\lim_{x \to a}[f(x) + g(x)] = b + c$.

(ii) Note that for $x \in D_f \cap D_g$,

$$|f(x)g(x) - bc| = |f(x)[g(x) - c] + [f(x) - b]c|.$$

Since $\lim_{x \to a} f(x) = b$, there exists $M > 0$ and $\delta_0 > 0$ such that

$$|f(x)| \leq M \quad \text{whenever} \quad x \in D_f, \, 0 < |x - a| < \delta_0.$$

Hence,

$$|f(x)g(x) - bc| \leq M|g(x) - c| + |f(x) - b||c| \tag{2.1}$$

whenever $x \in D_f \cap D_g$ with $0 < |x - a| < \delta_0$. Now, since $\lim_{x \to a} f(x) = b$ and $\lim_{x \to a} g(x) = c$, there exist $\delta_1 > 0, \delta_2 > 0$ such that

$$M|g(x) - c| < \varepsilon/2 \quad \text{whenever} \quad x \in D_g, \, 0 < |x - a| < \delta_1, \tag{2.2}$$

$$|c||f(x) - b| < \varepsilon/2 \quad \text{whenever} \quad x \in D_f, \, 0 < |x - a| < \delta_2. \tag{2.3}$$

From (2.1), (2.2), (2.3), we obtain

$$|f(x)g(x) - bc| < \varepsilon$$

whenever $x \in D_f \cap D_g$ with $0 < |x - a| < \delta := \min\{\delta_0, \delta_1, \delta_2\}$. Thus, $\lim_{x \to a} f(x)$
$g(x) = bc$.

(iii) Since $\lim_{x \to a} g(x) = c \neq 0$, by Theorem 2.1.3, there exists $\delta_0 > 0$ such that
$|g(x)| \geq |c|/2$ for all $x \in D_g$ with $0 < |x - a| < \delta_0$. Hence for $x \in D_f \cap D_g$ with
$0 < |x - a| < \delta_0$, we have

$$\begin{aligned}
\left| \frac{f(x)}{g(x)} - \frac{b}{c} \right| &= \frac{|cf(x) - bg(x)|}{|cg(x)|} \\
&\leq \frac{|c[f(x) - b] + [c - g(x)]b|}{|cg(x)|} \\
&\leq \frac{|c| |f(x) - b| + |c - g(x)| |b|}{(|c|^2/2)}
\end{aligned}$$

Now, since $\lim_{x \to a} f(x) = b$ and $\lim_{x \to a} g(x) = c$, from the above we can conclude
that there exists $\delta > 0$ such that

$$\left| \frac{f(x)}{g(x)} - \frac{b}{c} \right| < \varepsilon \quad \text{whenever} \quad x \in D_f \cap D_g, \ 0 < |x - a| < \delta.$$

Thus, $\lim_{x \to a} [f(x)/g(x)] = b/c$. ∎

***Proof* Alternative proof using sequences** Let (x_n) be any sequence in $D_f \cap D_g$
such that $x_n \to a$. Hence, by Theorem 2.1.6, we have $\lim_{n \to \infty} f(x_n) = b$ and
$\lim_{n \to \infty} g(x_n) = c$. Hence, by Theorem 1.1.3 (i) and Theorem 1.1.8 (i), we have

$$\lim_{n \to \infty} [f(x_n) + g(x_n)] = b + c, \quad \lim_{n \to \infty} f(x_n)g(x_n) = bc.$$

Also, since $c \neq 0$ and $\lim_{n \to \infty} g(x_n) = c$, there exists $N \in \mathbb{N}$ such that $g(x_n) \neq 0$
for all $n \geq N$. Therefore, by Theorem 1.1.8 (ii), we have

$$\lim_{n \to \infty} \frac{f(x_n)}{g(x_n)} = \frac{b}{c}.$$

Therefore, by Theorem 2.1.7, we have

$$\lim_{x \to a} [f(x) + g(x)] = b + c, \quad \lim_{x \to a} f(x)g(x) = bc, \quad \lim_{x \to a} f(x)/g(x) = b/c.$$

This completes the proof. ∎

Theorem 2.1.9 (Sandwich theorem) *If f and g have the same limit b as x approaches a, and if h is a function such that $f(x) \leq h(x) \leq g(x)$ for all x in a deleted neighbourhood of a, then $\lim_{x \to a} h(x) = b$.*

Proof Suppose $\lim_{x \to a} f(x) = b = \lim_{x \to a} g(x)$. Let $\varepsilon > 0$ be given. Then there exists $\delta_0 > 0$ such that if $x \in D_f \cap D_g$ and $0 < |x - a| < \delta_0$, then

$$f(x) \in (b - \varepsilon, b + \varepsilon) \quad \text{and} \quad g(x) \in (b - \varepsilon, b + \varepsilon).$$

Since $f(x) \leq h(x) \leq g(x)$ for all $x \in D_f \cap D_g \cap D_h$ with $0 < |x - a| < \delta_1$ for some δ_1, we obtain that $h(x) \in (b - \varepsilon, b + \varepsilon)$ whenever $x \in D_f \cap D_g \cap D_h$ with $0 < |x - a| < \delta := \min\{\delta_0, \delta_1\}$. Thus, $\lim_{x \to a} h(x) = b$. ∎

Proof Alternative proof using sequences Let (x_n) be any sequence in $D_f \cap D_g \cap D_h$ such that $x_n \to a$. By assumption $\lim_{x \to a} f(x) = b = \lim_{x \to a} g(x)$. Hence, by Theorem 2.1.6, we have $\lim_{n \to \infty} f(x_n) = b = \lim_{n \to \infty} g(x_n)$. Again by assumption, we have $f(x_n) \leq h(x_n) \leq g(x_n)$ for all $n \in \mathbb{N}$. Therefore, by the Sandwich theorem Theorem, 1.1.3 (iv), for sequences, we have $\lim_{x \to a} h(x) = b$. Hence, by Theorem 2.1.7, $\lim_{x \to a} h(x) = b$. ∎

Theorem 2.1.10 *Suppose $\lim_{x \to a} f(x) = b$, $\lim_{x \to a} g(x) = c$ and $f(x) \geq g(x)$ for all x in a deleted neighbourhood of a. Then $b \geq c$.*

Proof Suppose $b < c$. Since $\lim_{x \to a} f(x) = b$ and $\lim_{x \to a} g(x) = c$, for $\varepsilon > 0$ small enough, say $0 < \varepsilon < (c - b)/2$, there exists $\delta > 0$ such that

$$f(x) \in (b - \varepsilon, b + \varepsilon) \quad \text{and} \quad g(x) \in (c - \varepsilon, c + \varepsilon)$$

for all $x \in D_f \cap D_g$ with $0 < |x - a| < \delta_0$. Hence, $f(x) < g(x)$ for all $x \in D_f \cap D_g$ with $0 < |x - a| < \delta_0$, which is a contradiction to the assumption in the theorem. Hence, $b \geq c$. ∎

Exercise 2.1.3 Provide an alternative proof for the above theorem using sequences.◄

Theorem 2.1.11 *Suppose $\lim_{x \to a} f(x) = b$ and $\lim_{y \to b} g(y) = c$. Further, assume that $f(x) \in D_g \setminus \{b\}$ for every $x \in D_f \setminus \{a\}$. Then*

$$\lim_{x \to a} (g \circ f)(x) = c.$$

Proof Let $\varepsilon > 0$ be given. Then there exists $\delta_1 > 0$ such that

$$0 < |y - b| < \delta_1 \quad \Rightarrow \quad |g(y) - c| < \varepsilon.$$

Also, let $\delta_2 > 0$ be such that

$$0 < |x - a| < \delta_2 \quad \Rightarrow \quad |f(x) - b| < \delta_1.$$

Hence, along with the given condition that $f(x) \in D_2 \setminus \{b\}$ for every $x \in D_1 \setminus \{a\}$,

$$0 < |x - a| < \delta_2 \quad \Rightarrow \quad 0 < |f(x) - b| < \delta_1 \quad \Rightarrow \quad |g(f(x)) - c| < \varepsilon.$$

This completes the proof. ∎

Proof **Alternative proof using sequences** By Theorem 2.1.7, it is enough to prove that for any sequence (x_n) in $D_g \setminus \{a\}$ which converges to a, the sequence $(g(f(x_n)))$ converges to c. So, let (x_n) be in $D_f \setminus \{0\}$ such that $x_n \to a$. Since $\lim_{x \to a} f(x) = b$, by Theorem 2.1.6, $f(x_n) \to b$. Let $y_n = f(x_n)$, $n \in \mathbb{N}$. By the assumption, $y_n \in D_g \setminus \{b\}$ for all $n \in \mathbb{N}$. Since $\lim_{y \to b} g(y) = c$ and $y_n \to b$, again by Theorem 2.1.6, $g(y_n) \to c$. Thus we obtained $g(f(x_n)) \to c$, completing the proof. ∎

Example 2.1.8 If $f(x)$ is a polynomial, say $f(x) = a_0 + a_1 x + \ldots + a_k x^k$, then for any $a \in \mathbb{R}$,

$$\lim_{x \to a} f(x) = f(a).$$

This follows from the results in Theorem 2.1.8 by making use of the fact that $\lim_{x \to a} x = a$. ◇

Example 2.1.9 Let $f(x) = \frac{x^2 - 4}{x - 2}$ for $x \in D = \mathbb{R} \setminus \{2\}$. Then

$$\lim_{x \to 2} f(x) = 4.$$

Note that, for $x \neq 2$,

$$f(x) = \frac{(x + 2)(x - 2)}{x - 2} = (x + 2).$$

Hence, $\lim_{x \to 2} f(x) = \lim_{x \to 2}(x + 2) = 4.$ ◇

Example 2.1.10 We show that

$$\lim_{x \to 0} \sin(x) = 0 \quad \text{and} \quad \lim_{x \to 0} \cos(x) = 1.$$

From the graph of the function $\sin x$, it is clear that

$$0 < |x| < \frac{\pi}{2} \quad \Rightarrow \quad 0 < |\sin x| < |x|.$$

Hence, from Theorem 2.1.10, we have $\lim_{x \to 0} |\sin x| = 0$, so that we also have $\lim_{x \to 0} \sin(x) = 0$ (Figs. 2.4 and 2.5).

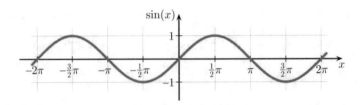

Fig. 2.4 Graph of sin x

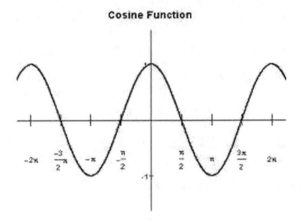

Fig. 2.5 Graph of cos x

Also, since $\cos x = 1 - 2\sin^2(x/2)$ and $\lim_{x\to 0} \sin(x/2) = 0$, Theorem 2.1.8 implies $\lim_{x\to 0} \cos x = 1$ (Fig. 2.6). ◇

Example 2.1.11 We show that

$$\lim_{x\to 0} \frac{\sin x}{x} = 1.$$

Note that $\lim_{x\to 0} \sin x = 0$ and $\lim_{x\to 0} x = 0$. Hence, we cannot apply Theorem 2.1.8(iii). So, we have to look for alternate arguments.

From the graph of $\sin x$, we see that

$$0 < x < \frac{\pi}{2} \Rightarrow \sin x < x < \tan x.$$

Hence,

$$0 < x < \frac{\pi}{2} \Rightarrow \cos x < \frac{\sin x}{x} < 1.$$

Since $\frac{\sin(-x)}{-x} = \frac{\sin x}{x}$ and $\cos(-x) = \cos x$, it follows that

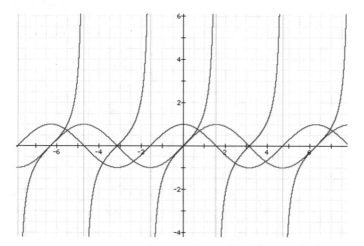

Fig. 2.6 Graph of $\sin x$, $\cos x$ and $\tan x$

Fig. 2.7 $\mathrm{sinc}(x) := \dfrac{\sin x}{x}$, $x \neq 0$

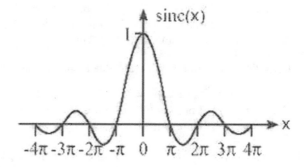

$$0 < |x| < \frac{\pi}{2} \Rightarrow \cos x < \frac{\sin x}{x} < 1.$$

Since $\lim_{x\to 0} \cos x = 1$, by Theorem 2.1.10, we have $\lim_{x\to 0} \dfrac{\sin x}{x} = 1$ (Fig. 2.7).◊

Remark 2.1.3 In the above two examples we have used some familiar properties of the trigonometric functions $\sin x$, $\cos x$ and $\tan x$, though we have not defined these functions formally. We shall define these functions formally in the due course. ◊

Exercise 2.1.4 Let $f : \mathbb{R} \to \mathbb{R}$ be such that $f(x + y) = f(x) + f(y)$. Suppose $\lim_{x\to 0} f(x)$ exists. Prove that $\lim_{x\to 0} f(x) = 0$ and for any $c \in \mathbb{R}$, $\lim_{x\to c} f(x) = f(c)$. ◁

Exercise 2.1.5 Suppose f is a function defined in a neighbourhood of x_0 and $\lim_{x\to x_0} f(x)$ exists. Let φ be a function defined in a neighbourhood I_0 of x_0 such that for every $r > 0$, $|\varphi(x) - x_0| < r$ whenever $x \in I_0$ and $|x - x_0| < r$. Prove that $\lim_{x\to x_0} f(\varphi(x))$ exists and $\lim_{x\to x_0} f(\varphi(x)) = \lim_{x\to x_0} f(x)$. ◁

Fig. 2.8 Limit does not exist
at 0, but left and right limits
exist

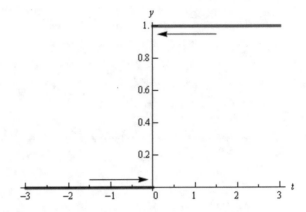

2.1.5 Left Limit and Right Limit

It can happen that $\lim_{x \to a} f(x)$ does not exist, but $f(x)$ can approach a particular value as x approaches a either from left or from right, as in the following figure corresponding to Example 2.1.3 (Fig. 2.8).

Definition 2.1.5 Let f be a real valued function defined on a set $D \subseteq \mathbb{R}$.

(1) Suppose $D \cap (-\infty, a) \neq \emptyset$ and a is a limit point of $D \cap (-\infty, a)$. Then we say that $f(x)$ has the **left limit** $b \in \mathbb{R}$ as x approaches $a \in \mathbb{R}$ from left if for every $\varepsilon > 0$, there exists $\delta > 0$ such that

$$|f(x) - b| < \varepsilon \quad \forall x \in D \cap (a - \delta, a),$$

and in that case we write $\lim_{x \to a^-} f(x) = b$. $\Diamond$

(2) Suppose $D \cap (a, \infty) \neq \emptyset$ and a is a limit point of $D \cap (a, \infty)$. Then we say that $f(x)$ has the **right limit** $b \in \mathbb{R}$ as x approaches $a \in \mathbb{R}$ from right if for every $\varepsilon > 0$, there exists $\delta > 0$ such that

$$|f(x) - b| < \varepsilon \quad \forall x \in D \cap (a, a + \delta),$$

and in that case we write $\lim_{x \to a^+} f(x) = b$.

Notation 2.1.3 We shall use the notations:

$$f(a-) := \lim_{x \to a^-} f(x), \qquad f(a+) := \lim_{x \to a^+} f(x)$$

whenever the above limits exist. $\Diamond$

We have the following characterizations in terms of sequences (*Verify*):

1. $\lim_{x \to a^-} f(x) = b$ if and only if for every sequence (x_n) in $D \setminus \{a\}$,

$$x_n < a \quad \forall n \in \mathbb{N}, \quad x_n \to a \quad \Rightarrow \quad f(x_n) \to b.$$

2. $\lim_{x \to a^+} f(x) = b$ if and only if for every sequence (x_n) in $D \setminus \{a\}$,

$$x_n > a \quad \forall n \in \mathbb{N}, \quad x_n \to a \quad \Rightarrow \quad f(x_n) \to b.$$

The proof of the following theorem is left as an exercise.

Theorem 2.1.12 *Let f be a real valued function defined on a set $D \subseteq \mathbb{R}$, and let $a \in \mathbb{R}$ be a limit point of D. Then $\lim_{x \to a} f(x)$ exists if and only if $\lim_{x \to a^-} f(x)$ and $\lim_{x \to a^+} f(x)$ exist and $\lim_{x \to a^-} f(x) = \lim_{x \to a^+} f(x)$, and in that case*

$$\lim_{x \to a} f(x) = \lim_{x \to a^-} f(x) = \lim_{x \to a^+} f(x).$$

$\lim_{x \to a} f(x)$ *does not exist in the following cases:*
(i) $\lim_{x \to a^-} f(x)$ *does not exist*
(ii) $\lim_{x \to a^+} f(x)$ *does not exist,*
(iii) $\lim_{x \to a^-} f(x)$ *and $\lim_{x \to a^+} f(x)$ exist, but they are not equal*

We may observe that

1. $\lim_{x \to a^-} f(x) = b \iff \lim_{h \to 0^+} f(a - h) = b$,
2. $\lim_{x \to a^+} f(x) = b \iff \lim_{h \to 0^+} f(a + h) = b$.

Exercise 2.1.6 Verify the above two results. ◁

Example 2.1.12 We consider a few examples to illustrate Theorem 2.1.12.

(i) Let $f : \mathbb{R} \to \mathbb{R}$ be defined by

$$f(x) = \begin{cases} 1/x, & x > 0, \\ 1, & x \le 0. \end{cases}$$

Then, $\lim_{x \to 0^-} f(x) = 1$, but $\lim_{x \to 0^+} f(x)$ does not exist.

(ii) Let $f : \mathbb{R} \to \mathbb{R}$ be defined by

$$f(x) = \begin{cases} 1/x, & x < 0, \\ 1, & x \ge 0. \end{cases}$$

Then, $\lim_{x \to 0^+} f(x) = 1$, but $\lim_{x \to 0^-} f(x)$ does not exist.

(iii) Let $f : \mathbb{R} \to \mathbb{R}$ be defined by

$$f(x) = \begin{cases} 1/x, & x \neq 0, \\ 1, & x = 0. \end{cases}$$

Then, $\lim_{x \to 0^+} f(x)$ and $\lim_{x \to 0^-} f(x)$ do not exist.

(iv) Let f be as in Example 2.1.3, that is, $f : [-1, 1] \to \mathbb{R}$ defined by

$$f(x) = \begin{cases} 0, & -1 \leq x \leq 0, \\ 1, & 0 < x \leq 1. \end{cases}$$

Then, $\lim_{x \to 0^-} f(x)$ and $\lim_{x \to 0^+} f(x)$ exist, but $\lim_{x \to 0} f(x)$ does not exist. $\Diamond$

2.1.6 Limit at $\pm\infty$ and Limit $\pm\infty$

Definition 2.1.6 Let f be a real valued function defined on $D_f \subseteq \mathbb{R}$.

(1) If D_f contains (a, ∞) for some $a \in \mathbb{R}$, then f **is said to have the limit** b **as** $x \to \infty$, if for every $\varepsilon > 0$, there exists $M > a$ such that

$$|f(x) - b| < \varepsilon \quad \text{whenever} \quad x > M,$$

and in that case we write $\lim_{x \to \infty} f(x) = b$.

(2) If D_f contains $(-\infty, a)$ for some $a \in \mathbb{R}$, then f **is said to have the limit** b **as** $x \to -\infty$, if for every $\varepsilon > 0$, there exits $M < a$ such that

$$|f(x) - b| < \varepsilon \quad \text{whenever} \quad x < M,$$

and in that case we write $\lim_{x \to -\infty} f(x) = b$. $\Diamond$

Notation 2.1.4 When we write $\lim_{x \to \infty} f(x) = b$ we mean that D_f contains an interval of the form (a, ∞) for some $a \in \mathbb{R}$ and the limit is in the sense of Definition 2.1.6(1). Similarly, when we write we $\lim_{x \to -\infty} f(x) = b$, it is assumed that D_f contains an interval of the form $(-\infty, a)$ for some $a \in \mathbb{R}$, and the limit is in the sense of Definition 2.1.6(2).

Also, for a sequence (x_n) in $\mathbb{R}$, if we write $f(x_n) \to b$ for some $b \in \mathbb{R}$, we assume that x_n belongs to D_f, and the limit of $(f(x_n))$ is b.

Now, we give the sequential characterizations of limits as given in Definition 2.1.6. $\Diamond$

Theorem 2.1.13 *The following hold.*

(i) $\lim_{x \to \infty} f(x) = b$ *if and only if for every sequence* (x_n), $x_n \to \infty$ *implies* $f(x_n) \to b$.

(ii) $\lim_{x \to -\infty} f(x) = b$ *if and only if for every sequence* (x_n), $x_n \to -\infty$ *implies* $f(x_n) \to b$.

Proof Suppose $\lim_{x \to \infty} f(x) = b$, and let (x_n) be such that $x_n \to \infty$. Let $\varepsilon > 0$ be given. To show that there exists $N \in \mathbb{N}$ such that $|f(x_n) - b| < \varepsilon$ for all $n \geq N$. Since $\lim_{x \to \infty} f(x) = b$, there exists $M > 0$ such that

$$x \in D_f, \ x > M \quad \Rightarrow \quad |f(x) - b| < \varepsilon. \tag{2.1}$$

Since $x_n \to \infty$, there exists $n_0 \in \mathbb{N}$ such that

$$n \geq n_0 \quad \Rightarrow \quad x_n > M. \tag{2.2}$$

From (2.1) and (2.2) above we have

$$n \geq n_0 \quad \Rightarrow \quad |f(x_n) - b| < \varepsilon.$$

Conversely, suppose that for every (x_n) with $x_n \to \infty$, we have $f(x_n) \to b$. Assume for a moment that $\lim_{x \to \infty} f(x) \neq b$. Then there exists $\varepsilon_0 > 0$ such that for every $n \in \mathbb{N}$, there exists $x_n > n$ and $|f(x_n) - b| \geq \varepsilon_0$. Thus, we obtained a sequence (x_n) with $x_n \to \infty$, but $f(x_n) \not\to b$. This is a contradiction to the hypothesis. Thus, we have proved (i). Proof of (ii) follows by similar arguments. ∎

The following results can be verified as in the case of limits at points in $\mathbb{R}$.

1. If $\lim_{x \to \infty} f(x) = b$ and $\lim_{x \to \infty} g(x) = c$, then

$$\lim_{x \to \infty} [f(x) + g(x)] = b + c, \quad \lim_{x \to \infty} f(x)g(x) = bc.$$

2. If $\lim_{x \to \infty} f(x) = b$, $\lim_{x \to \infty} g(x) = c$ and $c \neq 0$, then there exists $M_0 > 0$ such that $g(x) \neq 0$ for all $x > M_0$ and

$$\lim_{x \to \infty} \frac{f(x)}{g(x)} = \frac{b}{c}.$$

Exercise 2.1.7 Verify the above results. ◁

Example 2.1.13 (i) For any $k \in \mathbb{N}$,

$$\lim_{x \to \infty} \frac{1}{x^k} = 0, \quad \lim_{x \to -\infty} \frac{1}{x^k} = 0.$$

From the results on sequences, for any sequence (x_n) of nonzero numbers, we have

$$x_n \to \infty \iff \frac{1}{x_n^k} \to 0 \quad \text{and} \quad x_n \to -\infty \iff \frac{1}{x_n^k} \to 0.$$

Hence, we obtain the required limits, by using Theorem 2.1.13.

(ii) $\lim_{x\to\infty} \dfrac{1+x}{1+x^2} = 0$: Let $f(x) = \dfrac{1+x}{1+x^2}$ for $x \in \mathbb{R}$. By (i),

$$f(x) = \frac{1+x}{1+x^2} = \frac{1/x^2 + 1/x}{1/x^2 + 1} \to \frac{0}{1} = 0.$$

(iii) $\lim_{x\to\infty} \dfrac{1+x}{1-x} = -1$: Let $f(x) = \dfrac{1+x}{1-x}$ for $x \neq 1$. By (i),

$$f(x) = \frac{1+x}{1-x} = \frac{1/x + 1}{1/x - 1} \to \frac{1}{-1} = -1.$$

(iv) $\lim_{x\to\infty} \dfrac{1+2x}{1+3x} = \dfrac{2}{3}$: Let $f(x) = \dfrac{1+2x}{1+3x}$ for $x \neq -1/3$. By (i),

$$f(x) = \frac{1+2x}{1+3x} = \frac{1/x + 2}{1/x + 3} \to \frac{2}{3}. \qquad\qquad \Diamond$$

Exercise 2.1.8 Prove the limits in Example 2.1.13(i) by using ε, M arguments, without using Theorem 2.1.13.   ◁

Definition 2.1.7 We define the following:

(1) $\lim_{x\to a} f(x) = \infty$ if for every $M > 0$, there exists $\delta > 0$ such that

$$0 < |x - a| < \delta \quad \Rightarrow \quad f(x) > M.$$

(2) $\lim_{x\to a} f(x) = -\infty$ if for every $M > 0$, there exists $\delta > 0$ such that

$$0 < |x - a| < \delta \quad \Rightarrow \quad f(x) < -M.$$

(3) $\lim_{x\to+\infty} f(x) = \infty$ if for every $M > 0$, there exists $\alpha > 0$ such that

$$x > \alpha \quad \Rightarrow \quad f(x) > M.$$

(4) $\lim_{x\to+\infty} f(x) = -\infty$ if for every $M > 0$, there exists $\alpha > 0$ such that

$$x > \alpha \quad \Rightarrow \quad f(x) < -M.$$

(5) $\lim_{x\to-\infty} f(x) = \infty$ if for every $M > 0$, there exists $\alpha > 0$ such that

$$x < -\alpha \quad \Rightarrow \quad f(x) > M.$$

(6) $\lim_{x\to-\infty} f(x) = -\infty$ if for every $M > 0$, there exists $\alpha > 0$ such that

$$x < -\alpha \quad \Rightarrow \quad f(x) < -M. \qquad \diamond$$

It can be easily shown that

1. $\lim_{x \to a} f(x) = \infty \iff \lim_{x \to a}[-f(x)] = -\infty$,
2. $\lim_{x \to +\infty} f(x) = \infty \iff \lim_{x \to +\infty}[-f(x)] = -\infty$,
3. $\lim_{x \to -\infty} f(x) = \infty \iff \lim_{x \to -\infty}[-f(x)] = -\infty$.

Exercise 2.1.9 Verify the above statements. ◁

Example 2.1.14 f(i) $\lim_{x \to 0} \dfrac{1}{x^2} = \infty$: Taking $f(x) = \dfrac{1}{x^2}$ for $x \neq 0$ and $M > 0$, we observe that

$$f(x) > M \iff \frac{1}{x^2} > M \iff |x| < \frac{1}{\sqrt{M}}.$$

Hence, for $0 < \delta < 1/\sqrt{M}$,

$$|x| < \delta \quad \Rightarrow \quad |x| < \frac{1}{\sqrt{M}} \quad \Rightarrow \quad f(x) > M.$$

Thus, $\lim_{x \to 0} \dfrac{1}{x^2} = \infty$.

(ii) $\lim_{x \to 1} \left| \dfrac{1+x}{1-x} \right| = \infty$: Let $f(x) = \left| \dfrac{1+x}{1-x} \right|$ for $x \neq 1$. Then for $M > 0$,

$$f(x) = \left| \frac{1+x}{1-x} \right| > M \iff |1 - x| < \frac{|1+x|}{M}.$$

Note that

$$\frac{|1+x|}{M} = \frac{|2 - (1-x)|}{M} \geq \frac{2 - |1-x|}{M}.$$

Hence,

$$|x - 1| < \frac{2 - |1-x|}{M} \quad \Rightarrow \quad f(x) > M,$$

i.e.,

$$|x - 1| < \frac{2}{M+1} \quad \Rightarrow \quad f(x) > M.$$

Thus, $\lim_{x \to 1} \left| \dfrac{1+x}{1-x} \right| = \infty$.

(iii) Let $f(x) = x^2$, $x \in \mathbb{R}$. For $M > 0$,

$$f(x) = x^2 > M \iff |x| > \sqrt{M}.$$

Thus,

$$x > \sqrt{M} \quad \Rightarrow f(x) > M$$

and

$$x < -\sqrt{M} \quad \Rightarrow f(x) > M.$$

Thus, $\lim_{x \to \infty} f(x) = \infty$ and $\lim_{x \to -\infty} f(x) = \infty$.   ◊

Example 2.1.15 (The number e) We have seen in Example 1.1.26 that $\lim_{n \to \infty} \left(1 + \frac{1}{n}\right)^n$ exists, and we denoted its value by e. Now we show that

$$\lim_{x \to \infty} \left(1 + \frac{1}{x}\right)^x = e.$$

Let $\varepsilon > 0$ be given. We have to find an $M > 0$ such that

$$e - \varepsilon < \left(1 + \frac{1}{x}\right)^x < e + \varepsilon \quad \text{whenever} \quad x > M.$$

Note that, for every $n \in \mathbb{N}$, if $x \in \mathbb{R}$ is such that $n \le x \le n + 1$, then

$$1 + \frac{1}{n + 1} \le 1 + \frac{1}{x} \le 1 + \frac{1}{n}$$

so that

$$\left(1 + \frac{1}{n + 1}\right)^n \le \left(1 + \frac{1}{x}\right)^x \le \left(1 + \frac{1}{n}\right)^{n+1},$$

i.e.,

$$\alpha_n \le \left(1 + \frac{1}{x}\right)^x \le \beta_n,$$

where

$$\alpha_n = \left(1 + \frac{1}{n + 1}\right)^n = \left(1 + \frac{1}{n + 1}\right)^{-1} \left(1 + \frac{1}{n + 1}\right)^{n+1},$$

$$\beta_n = \left(1 + \frac{1}{n}\right)^{n+1} = \left(1 + \frac{1}{n}\right)^n \left(1 + \frac{1}{n}\right).$$

Since $\left(1 + \frac{1}{n+1}\right)^{-1} \to 1$ and $\left(1 + \frac{1}{n}\right) \to 1$, we have $\alpha_n \to e$ and $\beta_n \to e$. Therefore, there exists $N \in \mathbb{N}$ such that

$$e - \varepsilon < \alpha_n < e + \varepsilon, \qquad e - \varepsilon < \beta_n < \varepsilon + \varepsilon$$

for all $n \geq N$. Now, for $x > N$, let $n \geq N$ be such that $n \leq x \leq n + 1$. Then we have

$$e - \varepsilon < \alpha_n \leq \left(1 + \frac{1}{x}\right)^x \leq \beta_n < e + \varepsilon.$$

Thus,

$$e - \varepsilon < \left(1 + \frac{1}{x}\right)^x < e + \varepsilon \quad \text{whenever} \quad x > N.$$

Thus, we have proved $\lim_{x \to \infty} \left(1 + \frac{1}{x}\right)^x = e$.   ◇

$$\lim_{x \to \infty} \left(1 + \frac{1}{x}\right)^x = e = \lim_{n \to \infty} \left(1 + \frac{1}{n}\right)^n.$$

Exercise 2.1.10 Suppose (α_n) and (β_n) are sequences of positive real numbers and f is a (real valued) function defined on $(0, \infty)$ having the following property: For $n \in \mathbb{N}$, $x \in \mathbb{R}$,

$$n < x < n + 1 \implies \alpha_n \leq f(x) \leq \beta_n.$$

If (α_n) and (β_n) converge to the same limit, say b, then $\lim_{x \to \infty} f(x) = b$. (*Hint:* Use the arguments used in the Example 2.1.15.)   ◁

2.2 Continuity of a Function

Suppose f is defined on an interval I and $x_0 \in I$. If $\lim_{x \to x_0} f(x)$ exists, then a natural question would be whether the limit is equal to $f(x_0)$.

2.2.1 Definition and Some Basic Results

Definition 2.2.1 Let f be a real valued function defined on a set $D \subseteq \mathbb{R}$. Then f is said to be **continuous at a point** $x_0 \in D$ if for every $\varepsilon > 0$, there exists a $\delta > 0$ such that

$$|f(x) - f(x_0)| < \varepsilon \quad \text{whenever} \quad x \in D, \ |x - x_0| < \delta.$$

The function f is said to be **continuous** on D if it is continuous at every point in D.   ◇

Note that, in the above definition, we did not assume that x_0 is a limit point of D. However, if we assume that x_0 is a limit point of D, then we have the following characterization of continuity.

Fig. 2.9 Limit does not exist
at 0

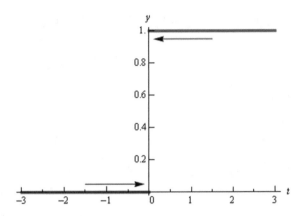

Theorem 2.2.1 *Let $x_0 \in D$ be a limit point of D. Then, for a function $f : D \to \mathbb{R}$, the following are equivalent.*

(i) f is continuous at x_0.
(ii) $\lim_{x \to x_0} f(x)$ exists and it is equal to $f(x_0)$.
(iii) For every sequence (x_n) in D, $x_n \to x_0$ implies $f(x_n) \to f(x_0)$.

Recall that a point x_0 in an interval I is a limit point of I if and only if either $x_0 \in I$ or if x_0 is an endpoint of I. In this book, we shall consider continuity of functions which are defined on intervals.

Convention: When we say that f is continuous at a point $x_0 \in \mathbb{R}$, we mean that f is defined on an interval containing x_0 and f is continuous at x_0.

Example 2.2.1 Let $f(x) = a_0 + a_1 x + \cdots a_k x^k$ for some $a_0, a_1, \ldots, a_k$ in $\mathbb{R}$ and $k \in \mathbb{N}$. We know (cf. Example 2.1.8) that, for any $x_0 \in \mathbb{R}$,

$$\lim_{x \to x_0} f(x) = f(x_0).$$

Hence, by Theorem 2.2.1, f is continuous at every $x_0 \in \mathbb{R}$ (Fig. 2.9).   ◊

Example 2.2.2 Let $f : [-1, 1] \to \mathbb{R}$ be as in Example 2.1.3, i.e.,

$$f(x) = \begin{cases} 0, & -1 \le x \le 0, \\ 1, & 0 < x \le 1. \end{cases}$$

We have seen that $\lim_{x \to 0} f(x)$ does not exist. Hence, by Theorem 2.2.1, f is not continuous at 0.   ◊

Example 2.2.3 We have seen in Examples 2.1.10 and 2.1.11 that

$$\lim_{x \to 0} \sin x = 0, \quad \lim_{x \to 0} \cos x = 1, \quad \lim_{x \to 0} \frac{\sin x}{x} = 1.$$

Fig. 2.10 $\lim_{x \to 0} \dfrac{\sin x}{x} = 1$

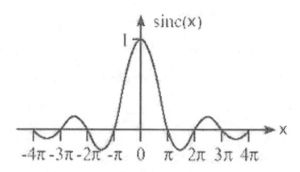

Hence, by Theorem 2.2.1, the functions

$$f(x) := \sin x, \quad g(x) := \cos x, \quad h(x) := \begin{cases} \frac{\sin x}{x}, & x \neq 0, \\ 1, & x = 0 \end{cases}$$

are continuous at 0 (Fig. 2.10). $\Diamond$

The following theorem is a consequence of Theorems 2.1.8 and 2.2.1.

Theorem 2.2.2 *Suppose f and g are defined on an interval I and both f and g are continuous at $x_0 \in I$. Then $f + g$ and fg are continuous at x_0.*

The following Theorem is analogous to Theorem 2.1.11.

Theorem 2.2.3 *Suppose $f : I \to \mathbb{R}$ is continuous at a point $x_0 \in I$ and $g : J \to \mathbb{R}$ is continuous at the point $y_0 := f(x_0)$, where J is an interval such that $f(I) \subseteq J$. Then $g \circ f : I \to \mathbb{R}$ is continuous at x_0.*

Proof Let (x_n) be any sequence in I such that $x_n \to x_0$. Since f is continuous at x_0, we have $f(x_n) \to f(x_0)$. Let $y_n = f(x_n)$, $n \in \mathbb{N}$. Since g is continuous at $y_0 := f(x_0)$, $g(y_n) \to g(y_0)$. Thus, we have proved that for every sequence (x_n) in I with $x_n \to x_0$, $(g \circ f)(x_n) \to (g \circ f)(x_0)$. Hence, by Theorem 2.2.1, $g \circ f$ is continuous at x_0. ∎

Here is another characterization of continuity at a point.

Theorem 2.2.4 *A function $f : I \to \mathbb{R}$ is continuous at a point $x_0 \in I$ if and only if for every open interval J containing $f(x_0)$, there exists an open interval I_0 containing x_0 such that*
$$f(x) \in J \quad \text{whenever} \quad x \in I_0 \cap I.$$

Proof Suppose f is continuous at x_0 and J be such that $f(x_0) \in J$. For $\varepsilon > 0$, let $\delta > 0$ be such that

$$x \in I, \quad |x - x_0| < \delta \quad \Rightarrow \quad |f(x) - f(x_0)| < \varepsilon,$$

i.e., taking $I_0 = (x_0 - \delta, x_0 + \delta)$,

$$x \in I_0 \cap I \quad \Rightarrow \quad f(x) \in (f(x_0) - \varepsilon, f(x_0) + \varepsilon).$$

Choosing $\varepsilon > 0$ small enough such that $(f(x_0) - \varepsilon, f(x_0) + \varepsilon) \subseteq J$, we obtain

$$x \in I_0 \cap I \quad \Rightarrow \quad f(x) \in J.$$

Conversely, suppose that for every open interval J containing $f(x_0)$, there exists an open interval I_0 containing x_0 such that $x \in I_0 \cap I$ implies $f(x) \in J$. So, given $\varepsilon > 0$, we may take $J = (f(x_0) - \varepsilon, f(x_0) + \varepsilon)$. Then, by taking $\delta > 0$ such that $(x_0 - \delta, x_0 + \delta) \subseteq I_0$, we obtain

$$x \in I, \quad |x - x_0| < \delta \quad \Rightarrow \quad |f(x) - f(x_0)| < \varepsilon.$$

Thus, f is continuous at x_0. ∎

Theorem 2.2.5 *Suppose f is a continuous function defined on an interval I and $x_0 \in I$. Suppose $f(x_0) \neq 0$. Then there exists an open interval I_0 containing x_0 such that $f(x) \neq 0$ for every $x \in I_0 \cap I$. Further, the function $g : I_0 \cap I \to \mathbb{R}$ defined by $g(x) = 1/f(x)$ is continuous at x_0.*

Proof Suppose $f(x_0) \neq 0$. Let $J = (a, b)$ be an open interval containing $f(x_0)$ such that $0 \notin J$. Then by Theorem 2.2.4, there exists an open interval I_0 containing x_0 such that $f(x) \in J$ whenever $x \in I_0 \cap I$. In particular, $f(x) \neq 0$ for all $x \in I_0 \cap I$ and $g(x) = 1/f(x)$ is defined on $I_0 \cap I$.

Next, we observe that for every $x \in I_0 \cap I$,

$$\frac{1}{f(x)} - \frac{1}{f(x_0)} = \frac{f(x_0) - f(x)}{f(x) f(x_0)}.$$

Since $f(x) \neq 0$ for all $x \in I_0 \cap I$ we have $|f(x)| > c := \min\{|a|, |b|\}$ for all $x \in I_0 \cap I$. Therefore,

$$\left| \frac{1}{f(x)} - \frac{1}{f(x_0)} \right| = \frac{|f(x_0) - f(x)|}{|f(x) f(x_0)|} \leq \frac{|f(x_0) - f(x)|}{c^2}$$

for all $x \in I_0 \cap I$. Now, by continuity of f at x_0, for every $\varepsilon > 0$, there exists $\delta > 0$ such that

$$|f(x_0) - f(x)| < c^2 \varepsilon \quad \text{whenever} \quad x \in I_0 \cap I, \quad |x - x_0| < \delta.$$

Hence,

$$\left| \frac{1}{f(x)} - \frac{1}{f(x_0)} \right| < \varepsilon \quad \text{whenever} \quad x \in I_0 \cap I, \quad |x - x_0| < \delta.$$

Thus, $1/f$ is continuous at x_0. ∎

Theorems 2.2.2 and 2.2.5 imply the following theorem.

Theorem 2.2.6 *Suppose* $f : I \to \mathbb{R}$ *and* $g : I \to \mathbb{R}$ *are continuous at a point* $x_0 \in I$ *and* $g(x_0) \neq 0$. *Then there exists an open interval* I_0 *containing* x_0 *such that* f/g *is well defined on* $I_0 \cap I$ *and* f/g *is continuous at* x_0.

Exercise 2.2.1 Suppose f is a continuous function defined on an interval I and $x_0 \in I$. Prove the following.

1. If $\alpha \geq 0$ is such that $|f(x_0)| > \alpha$, then there exists a subinterval I_0 of I containing x_0 such that $|f(x)| > \alpha$ for all $x \in I_0$.
2. If $f(x_0) > 0$, then there exists a subinterval I_0 of I containing x_0 such that $f(x) \geq f(x_0)/2$ for all $x \in I_0$.
3. If $f(x_0) < 0$, then there exists a subinterval I_0 of I containing x_0 such that $f(x) \leq f(x_0)/2$ for all $x \in I_0$. ◁

2.2.2 Some More Examples

In the following examples a particular procedure is adopted to show continuity or discontinuity of a function. The reader may adopt any other alternate procedure, for instance, any one of the characterizations in Theorem 2.2.1.

Example 2.2.4 For given $x_0 \in \mathbb{R}$, let $f(x) = |x - x_0|$, $x \in \mathbb{R}$. Then f is continuous on $\mathbb{R}$. To see this, note that, for $a \in \mathbb{R}$,

$$|f(x) - f(a)| = ||x - x_0| - |a - x_0|| \leq |(x - x_0) - (a - x_0)| = |x - a|.$$

Hence, for every $\varepsilon > 0$, we have

$$|x - a| < \varepsilon \Rightarrow |f(x) - f(a)| < \varepsilon. \qquad \Diamond$$

Example 2.2.5 Recall from Example 2.2.9 that

$$\lim_{x \to 2} \frac{x^2 - 4}{x - 2} = 4.$$

Hence, f defined by

$$f(x) := \begin{cases} \frac{x^2 - 4}{x - 2}, & x \neq 2, \\ 4, & x = 2 \end{cases}$$

is continuous at 2. However, the function

$$g(x) := \begin{cases} \frac{x^2-4}{x-2}, & x \neq 2, \\ \alpha, & x = 2 \end{cases}$$

is not continuous at 2 for any $\alpha \neq 4$.

Also, if $a \neq 2$, then $x - 2$ is nonzero in a neighbourhood of a, and the functions $x - 2$ and $x^2 - 4$ are continuous on $\mathbb{R}$. Hence, by Theorem 2.2.6, f is continuous at every $a \neq 2$ as well. By similar arguments, g is continuous at every point $a \neq 2$. $\Diamond$

Example 2.2.6 We already observed in Example 2.2.3 that the functions f, g, h defined by

$$f(x) = \sin x, \quad g(x) = \cos x, \quad h(x) = \begin{cases} \frac{\sin x}{x}, & x \neq 0, \\ 1, & x = 0 \end{cases}$$

are continuous at 0. Now, we show that they are continuous at every point in $\mathbb{R}$.

Note that for $x, y \in \mathbb{R}$,

$$\sin x - \sin y = 2 \sin \left(\frac{x-y}{2}\right) \cos \left(\frac{x+y}{2}\right)$$

so that

$$|\sin x - \sin y| \leq |x - y| \quad \forall x, y \in \mathbb{R}.$$

Hence, for every $\varepsilon > 0$ and for every $x_0 \in \mathbb{R}$,

$$x \in \mathbb{R}, \quad |x - x_0| < \varepsilon \quad \Rightarrow \quad |\sin x - \sin x_0| < \varepsilon.$$

Thus, f is continuous at every point in $\mathbb{R}$. Since $\cos x = 1 - 2\sin^2(x/2)$, $x \in \mathbb{R}$, it also follows that g is continuous at every point in $\mathbb{R}$. Now, let $x_0 \neq 0$. Then the continuity of h at x_0 follows from Theorem 2.2.6, since $h = f/f_0$ where $f_0(x) = x$ in nonzero in a neighbourhood of x_0. $\Diamond$

Exercise 2.2.2 Prove continuity of the function $f(x) := \frac{\sin x}{x}$, $x \neq 0$ at every nonzero point in $\mathbb{R}$ by using $\varepsilon - \delta$ arguments. $\triangleleft$

Example 2.2.7 Let f be defined by $f(x) = 1/x$ on $(0, 1]$. Then there does not exist a continuous function g on $[0, 1]$ such that $g(x) = f(x)$ for all $x \in (0, 1]$:

Suppose there is a function g defined on $[0, 1]$ such that $g(x) = f(x)$ for all $x \in (0, 1]$. Then we have $1/n \to 0$ but $g(1/n) = f(1/n) = n \to \infty$. Thus, $g(1/n) \not\to g(0)$. $\Diamond$

Exercise 2.2.3 Show by $\varepsilon - \delta$ arguments that f defined by $f(x) = 1/x$, $x \neq 0$, is continuous at every $x_0 \neq 0$. $\triangleleft$

Example 2.2.8 The function f defined by $f(x) = \sqrt{x}, x \geq 0$ is continuous at every $x_0 \geq 0$:

Let $\varepsilon > 0$ be given. First consider the point $x_0 = 0$. Then we have

$$|f(x) - f(x_0)| = \sqrt{x} < \varepsilon \quad \text{whenever} \quad |x| < \varepsilon^2.$$

Thus, f is continuous at $x_0 = 0$. Next assume that $x_0 > 0$. Since $|x - x_0| = (\sqrt{x} + \sqrt{x_0})|\sqrt{x} - \sqrt{x_0}|$, we have

$$|\sqrt{x} - \sqrt{x_0}| = \frac{|x - x_0|}{\sqrt{x} + \sqrt{x_0}} \leq \frac{|x - x_0|}{\sqrt{x_0}}.$$

Thus,

$$|\sqrt{x} - \sqrt{x_0}| < \varepsilon \quad \text{whenever} \quad |x - x_0| < \delta := \varepsilon\sqrt{x_0}.$$

More generally, we have the following example. ◇

Example 2.2.9 Let $k \in \mathbb{N}$. Then the function f defined by $f(x) = x^{1/k}, x \geq 0$ is continuous at every $x_0 \geq 0$:

Let $\varepsilon > 0$ be given. First consider the point $x_0 = 0$. Then we have

$$|f(x) - f(x_0)| = x^{1/k} < \varepsilon \quad \text{whenever} \quad |x| < \varepsilon^k.$$

Thus, f is continuous at $x_0 = 0$. Next assume that $x_0 > 0$. Let $y = x^{1/k}$ and $y_0 = x_0^{1/k}$. Since

$$y^k - y_0^k = (y - y_0)(y^{k-1} + y^{k-2}y_0 + \cdots + yy_0^{k-2} + y_0^{k-1}),$$

so that

$$x - x_0 = (x^{1/k} - x_0^{1/k})(y^{k-1} + y^{k-2}y_0 + \cdots + yy_0^{k-2} + y_0^{k-1}).$$

Hence,

$$|x^{1/k} - x_0^{1/k}| = \frac{|x - x_0|}{y^{k-1} + y^{k-2}y_0 + \ldots + yy_0^{k-2} + y_0^{k-1}} \leq \frac{|x - x_0|}{y_0^{k-1}}.$$

Thus,

$$|x^{1/k} - x_0^{1/k}| < \varepsilon \quad \text{whenever} \quad |x - x_0| < \delta := \varepsilon y_0^{k-1} = \varepsilon x_0^{1-1/k}.$$

Thus, f is continuous at every $x_0 > 0$. ◇

Remark 2.2.1 In the above two examples we assumed the knowledge of the square-root and the kth-root of a positive number for $k \in \mathbb{N}$. In school, for $a > 0$ and $k \in \mathbb{N}$, the number $a^{1/k}$ is defined as a number whose k^{th} power is a, that is, $(a^{1/k})^k = a$.

However, existence of such a number is not established. In the next section, we shall prove that the function $f_k : (0, \infty) \rightarrow (0, \infty)$ defined by

$$f_k(x) = x^k, \quad x \in (0, \infty)$$

is bijective so that for every $a > 0$, there exists a unique number $b > 0$ such that $b^k = a$. Such a number b is denoted by $a^{1/k}$, and called the kth-*root* of a.   ◊

Example 2.2.10 For a rational number r, let $f(x) = x^r$ for $x > 0$. Then using Example 2.2.9 together with Theorem 2.2.3, we see that f is continuous at every $x_0 > 0$.   ◊

Remark 2.2.2 We know that for any given $r \in \mathbb{R}$, there exists a sequence (r_n) of rational numbers such that $r_n \rightarrow r$. For $n \in \mathbb{N}$, let $f_n(x) = x^{r_n}$, $x > 0$. Since each f_n is continuous on $[0, \infty)$ one may ask whether

$$\lim_{n \rightarrow \infty} f_n(x)$$

exists for each $x \in [0, \infty)$, and if this limit exists, then one may define x^r as

$$x^r = \lim_{n \rightarrow \infty} f_n(x), \quad x \in [0, \infty).$$

Also, one may want to know whether the function $x \mapsto x^r$ continuous. We shall discuss this issue in a latter section, where we shall introduce two important classes of functions, namely *exponential* and *logarithm functions*. In fact, our discussion will also include, as special cases, the Examples 2.2.8–2.2.10.   ◊

In all the examples given above, either the function is continuous everywhere on the domain of the function or discontinuous at a one point in the domain of definition. Functions with only a finite number of discontinuities can be easily constructed. For instance, the function

$$f(x) = \begin{cases} 4x, & 0 \leq x < 1/4 \\ 4x - 1, & 1/4 \leq x < 1/2, \\ 4x - 2, & 1/2 \leq x < 3/4, \\ 4x - 3, & 3/4 \leq x \leq 1 \end{cases}$$

defined on the interval $[0, 1]$ has three points of discontinuity. The discontinuities are at $1/4$, $1/2$ and $3/4$.

What about functions with infinite number of discontinuities?

Look at the following examples:

Example 2.2.11 Let $f : [0, \infty) \to \mathbb{R}$ be defined by

$$f(x) = x - n \quad \text{whenever} \quad n \leq x < n + 1.$$

Then f has discontinuities at every positive integer.   ◊

Example 2.2.12 Let $J = \{1, 1/2, \ldots\}$ and $f : [0, 1] \to \mathbb{R}$ be defined by

$$f(x) = \begin{cases} 0, & x \notin J, \\ x, & x \in J, \end{cases}$$

that is, $f(1/n) = 1/n$ for all $n \in \mathbb{N}$ and $f(x) = 0$ for $x \notin \{1, 1/2, \ldots\}$. Clearly, this function is not continuous at $1/n$ for any $n \in \mathbb{N}$, and it is continuous at every $x \in (0, 1] \setminus J$. Also, it is continuous at 0.

To see the last statement, consider a sequence (x_n) in $[0, 1]$ such that $x_n \to 0$. Now, for a given $\varepsilon > 0$, let $k \in \mathbb{N}$ be such that $1/k < \varepsilon$, and let $N_k \in \mathbb{N}$ be such that $0 \leq x_n < 1/k$ for all $n \geq N_k$. Then we have

$$|f(x_n)| \leq 1/k < \varepsilon \quad \forall n \geq N_k.$$

Thus, $f(x_n) \to 0$ as $n \to \infty$.   ◊

Example 2.2.13 Let I be an interval, $S := \{a_n : n \in \mathbb{N}\} \subseteq I$ and let $f : I \to \mathbb{R}$ be defined by $f(a_n) = 1/n$ for all $n \in \mathbb{N}$ and $f(x) = 0$ for $x \notin S$. Clearly, this function is not continuous at every $x \in S$. We show that f is continuous at every $x \in I \setminus S$. Let $x_0 \in I \setminus S$. Then $f(x_0) = 0$. For $\varepsilon > 0$, we have to find a $\delta > 0$ such that $|x - x_0| < \delta$ implies $|f(x)| < \varepsilon$.

For $\delta > 0$, let $J_\delta := (x_0 - \delta, x_0 + \delta) \cap I$. For $\varepsilon > 0$, let $k \in \mathbb{N}$ be such that $1/k < \varepsilon$. Choose $\delta > 0$ such that

$$a_1, a_2, \ldots, a_k \notin J_\delta.$$

For instance, we may choose $0 < \delta < \min\{|x_0 - a_i| : i = 1, \ldots, k\}$. Then we have

$$J_\delta \cap \{a_1, a_2, \ldots\} = \{a_{k+1}, a_{k+2}, \ldots\}.$$

Hence, for $x \in J_\delta$, we have either $f(x) = 0$ or $f(x) = 1/n$ for some $n > k$. Thus,

$$|f(x)| \leq \frac{1}{k} < \varepsilon.$$

Thus we have proved that f is continuous at x_0.

If we take S to be the set of all rational numbers in I, then the corresponding function is continuous at every irrational number in I and discontinuous at every rational number in I.   ◊

2.2.3 Some Properties of Continuous Functions

Recall that a subset S of $\mathbb{R}$ is said to be *bounded* if there exists $M > 0$ such that $|s| \le M$ for all $s \in S$, and a set which is not bounded is called an *unbounded set*. Recall also that if S is a bounded subset of $\mathbb{R}$, then S has the infimum and the supremum, not necessarily in S.

For $S \subseteq \mathbb{R}$, we have the following:

1. Suppose S is bounded, and say $\alpha := \inf S$ and $\beta := \sup S$. Then there exist sequences (s_n) and (t_n) in S such that $s_n \to \alpha$ and $t_n \to \beta$.
2. S is unbounded if and only if there exists a sequence (s_n) in S which is unbounded.
3. S is unbounded if and only if there exists a sequence (s_n) in S such that $|s_n| \to \infty$ as $n \to \infty$.
4. If (s_n) is a sequence in S which is unbounded, then there exists a subsequence (s_{k_n}) of (s_n) such that $|s_{k_n}| \to \infty$ as $n \to \infty$.
5. If (s_n) is a sequence in S such that $|s_n| \to \infty$ as $n \to \infty$, and if (s_{k_n}) is a subsequence of (s_n), then $|s_{k_n}| \to \infty$ as $n \to \infty$.

Exercise 2.2.4 Prove the above statements.   ◁

Theorem 2.2.7 *Suppose f is a real valued continuous function defined on a closed and bounded interval $[a, b]$. Then f is a bounded function.*

Proof Assume for a moment that f is not a bounded function. Then, there exists a sequence (x_n) in $[a, b]$ such that $|f(x_n)| \to \infty$. Since (x_n) is a bounded sequence, by Bolzano–Weierstrass theorem (Theorem 1.1.13), there exists a subsequence (x_{k_n}) of (x_n) such that $x_{k_n} \to x$ for some $x \in [a, b]$. Therefore, by the continuity of f, $f(x_{k_n}) \to f(x)$. In particular, $(f(x_{k_n}))$ is a bounded sequence. This is a contradiction to the fact that $|f(x_n)| \to \infty$. Thus, we have proved that f cannot be unbounded. ∎

Remark 2.2.3 The conditions in Theorem 2.2.7 are only sufficient conditions; they are not necessary conditions. To see this consider the function

$$f(x) = \begin{cases} 1, & 0 < x \le 1, \\ 2, & 1 < x < \infty. \end{cases}$$

Then f defined on $I = (0, \infty)$ is not continuous and I is neither closed nor bounded, but f is a bounded function.

It is also true that, if we drop any of the conditions in the theorem, then the conclusion need not be true. To see this, consider the unbounded functions in the following examples:

1. Let

$$f(x) = \begin{cases} 1/x, & x \in (0, 1], \\ 1, & x = 0. \end{cases}$$

In this case f is not continuous, though it is defined on a closed and bounded interval $[0, 1]$.

2. Let $f(x) = 1/x$, $x \in (0, 1]$. In this case f is continuous, but its domain $(0, 1]$ is not a closed set.
3. Let $f(x) = x$, $x \in [0, \infty)$. In this case f is continuous, but its domain $[0, \infty)$ is not bounded.   ◊

Maximum and Minimum

Suppose f is a bounded function defined on an interval I. Then, we know that

$$\inf_{x \in I} f(x) := \inf\{f(x) : x \in I\} \quad \text{and} \quad \sup_{x \in I} f(x) := \sup\{f(x) : x \in I\}$$

exist.

Definition 2.2.2 A function f defined on an interval I is said to attain

(1) **maximum** at a point $x_1 \in I$ if $f(x_1) = \sup_{x \in I} f(x)$,
(2) **minimum** at a point $x_2 \in I$ if $f(x_2) = \inf_{x \in I} f(x)$.

The function f is said to attain **extremum** at a point $x_0 \in I$ if f attains either maximum or minimum at x_0.   ◊

If f attains maximum (respectively, minimum) at a point in I,then we write $\max_{x \in I} f(x)$ for $\sup_{x \in I} f(x)$ (respectively, $\min_{x \in I} f(x)$ for $\inf_{x \in I} f(x)$). A natural question would be the following:

Under what conditions on f and I does the function attain maximum and minimum at some points in I?

In this regard, we have the following theorem.

Theorem 2.2.8 *Suppose f is a continuous function defined on a closed and bounded interval $[a, b]$. Then there exists x_0, y_0 in $[a, b]$ such that*

$$f(x_0) = \inf_{a \le x \le b} f(x) \quad and \quad f(y_0) = \sup_{a \le x \le b} f(x).$$

Proof Let $\alpha = \inf_{a \le x \le b} f(x)$. By the definition of the infimum of a set, there exists a sequence (x_n) in $[a, b]$ such that $f(x_n) \to \alpha$. Since (x_n) is a bounded sequence, by Bolzano–Weierstrass theorem, there exists a subsequence (x_{k_n}) such that $x_{k_n} \to x_0$ for some $x_0 \in [a, b]$. By continuity of f, $f(x_{k_n}) \to f(x_0)$. But, we already have $f(x_{k_n}) \to \alpha$. Hence, $\alpha = f(x_0)$, that is, $f(x_0) = \inf_{a \le x \le b} f(x)$.

Similarly, using the definition of supremum, it can be shown that there exists $y_0 \in [a, b]$ such that $f(y_0) = \sup_{a \le x \le b} f(x)$.   ∎

Every continuous function $f : [a, b] \to \mathbb{R}$ attain maximum and Minimum at some points in $[a, b]$

Remark 2.2.4 The conclusion in Theorem 2.2.8 need not hold if the domain of the function is not of the form $[a, b]$ or if f is not continuous. For example, $f : (0, 1] \to \mathbb{R}$ defined by $f(x) = 1/x$ for $x \in (0, 1]$ is continuous, but does not attain supremum. Same is the case if $g : [0, 1] \to \mathbb{R}$ is defined by

$$g(x) = \begin{cases} 1/x, & x \in (0, 1], \\ 1, & x = 0. \end{cases}$$

Thus, neither continuity nor the fact that the domain is a closed and bounded interval can be dropped. This does not mean that the conclusion in the theorem does not hold for all those functions which are either discontinuous or their domains are not closed intervals. For example, consider $f : [0, 1) \to \mathbb{R}$ defined by

$$f(x) = \begin{cases} 0, & x \in [0, 1/2), \\ 1, & x \in [1/2, 1). \end{cases}$$

Then we see that neither f is continuous, nor its domain is of the form $[a, b]$. But, f attains both its maximum and minimum.   ◇

Intermediate Value Theorem

Suppose f is a continuous real valued function defined on an interval. We ask the following question:

Given any number γ lying between any two values α and β of the function f, does f attain the value γ at some point in the domain of definition of f?

The answer is in the affirmative. This result is known as the *Intermediate value theorem*.

Theorem 2.2.9 (Intermediate value theorem (IVT)) *Suppose f is a continuous function defined on a closed and bounded interval $[a, b]$. Let γ be a number between $f(a)$ and $f(b)$. Then there exists $c \in [a, b]$ such that $f(c) = \gamma$ (Fig. 2.11).*

Before giving its proof, let us look at the interpretations of the theorem geometrically and algebraically.
Geometric interpretation:

Consider the curve C represented by the equation $y = f(x)$, where f is a continuous function on $[a, b]$. If γ lies between the values $f(a)$ and $f(b)$, then the curve C intersects with the straight line represented by the equation $y = \gamma$

Algebraic interpretation:

Fig. 2.11 Intermediate
Value Property

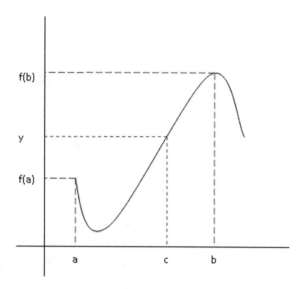

If f is a continuous function defined on $[a, b]$ and γ lies between the values $f(a)$ and $f(b)$, then the equation $f(x) = \gamma$ has at least one solution in $[a, b]$

Theorem 2.2.9 ensures that if γ lies between $f(a)$ and $f(b)$, then there exists at least one point c between a and b such that $f(c) = \gamma$. There can be more than one such points. Obviously, if f is a constant function, then every point in $[a, b]$ is having this property. Even if the function is not constant in any subinterval, then also there can be more than one such points. In this regard, we may look at the Fig. 2.12. Also, for the function f given by

$$f(x) = x^2, \quad x \in [-1, 1],$$

we have $f(-1/2) = 1/4 = f(1/2)$.

Also, a function can take all intermediate values without being continuous. For example, the function $f : [-1, 1] \to \mathbb{R}$ defined by

$$f(x) = \begin{cases} x, & x \in [-1, 0], \\ 1 - x, & x \in (0, 1] \end{cases}$$

is not continuous at 0, but it takes all intermediate values between any two values.

Now, we give the proof of Theorem 2.2.9 adapted from Ghorpade and Limaye [6].

Proof of Theorem 2.2.9 Without loss of generality, assume that $f(a) < f(b)$. If γ is either $f(a)$ or $f(b)$, then there is nothing to prove. So, let $f(a) < \gamma < f(b)$. Let

Fig. 2.12 Intermediate value property

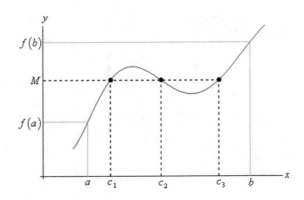

$$S = \{x \in [a, b] : \ f(x) < \gamma\}.$$

Note that S is non-empty (since $a \in S$) and bounded above (since $x \leq b$ for all $x \in S$). Let

$$u := \sup S.$$

Then $u \in [a, b]$ and there exists a sequence (u_n) in S such that $u_n \to u$. Hence, by continuity of f, $f(u_n) \to f(u)$. Since $f(u_n) < \gamma$ for all $n \in \mathbb{N}$, we have $f(u) \leq \gamma$. Note that $u \neq b$, since $f(u) \leq \gamma < f(b)$.

Now, let (v_n) be a sequence in (u, b) such that $v_n \to u$. Then, again by continuity of f, $f(v_n) \to f(u)$. Since $v_n > u, v_n \notin S$ and hence $f(v_n) \geq \gamma$. Therefore, $f(u) \geq \gamma$. Thus, $a \leq u \leq b$ and $f(u) = \gamma$.   ∎

Example 2.2.14 Consider the function

$$f(x) = x^5 + 2x^3 + 1, \quad x \in \mathbb{R}.$$

Does there exists $x_0 \in \mathbb{R}$ such that $f(x_0) = 0$? The answer is in the affirmative by IVT, since $f(-1) = -2$ and $f(1) = 4$. Not only that, since

$$f(x) \to -\infty \quad \text{as} \quad x \to -\infty \quad \text{and} \quad f(x) \to \infty \quad \text{as} \quad x \to \infty,$$

f attain every real value, that is, for every $y \in \mathbb{R}$, there exists $x \in \mathbb{R}$ such that $f(x) = y$. In other words, f is an onto function.   ◊

Example 2.2.15 Consider the function

$$f(x) = x^3 + \sin x + 1, \quad x \in \mathbb{R}.$$

Does there exists $x_0 \in \mathbb{R}$ such that $f(x_0) = 0$? Note that

$$f(0) = 1 \quad \text{and} \quad f(-2) = -8 + \sin(-2) + 1 < 0.$$

Hence, by IVT, there exists $x_0 \in \mathbb{R}$ such that $f(x_0) = 0$. As in the case of the last example, we see that f attain every real value. ◊

The following two corollaries are immediate consequences of IVT.

Corollary 2.2.10 *Let f be a continuous function defined on an interval. Then range of f is an interval.*

Corollary 2.2.11 *Suppose f is a continuous function defined on an interval I. If $a, b \in I$ are such that $f(a)$ and $f(b)$ have opposite signs, then there exists $x_0 \in I$ such that $f(x_0) = 0$.*

Now, we derive another important property of continuous functions.

Theorem 2.2.12 *Suppose f is a continuous function defined on a closed and bounded interval I. Then its range is a closed and bounded interval.*

Proof We know, by Theorem 2.2.7 Corollary 2.2.10, that range of f is a bounded interval, say J. Hence, it is enough to show that J contains its endpoints. For this, let c be an end point of J. Let (y_n) in J such that $y_n \to c$. Let $x_n \in I$ be such that $f(x_n) = y_n$, $n \in \mathbb{N}$. Since I is closed and bounded, (x_n) has a subsequence (x_{k_n}) which converges to some point $x_0 \in I$. By continuity of f, $y_{k_n} = f(x_{k_n}) \to f(x_0)$. Thus, we obtain $c = f(x_0) \in J$. This completes the proof. ∎

Continuity of the Inverse of a Function

Suppose f is defined on a set $D \subseteq \mathbb{R}$. We may recall the following from elementary set theory:

If f is injective, i.e., one-one, then we know that a function g can be defined on the range $E := f(D)$ of f by $g(y) = x$ for $y \in E$, where $x \in D$ is the unique element in D such that $f(x) = y$. The above function g is called the **inverse** of f. Note that the domain of the inverse of f is the range of f.

By Corollary 2.2.10, we know that range of a continuous function defined on an interval I is also an interval. Suppose f is also injective. Then a natural question one would like to ask is whether its inverse is also continuous. First we answer this question affirmatively by assuming that the domain of the function is a closed and bounded interval.

Theorem 2.2.13 (Inverse function theorem (IFT)) *Let f be a continuous injective function defined on a closed and bounded interval I. Then its inverse from its range is continuous.*

Proof Suppose $J = f(I)$, the range of f. Let $y_0 \in J$ and (y_n) be an arbitrary sequence in J which converges to y_0. Let $x_n = f^{-1}(y_n)$, $n \in \mathbb{N}$ and $x_0 = f^{-1}(y_0)$. We have to show that $x_n \to x_0$.

Suppose, on the contrary, that $x_n \not\to x_0$. Then there exists $\varepsilon_0 > 0$ and a subsequence (u_n) of (x_n) such that $u_n \notin (x_0 - \varepsilon_0, x_0 + \varepsilon_0)$ for all $n \in \mathbb{N}$. Since I is a bounded interval, (u_n) is a bounded sequence. Hence, (u_n) has a subsequence (v_n)

which converges to some $v \in \mathbb{R}$. Since I is a closed interval, $v \in I$. Now, continuity of f implies that $f(v_n) \to f(v)$. But, since $(f(v_n))$ is a subsequence of (y_n), and since $y_n \to y_0$, we have $f(v) = y_0 = f(x_0)$. Now, since f is injective, $v = x_0$. Thus we have proved that $v_n \to x_0$. This is a contradiction to the fact that $v_n \notin (x_0 - \varepsilon_0, x_0 + \varepsilon_0)$ for all $n \in \mathbb{N}$.   ∎

Next we shall prove the conclusion in the last theorem by dropping the condition that I is closed and bounded, but assuming an additional condition on f, namely that it is *strictly monotonic*.

Definition 2.2.3 Let f be defined on an interval I. Then

(1) f is said to be **monotonically increasing** on I if

$$x, y \in I, \quad x < y \quad \Rightarrow \quad f(x) \leq f(y),$$

(2) f is said to be **strictly monotonically increasing** on I if

$$x, y \in I, \quad x < y \quad \Rightarrow \quad f(x) < f(y),$$

(3) f is said to be **monotonically decreasing** on I if

$$x, y \in I, \quad x < y \quad \Rightarrow \quad f(x) \geq f(y).$$

(4) f is said to be **strictly monotonically decreasing** on I if

$$x, y \in I, \quad x < y \quad \Rightarrow \quad f(x) > f(y).$$

If f is either monotonically increasing (respectively, strictly monotonically increasing) or monotonically decreasing (respectively, strictly monotonically decreasing) on I, then it is called a **monotonic (respectively, strictly monotonic)** function. ◊

Observe:

$$f \text{ is strictly monotonic on } I \quad \Rightarrow \quad f \text{ is injective on } I.$$

The converse of the above statement is not true. For example, the function

$$f(x) = \begin{cases} x, & -1 \leq x \leq 0, \\ 1 - x, & 0 < x \leq 1, \end{cases}$$

is injective but not strictly monotonic on $[-1, 1]$.

Sometimes, the terminology increasing, decreasing, strictly increasing, strictly decreasing, are used in place of monotonically increasing, monotonically decreasing, strictly monotonically increasing, and strictly monotonically decreasing, respectively.

Example 2.2.16 We observe the following.

(i)   The function $f(x) = x$ is strictly increasing on $\mathbb{R}$.

(ii) The function $f(x) = -x$ is strictly decreasing on $\mathbb{R}$.

(iii) The function $f(x) = x^2$ is strictly increasing for $x \geq 0$ and strictly decreasing for $x \leq 0$.

(iv) The function $f(x) = x^3$ is strictly increasing on $\mathbb{R}$.

(v)   For $k \in \mathbb{N}$, $f(x) = x^k$ is strictly increasing on $[0, \infty)$.

(vi) The function $f(x) = \sin x$ is strictly increasing on $[0, \pi/2]$ and strictly decreasing on $[\pi/2, \pi]$.

(vii) The function $f(x) = \cos x$ is strictly decreasing on $[0, \pi]$.   ◊

Now, we prove a companion result to Theorem 2.2.13 without assuming that the domain of the function is closed and bounded, but assuming that the function is strictly monotonic.

Theorem 2.2.14 (Inverse function theorem (IFT)) *Let f be a continuous function defined on an interval I. Suppose f is strictly monotonic on I. Then f is injective and its inverse from its range is continuous.*

Proof We assume that f is strictly monotonically increasing. The case when f is strictly monotonically decreasing will follow by similar arguments.

Since f is continuous, its range is also an interval, say J. By the assumption, for $x_1, x_2 \in J$, $x_1 < x_2 \Rightarrow f(x_1) < f(x_2)$. Hence, f is injective. Let g be its inverse from the range J. Let $y_0 \in J$ and (y_n) in J be such that $y_n \to y_0$. Let $x_n = g(y_n)$, $n \in \mathbb{N}$ and $x_0 = g(y_0)$ for that $y_n = f(x_n)$ and $y_0 = f(x_0)$ for all $n \in \mathbb{N}$. We have to show that $x_n \to x_0$.

Suppose $x_n \not\to x_0$. Then there exists $\varepsilon > 0$ and a subsequence (x_{k_n}) of (x_n) such that

$$x_{k_n} \notin (x_0 - \varepsilon, x_0 + \varepsilon) \quad \forall n \in \mathbb{N}.$$

Hence, for each $n \in \mathbb{N}$, either $x_{k_n} < x_0 - \varepsilon$ or $x_{k_n} > x_0 + \varepsilon$. Hence, using the fact that f is strictly monotonically increasing, we have the following, for each $n \in \mathbb{N}$:

$$x_{k_n} < x_0 - \varepsilon \quad \Rightarrow \quad x_0 - \varepsilon \in I \text{ and } f(x_{k_n}) < f(x_0 - \varepsilon) < f(x_0),$$

$$x_{k_n} > x_0 + \varepsilon \quad \Rightarrow \quad x_0 + \varepsilon \in I \text{ and } f(x_{k_n}) > f(x_0 + \varepsilon) > f(x_0).$$

Thus, $(f(x_0 - \varepsilon), f(x_0 + \varepsilon))$ is an open interval containing $f(x_0)$ such that $f(x_{k_n}) \notin (f(x_0 - \varepsilon), f(x_0 + \varepsilon))$ for every $n \in \mathbb{N}$. Hence, we can conclude that $y_{k_n} \not\to y_0$, which is a contradiction. Thus, we have proved that g is continuous. ∎

Theorem 2.2.15 (Root function) *For every $y \in [0, \infty)$ and $k \in \mathbb{N}$, there exists a unique $x \in [0, \infty)$ such that $x^k = y$.*

Proof We have observed in Example 2.2.16 that for $k \in \mathbb{N}$, the function

$$f(x) = x^k, \quad x \geq 0,$$

is strictly monotonically increasing. Also, $f(0) = 0$ and $f(x) \to \infty$ as $x \to \infty$. Hence, by IVT and IFT (Theorem 2.2.14), $f : [0, \infty) \to [0, \infty)$ is bijective and the inverse of f is continuous. In particular, for every $y \in [0, \infty)$, there exists a unique $x \in [0, \infty)$ such that $x^k = y$.   ∎

Definition 2.2.4 For $k \in \mathbb{N}$ and $y > 0$, the unique number $x > 0$ such that $x^k = y$ is called the kth-**root** of y, and it is denoted by $y^{1/k}$.   ◇

Remark 2.2.5 (i) In the proof of Theorem 2.2.14, the continuity of f is used only to assert that its range J is an interval so that its inverse f^{-1} is defined on an interval.

(ii) We know that strict monotonicity of a function implies that it is injective, but injectivity does not implies strict monotonicity. So, one may ask whether strict monotonicity assumption in Theorem 2.2.14 can be replaced by injectivity. The answer is in the affirmative as the following Exercise shows.   ◇

Exercise 2.2.5 Let f be an injective function defined on an interval I. Show that if f is continuous, then it is strictly monotonic on I.
[*Hint:* IVT].   ◁

2.2.4 Exponential and Logarithm Functions

We have already come across the expression a^b for $a > 0$ and $b \in \mathbb{R}$, though we have not defined it explicitly. However, from elementary arithmetic we know the definition of a^n for $n \in \mathbb{N}$, and the relation, $a^{m+n} = a^m a^n$ for $m, n \in \mathbb{N}$. Also, defining $a^{-n} = \frac{1}{a^n}$ for $n \in \mathbb{N}$ and using the convention $a^0 = 1$, we have

$$a^{m+n} = a^m a^n \quad \forall\, m, n \in \mathbb{Z}.$$

Also, we have defined $a^{1/n}$, the nth-root of a for any $a > 0$ and $n \in \mathbb{N}$ (cf. Theorem 2.2.15 and Definition 2.2.4). Thus, for any positive rational number $r = m/n$, we can define

$$a^r := (a^m)^{1/n}, \quad a^{-r} := \frac{1}{a^r}.$$

Thus, we have defined a^r for any $a > 0$ and $r \in \mathbb{Q}$. Now, the question is, can we have a meaningful definition of a^x for any $x \in \mathbb{R}$? To this end, we shall first define e^x, where e is the Euler constant defined by

$$e := \lim_{n \to \infty} \left(1 + \frac{1}{n}\right)^n \quad \text{or} \quad e := \sum_{n=0}^{\infty} \frac{1}{n!}.$$

We define e^x for $x \in \mathbb{R}$ in such a way that $e^1 = \sum_{n=0}^{\infty} \frac{1}{n!}$. More specifically, we define the concept of an *exponential function* $\exp(x)$, $x \in \mathbb{R}$, as

$$\exp(x) = \sum_{n=0}^{\infty} \frac{x^n}{n!}, \quad x \in \mathbb{R}.$$

Observe that the above series converges absolutely for every $x \in \mathbb{R}$. This is easily seen by using the ratio test. This series plays very significant roles in mathematics.

Definition 2.2.5 For $x \in \mathbb{R}$, the function

$$\exp(x) := \sum_{n=0}^{\infty} \frac{x^n}{n!}, \quad x \in \mathbb{R},$$

is called the **exponential function**. ◊

Clearly,

$$\exp(0) = 1, \quad \exp(1) = e.$$

Our first attempt is to show that

$$\exp(r) = e^r$$

for every rational number. In order to do that we have to derive some of the important properties of the function $\exp(x)$. For that purpose, we shall make use of the following result on convergence of series.

Theorem 2.2.16 *Suppose that $\sum_{n=0}^{\infty} a_n$ and $\sum_{n=0}^{\infty} b_n$ are absolutely convergent series, and*

$$c_n = \sum_{k=0}^{n} a_k b_{n-k}, \quad n \in \mathbb{N} \cup \{0\}.$$

Then, the series $\sum_{n=0}^{\infty} c_n$ converges absolutely and

$$\left(\sum_{n=0}^{\infty} a_n \right) \left(\sum_{n=0}^{\infty} b_n \right) = \sum_{n=0}^{\infty} c_n.$$

Proof Since $\sum_{n=0}^{\infty} a_n$ and $\sum_{n=0}^{\infty} b_n$ are absolutely convergent, they are convergent. Let their sums be A and B, respectively. Let

$$A_n = \sum_{i=0}^{n} a_i, \quad B_n = \sum_{i=0}^{n} b_i, \quad C_n = \sum_{i=0}^{n} c_i.$$

Then $A_n \to A$, $B_n \to B$ and $A_n B_n \to AB$. We have to prove that $C_n \to AB$.

First let us assume that the terms of the series are non-negative. Note that, if

$$\alpha_{ij} = a_i b_j, \quad i, j = 0, 1, \ldots, n,$$

then $A_n B_n$ is the sum of all entries of the matrix (α_{ij}) and C_n is the sum of the entries of the left upper triangular part of the matrix (α_{ij}), i.e.,

$$A_n B_n = \sum_{i=0}^{n} \sum_{j=0}^{n} \alpha_{ij}, \quad C_n = \sum_{i=0}^{n} \sum_{j=0}^{n-i} \alpha_{ij}.$$

Hence, it follows that

$$C_n \le A_n B_n \le C_{2n} \qquad\qquad (*)$$

for all $n \in \mathbb{N}$. Since $(A_n B_n)$ converges to AB and (C_n) is an increasing sequence of nonnegative terms, the relation $(*)$ implies that (C_n) is bounded, and hence it converges. Let $C_n \to C$. Since (C_{2n}) is a subsequence of (C_n), we also have the convergence $C_{2n} \to C$. Hence, by Sandwich theorem, $C_n \to AB$. This proves the case when the series are with nonnegative terms.

Next let us consider the general case. By what we have already proved, we have

$$\left(\sum_{i=0}^{\infty} |a_i| \right) \left(\sum_{i=0}^{\infty} |b_i| \right) = \sum_{k=0}^{\infty} \left(\sum_{i=0}^{k} |a_i|\, |b_{k-i}| \right).$$

Let

$$\hat{A}_n = \sum_{i=0}^{n} |a_i|, \quad \hat{B}_n = \sum_{i=0}^{n} |b_i|, \quad \hat{C}_n = \sum_{k=0}^{n} \left(\sum_{i=0}^{k} |a_i|\, |b_{k-i}| \right).$$

As in the last paragraph, we obtain $\hat{C}_n \le \hat{A}_n \hat{B}_n \le \hat{C}_{2n}$. Note that

$$|A_n B_n - C_n| \le \sum_{i=1}^{n} \sum_{j=n-i+1}^{n} |\alpha_{ij}| = \hat{A}_n \hat{B}_n - \hat{C}_n \le \hat{C}_{2n} - \hat{C}_n.$$

Since $(\hat{C}_n)$ converges, we obtain $\hat{C}_{2n} - \hat{C}_n \to 0$, and since $A_n B_n \to AB$, we have the convergence $C_n \to AB$. ∎

Definition 2.2.6 The series $\sum_{n=0}^{\infty} c_n$ with $c_n = \sum_{k=0}^{n} a_k b_{n-k}$ is called the **Cauchy product** of $\sum_{n=0}^{\infty} a_n$ and $\sum_{n=0}^{\infty} b_n$. ◇

It is to be observed that the Cauchy product of two (non-absolutely) convergent series need not be convergent. To see this one may consider the series $\sum_{n=1}^{\infty} a_n$ and $\sum_{n=1}^{\infty} b_n$ with $a_n = b_n = (-1)^{n+1}/\sqrt{n+1}$ for $n \in \mathbb{N}$. Since

$$c_n = \sum_{k=0}^{n} a_k b_{n-k} = (-1)^n \sum_{k=0}^{n} \frac{1}{\sqrt{k+1}\sqrt{n-k+1}}$$

and $\sqrt{(k+1)(n-k+1)} \le n+1$ we have $|c_n| \ge 1$ for all $n \in \mathbb{N}$. Hence, $\sum_{n=1}^{\infty} c_n$ is not convergent.

Properties of Exponential and Logarithm Functions

Recall from Definition 2.2.5 that for $x \in \mathbb{R}$,

$$\exp(x) := \sum_{n=0}^{\infty} \frac{x^n}{n!}, \quad x \in \mathbb{R}.$$

Theorem 2.2.17 *Let* $\exp(\cdot)$ *be the function as in Definition 2.2.5. Then the following results hold.*

(i) $\exp(x + y) = \exp(x)\exp(y) \quad \forall x, y \in \mathbb{R}$
(ii) $\exp(x) \neq 0 \quad \forall x \in \mathbb{R}.$
(iii) $\exp(-x) = \dfrac{1}{\exp(x)} \quad \forall x \in \mathbb{R}.$
(iv) $\exp(x) > 0 \quad \forall x \in \mathbb{R}.$
(v) $\exp(kx) = [\exp(x)]^k \quad \forall x \in \mathbb{R}, \ k \in \mathbb{Z}.$ *In particular,*

   (a) $\exp(k) = e^k, \quad \forall k \in \mathbb{Z},$
   (b) $[\exp(1/k)]^k = e \quad \forall k \in \mathbb{Z}.$
   (c) $\exp(m/n) = [\exp(1/n)]^m \quad \forall m, n \in \mathbb{Z} \text{ with } n \neq 0.$

(vi) $\exp(x) > 1 \iff x > 0 \ $ *and* $\ \exp(x) = 1 \iff x = 0.$
(vii) $x > y \iff \exp(x) > \exp(y).$
(viii) $\exp(x) \to \infty$ *as* $x \to \infty.$
(ix) $\exp(x) \to 0$ *as* $x \to -\infty.$

Proof Note that, for $x, y \in \mathbb{R}$,

$$\frac{(x+y)^n}{n!} = \frac{1}{n!}\sum_{k=0}^{n} \frac{n!}{k!(n-k)!}x^k y^{n-k} = \sum_{k=0}^{n} \frac{x^k}{k!}\frac{y^{n-k}}{(n-k)!}.$$

Hence, by Theorem 2.2.16 by taking $a_n = x^n/n!$ and $b_n = y^n/n!$, we have

$$\sum_{n=0}^{\infty} \frac{(x+y)^n}{n!} = \Big(\sum_{n=0}^{\infty} \frac{x^n}{n!}\Big)\Big(\sum_{n=0}^{\infty} \frac{y^n}{n!}\Big).$$

This proves (i). The results in (ii), (iii) and (v) follow from (i), and the result in (iv) follows from (iii), since $\exp(x) > 0$ for $x \geq 0$.

To see (vi), observe that $x > 0$ implies $\exp(x) > 1$. Next, suppose $x \leq 0$. If $x = 0$, then $\exp(x) = \exp(0) = 1$. If $x < 0$, then taking $y = -x$, we have $y > 0$, and hence $\exp(y) > 1$. Thus, $1/\exp(x) = \exp(-x) > 1$ so that $\exp(x) < 1$. Hence, $\exp(x) > 1 \iff x > 0$. From the above arguments, we also obtain $\exp(x) = 1 \iff x = 0$. The result in (vii) follows from the facts that

$$x > y \iff x - y > 0 \iff \exp(x - y) > 1$$

and the relation $\exp(x - y) = \exp(x)/\exp(y)$, which is a consequence of (i) and (iii).

The result in (viii) follows from the relation

$$\exp(x) = 1 + x + \sum_{n=2}^{\infty} \frac{x^n}{n!} \geq 1 + x \quad \forall x > 0,$$

and (ix) is a consequence of (iii) and (viii). ∎

In view of (v)(b) above, we may define

$$e^{1/k} := \exp(1/k) \quad \forall k \in \mathbb{N},$$

and hence by (v)(c),

$$e^{m/n} := [e^{1/n}]^m \quad \forall m, n \in \mathbb{N}.$$

Thus, for every rational number r, we can define

$$e^r := \exp(r)$$

which satisfies the usual index laws.

We know that every real number is a limit of a sequence of rational numbers. Thus, if $x \in \mathbb{R}$, there exists a sequence (x_n) of rational numbers that $x_n \to x$. So, we may define

$$e^x = \lim_{n \to \infty} e^{x_n}$$

provided the above limit exists. Thus, our next attempt is to show that the function $\exp(x)$, $x \in \mathbb{R}$, is continuous.

Theorem 2.2.18 *The function* $\exp(\cdot)$ *is continuous on* $\mathbb{R}$

Proof For brevity of expression, let us use the notation e^x for $\exp(x)$. Let x, $x_0 \in \mathbb{R}$. Then, by Theorem 2.2.17(i), we have

$$e^x - e^{x_0} = e^{x_0}(e^{x - x_0} - 1)$$

$$= e^{x_0} \sum_{n=1}^{\infty} \frac{(x - x_0)^n}{n!}$$

$$= e^{x_0}(x - x_0) \sum_{n=1}^{\infty} \frac{(x - x_0)^{n-1}}{n!}.$$

Thus, if $|x - x_0| \leq 1$, then

$$|e^x - e^{x_0}| \leq e^{x_0}|x - x_0| \sum_{n=1}^{\infty} \frac{1}{n!} = e^{x_0}(e - 1)|x - x_0|.$$

From this, we obtain that $\exp(\cdot)$ is a continuous function on $\mathbb{R}$. ∎

Notation 2.2.1 We know that for every $x \in \mathbb{R}$, there exists a sequence (x_n) of rational numbers such that $x_n \to x$. In view of Theorem 2.2.18,

$$e^{x_n} = \exp(x_n) \to \exp(x).$$

Hence, we shall use the notation e^x for $\exp(x)$ for every $x \in \mathbb{R}$, i.e.,

$$e^x := \sum_{n=0}^{\infty} \frac{x^n}{n!}, \quad x \in \mathbb{R}. \qquad \Diamond$$

With the above notation we have the identity:

$$e^{x+y} = e^x e^y \quad \forall x, y \in \mathbb{R}.$$

Theorem 2.2.19 *The function $\exp(\cdot)$ is bijective from $\mathbb{R}$ to $(0, \infty)$.*

Proof Using the notation $e^x := \exp(x)$, $x \in \mathbb{R}$, first we observe that $e^x > 0$ for every $x \in \mathbb{R}$, and for x_1, x_2 in $\mathbb{R}$

$$e^{x_2} - e^{x_1} = e^{x_1}[e^{x_2 - x_1} - 1].$$

Thus,

$$e^{x_2} = e^{x_1} \iff e^{x_2 - x_1} = 1 \iff x_1 = x_2,$$

showing that the function $x \mapsto e^x$ is one-one.

Next, we show that the function is onto. Let $y \in (0, \infty)$. Recall that (cf. Theorem 2.2.17)

$$e^x \to 0 \quad \text{as} \quad x \to -\infty, \qquad e^x \to \infty \quad \text{as} \quad x \to \infty.$$

Hence, there exists $M_1 > 0$ such that $e^x > y$ for all $x > M_1$, and there exists $M_2 > 0$ such that $e^x < y$ for all $x < -M_2$. Now, taking $x_1 > M_1$ and $x_2 < -M_2$, we obtain

$$e^{x_2} < y < e^{x_1}.$$

Hence, by the intermediate value property, there exists $x \in (x_2, x_1)$ such that $e^x = y$. ∎

Definition 2.2.7 For $b > 0$, the unique $a \in \mathbb{R}$ such that $e^a = b$ is called the **natural logarithm** of b, and it is denoted by $\ln b$. The function

$$\ln x, \quad x > 0,$$

is called the **natural logarithm function**.   ◊

Definition 2.2.8 For $a > 0$ and $x \in \mathbb{R}$, we define

$$a^x := e^{x \ln a}.$$ ◊

Remark 2.2.6 We note that $\ln e = 1$ so that if $a = e$, then the Definition 2.2.8 matches with Definition 2.2.5.   ◊

Theorem 2.2.20 *Let $a > 0$. Then the function $x \mapsto a^x$ is continuous and bijective from $\mathbb{R}$ to $(0, \infty)$.*

Proof By Definition 2.2.8, $a^x := e^{x \ln a}$ for $x \in \mathbb{R}$. Hence, the result is a consequence of Theorems 2.2.18 and 2.2.19, and using the fact that composition of two continuous functions is continuous.   ∎

The following corollary is immediate from the above theorem.

Corollary 2.2.21 *Let $a > 0$ be given. Then for every $x > 0$, there exists a unique $b \in \mathbb{R}$ such that $a^b = x$.*

Definition 2.2.9 Let $a > 0$. For $x > 0$, the unique $b \in \mathbb{R}$ such that $a^b = x$ is called the **logarithm** of x to the base a, and it is denoted by $\log_a x$. The function

$$\log_a x, \qquad x > 0,$$

is called the **logarithm function**.   ◊

Note that, for $x > 0$,
$$\log_e x = \ln x.$$

We observe that following.

1. For $y \in \mathbb{R}$, $\quad y = \ln x \iff e^y = x$.
2. For $a > 0$ and $y \in \mathbb{R}$, $\quad y = \log_a x \iff a^y = x$.
3. For $a > 0$ and $x > 0$, $\quad \log_a x = \dfrac{\ln x}{\ln a}$.
4. For $a > 0, b > 0$, show that $(\log_b a)(\log_a b) = 1$.

Exercise 2.2.6 Prove the above results.   ◁

Theorem 2.2.22 *For $a > 0$, the function $\log_a x$, $x > 0$, is continuous on $(0, \infty)$.*

Proof First we show that $\ln x$, $x > 0$ is continuous. Then the continuity of $\log_a x$, $x > 0$ will follow from the relation $\log_a x = \ln x / \ln a$.

Let x, x_0 belong to the interval $(0, \infty)$, and let $y = \ln x$ and $y_0 = \ln x_0$. Then we have $e^y = x$ and $e^{y_0} = x_0$. Assume, without loss of generality that $x > x_0$. Since $e^a > 1$ if and only if $a > 0$, we have $y > y_0$, and hence

$$x - x_0 = e^y - e^{y_0} = e^{y_0}(e^{y-y_0} - 1)$$
$$= e^{y_0} \sum_{n=1}^{\infty} \frac{(y - y_0)^n}{n!} \geq e^{y_0}(y - y_0).$$

Hence, $|y - y_0| \leq e^{-y_0}|x - x_0|$, that is,

$$|\ln x - \ln x_0| \leq e^{-y_0}|x - x_0|,$$

showing the continuity of $\ln x$, $x > 0$. ∎

Example 2.2.17 Using continuity of the function $x \mapsto \log x$, $x > 0$, we have

$$\lim_{n \to \infty} \frac{\log(1 + n)}{n} = 0.$$

Indeed, since $(1 + n)^{1/n} \to 1$, we have

$$\frac{\log(1 + n)}{n} = \log(1 + n)^{1/n} \to \log 1 = 0. \qquad \Diamond$$

Theorem 2.2.23 *For* $r \in \mathbb{R}$, *the function* $f : (0, \infty) \to \mathbb{R}$ *be defined by*

$$f(x) = x^r, \qquad x \in (0, \infty)$$

is continuous.

Proof For $r \in \mathbb{R}$ and $x > 0$, we have $x^r = e^{r \ln x}$. Hence, the result follows from Theorem 2.2.22 and Theorem 2.2.3. ∎

Notation 2.2.2 Often, the notation $\log x$ is used for the natural logarithm function in place $\ln x$. $\Diamond$

2.3 Differentiability of a Function

2.3.1 Definition and Examples

Let f be a continuous function defined on an open interval I and $x_0 \in I$. In the Cartesian plane, consider the curve represented by the equation

$$y = f(x), \qquad x \in I.$$

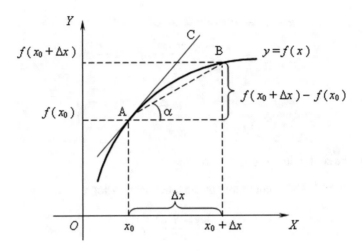

Fig. 2.13 Derivative

Then, for x in a neigbourhood of x_0, the quotient

$$\frac{f(x) - f(x_0)}{x - x_0}$$

is the slop of the line segment joining the points $X_0 = (x_0, f(x_0))$ and $X = (x, f(x))$. As $x \to x_0$, the point X approaches X_0, so that, if the limit

$$\lim_{x \to x_0} \frac{f(x) - f(x_0)}{x - x_0}$$

exists, then it can be thought of as the *slope of the tangent* to the curve at the point X_0.

Definition 2.3.1 Suppose f is a (real valued) function defined on an open interval I and $x_0 \in I$. Then f is said to be **differentiable at** x_0 if

$$\lim_{x \to x_0} \frac{f(x) - f(x_0)}{x - x_0}$$

exists, and in that case the value of the limit is called the **derivative** of f at x_0, and this value is denoted by $f'(x_0)$ (Fig. 2.13).   ◇

Notation 2.3.1 The number $f'(x_0)$, namely, the derivative of f at x_0 is also denoted by

$$\frac{\mathrm{d}f}{\mathrm{d}x}(x_0) \quad \text{or} \quad \frac{\mathrm{d}}{\mathrm{d}x} f(x)|_{x=x_0}.$$

The notation $\dfrac{df}{dx}$, introduced by Leibnitz, is useful in realizing that the expression $\dfrac{d}{dx}$ as an *operator* which associates each differentiable function f to the function f'. Writing the function f as an equation

$$y = f(x), \quad x \in I,$$

and denoting Δx a small number such that $x + \Delta x \in I$ whenever $x \in I$, the notation Δy is used for the difference $f(x + \Delta x) - f(x)$ so that

$$\frac{\Delta y}{\Delta x} = \frac{f(x + \Delta x) - f(x)}{\Delta x}.$$

The above expression is sometimes called as the *differential quotient* of f at x. In view of this, the notation $\frac{dy}{dx}$ is used for $f'(x)$, that is,

$$\frac{dy}{dx} = f'(x) = \lim_{\Delta x \to 0} \frac{\Delta y}{\Delta x}. \qquad \Diamond$$

A physical interpretation of the derivative: Suppose a particle is moving along a straight line. Suppose the particle is at points A and B at time t_0 and t, which are at distances $f(t_0)$ and $f(t)$, respectively, from a fixed point O on the line. Then the ratio

$$\frac{f(t) - f(t_0)}{t - t_0}$$

can be thought of as the average velocity of the particle while moving from A to B during the time interval $[t_0, t]$, and the limit

$$\lim_{t \to t_0} \frac{f(t) - f(t_0)}{t - t_0}$$

can be thought of as the *instantaneous velocity* at the point A.

Let f be a real valued function defined on an open interval I containing x_0. We observe the equivalence of the following statements.

1. f is differentiable at $x_0 \in I$.
2. The limit $\lim_{h \to 0} \dfrac{f(x_0 + h) - f(x_0)}{h}$ exists.
3. For every sequence (x_n) in $I \setminus \{x_0\}$ such that $x_n \to x_0$, the limit $\lim_{n \to \infty} \dfrac{f(x_n) - f(x_0)}{x_n - x_0}$ exists, and in that case

$$f'(x_0) = \lim_{n \to \infty} \frac{f(x_n) - f(x_0)}{x_n - x_0}.$$

Convention: Whenever we say that "a function f is differentiable at a point x_0", we mean that f is a real valued function defined on an open interval I containing x_0 and $f : I \to \mathbb{R}$ is differentiable at x_0.

Example 2.3.1 Let us look at the following simple examples.

(i) Derivative of a constant function is 0 at every point. Indeed, for $c \in \mathbb{R}$, if $f(x) = c$ for all $x \in \mathbb{R}$, then we have

$$\frac{f(x) - f(x_0)}{x - x_0} = 0$$

for every $x, x_0 \in \mathbb{R}$ with $x \neq x_0$. hence, $f'(x_0) = 0$.

(ii) Let $f(x) = x$, $x \in \mathbb{R}$. Then for any $x_0 \in \mathbb{R}$,

$$\frac{f(x) - f(x_0)}{x - x_0} = 1 \quad \forall x \neq x_0.$$

Hence $f'(x_0) = 1$.

More generally, let f be a polynomial function, that is,

$$f(x) = a_0 + a_1 x + \cdots + a_n x^n, \quad x \in \mathbb{R}$$

for some $n \in \mathbb{N}$. Then, for any $x_0 \in \mathbb{R}$, we have

$$f(x) - f(x_0) = a_1(x - x_0) + \cdots + a_n(x^n - x_0^n).$$

We know that, for any $k \in \mathbb{N}$,

$$x^k - x_0^k = (x - x_0) \sum_{i=1}^{k} x^{k-i} x_0^{i-1}.$$

Thus,

$$f(x) - f(x_0) = \sum_{k=1}^{n} a_k(x^k - x_0^k) = (x - x_0) g(x),$$

where

$$g(x) := \sum_{k=1}^{n} a_k \sum_{i=1}^{k} x^{k-i} x_0^{i-1}.$$

Thus,

$$\lim_{x \to x_0} \frac{f(x) - f(x_0)}{x - x_0} = \lim_{x \to x_0} g(x) = g(x_0) = \sum_{k=1}^{n} k a_k x_0^{k-1},$$

and therefore,

$$f'(x_0) = \sum_{k=1}^{n} k a_k x_0^{k-1}.$$

In particular, for $n \geq 2$,

$$f'(x_0) = a_1 + 2a_2 x_0 + \cdots + n a_n x_0^{n-1}.$$

(iii) Let $f(x) = \sin x$, $x \in \mathbb{R}$. Then for any $x, x_0 \in \mathbb{R}$ with $x \neq x_0$,

$$\frac{f(x) - f(x_0)}{x - x_0} = \frac{2 \cos \left(\frac{x+x_0}{2}\right) \sin \left(\frac{x-x_0}{2}\right)}{x - x_0}$$

$$= \cos \left(\frac{x + x_0}{2}\right) \frac{\sin \left(\frac{x-x_0}{2}\right)}{\frac{x-x_0}{2}}.$$

Then, using the composition rule for limits and the facts that

$$\lim_{y \to y_0} \cos y = \cos y_0 \quad \text{and} \quad \lim_{y \to 0} \frac{\sin y}{y} = 1,$$

we obtain

$$f'(x_0) = \lim_{x \to x_0} \frac{f(x) - f(x_0)}{x - x_0} = \cos x_0.$$

(iv) The function e^x is differentiable at every $x \in \mathbb{R}$ and

$$(e^x)' = e^x \quad \forall x \in \mathbb{R}.$$

To see this, first we note that for $h \neq 0$,

$$\frac{e^{x+h} - e^x}{h} - e^x = \frac{e^x}{h}(e^h - 1 - h) = \frac{e^x}{h} \sum_{n=2}^{\infty} \frac{h^n}{n!} = e^x h \sum_{n=2}^{\infty} \frac{h^{n-2}}{n!}.$$

Hence, $|h| \leq 1$ implies

$$\left| \frac{e^{x+h} - e^x}{h} - e^x \right| \leq e^x |h| \sum_{n=2}^{\infty} \frac{1}{n!} = e^x |h| (e - 2).$$

Thus, we have proved that the function e^x is differentiable at x and its derivative is e^x. ◊

Remark 2.3.1 In deriving the result in Example 2.3.1(iv), we used the following fact: If (a_n) is a sequence such that $\sum_{n=1}^{\infty} a_n$ converges absolutely, then $\sum_{n=1}^{\infty} a_n$ converges and

$$\left| \sum_{n=1}^{\infty} a_n \right| \le \sum_{n=1}^{\infty} |a_n|. \qquad\qquad \diamond$$

Many of the functions that occur in mathematics can be constructed with the help of the functions considered in the Example 2.3.1 using some properties of differentiation.

In Example 2.3.1(ii), we have seen that if f is a polynomial, then we can write

$$f(x) - f(x_0) = (x - x_0)g(x),$$

where g is a polynomial, and therefore, $f'(x_0) = g(x_0)$. Do we have similar situation for a general differentiable function? In this regard, we have the following theorem.

Theorem 2.3.1 (Caratheodory criterion[1]) *Suppose f is defined in an open interval I. Then f is differentiable at $x_0 \in I$ if and only if there exists a function $g : I \to \mathbb{R}$ which is continuous at x_0 and satisfies*

$$f(x) = f(x_0) + g(x)(x - x_0),$$

and in that case $f'(x_0) = g(x_0)$.

Proof Suppose f is differentiable at $x_0 \in I$. Then the function g defined on I by

$$g(x) = \begin{cases} \frac{f(x)-f(x_0)}{x-x_0}, & x \ne x_0, \\ f'(x_0), & x = x_0, \end{cases}$$

satisfies the requirements.

Conversely, suppose that there exists a function $g : I \to \mathbb{R}$ which is continuous at x_0 and satisfies $f(x) = f(x_0) + g(x)(x - x_0)$. Then we have

$$\lim_{x \to x_0} \frac{f(x) - f(x_0)}{x - x_0} = \lim_{x \to x_0} g(x) = g(x_0)$$

so that f is differentiable at $x_0 \in I$ and $f'(x_0) = g(x_0)$. $\blacksquare$

Let us illustrate the above theorem by some examples.

Example 2.3.2 Let $f(x) = \sqrt{x}$, $x > 0$. For $x_0, x > 0$, we have

$$f(x) - f(x_0) = \sqrt{x} - \sqrt{x_0} = \frac{x - x_0}{\sqrt{x} + \sqrt{x_0}} = (x - x_0)g(x),$$

where $g(x) = 1/(\sqrt{x} + \sqrt{x_0})$. Since g is continuous at x_0, we can infer from Theorem 2.3.1 that $f'(x_0) = g(x_0) = 1/(2\sqrt{x_0})$. $\diamond$

---

[1] Constantin Carathéodory (September 13, 1873 February 2,1950) was a Greek mathematician who made significant contributions to the theory of functions of a real variable, the calculus of variations, and measure theory.

Example 2.3.3 Let $f(x) = e^x := \sum_{n=0}^{\infty} \frac{x^n}{n!}$, $x \in \mathbb{R}$. Then we have

$$f(x) - f(x_0) = \sum_{n=1}^{\infty} \frac{x^n - x_0^n}{n!}$$

But,

$$x^n - x_0^n = (x - x_0) \sum_{i=1}^{n} x^{n-i} x_0^{i-1}.$$

Thus, formally,

$$f(x) - f(x_0) = (x - x_0) \sum_{n=1}^{\infty} \frac{1}{n!} \left(\sum_{i=1}^{n} x^{n-i} x_0^{i-1} \right) = (x - x_0) g(x),$$

with

$$g(x) = \sum_{n=1}^{\infty} \frac{1}{n!} \left(\sum_{i=1}^{n} x^{n-i} x_0^{i-1} \right).$$

If we can show that the series defining $g(x)$ actually converges and the limit is a continuous function, then we have $g(x_0) = e^{x_0}$. For showing the convergence of $\sum_{n=1}^{\infty} \frac{1}{n!} \left(\sum_{i=1}^{n} x^{n-i} x_0^{i-1} \right)$ to a continuous function, we need to use the concept of *uniform convergence of sequence of functions* which we shall do in a later chapter of this book.   ◊

2.3.2 Left and Right Derivatives

Recall that in the definition of continuity of a function we considered the domain of the function to be an interval, not necessarily an open interval, whereas in the definition of differentiability we took the interval to be an open interval. Even in the definition of differentiability we could have taken an arbitrary interval I and x_0 to be an end point of I if it belongs to that interval. In such case, we have the so called *right differentiability* or *left differentiability* at x_0 depending on whether x_0 is a right end point or left end point of I.

In fact right differentiability and left differentiability can be defined at an interior point as well. By an *interior point* of an interval I we mean those points in I which are not the end points. More generally:

Definition 2.3.2 A a point $a \in \mathbb{R}$ is said to be an **interior point** of a set $D \subseteq \mathbb{R}$ if D contains an open neighbourhood of a.   ◊

Definition 2.3.3 Let f be a real valued function defined on an interval I and $x_0 \in I$.

(1) If x_0 is such that $(x_0 - \delta_0, x_0] \subseteq I$ for some $\delta_0 > 0$, then f is said to be **left differentiable** at x_0 if

$$f'_-(x_0) := \lim_{x \to x_0-} \frac{f(x) - f(x_0)}{x - x_0} \quad \text{exists,}$$

and in that case, it is called the **left derivative** of f at x_0.

(2) If x_0 is such that $[x_0, x_0 + \delta_0) \subseteq I$ for some $\delta_0 > 0$, then f is said to be **right differentiable** at x_0 if

$$f'_+(x_0) := \lim_{x \to x_0+} \frac{f(x) - f(x_0)}{x - x_0} \quad \text{exists,}$$

and in that case, it is called the **right derivative** of f at x_0. $\Diamond$

Remark 2.3.2 In some of the books in calculus, one may find the notations $f'(x_0-)$ and $f'(x_0+)$ for left derivative and right derivative, respectively, at x_0. We preferred to use the notations $f'_-(x_0)$ and $f'_+(x_0)$ as the notations $f'(x_0-)$ and $f'(x_0+)$ can be confused with the left and right limits of the function f' at the point x_0.

Left derivative and right derivative at a point can be characterized as follows: Let f be a real valued function defined on an interval I and $x_0 \in I$.

(i) If $(x_0 - \delta_0, x_0] \subseteq I$ for some $\delta_0 > 0$, then $f'_-(x_0)$ exists if and only if for every sequence (x_n) in $(x_0 - \delta_0, x_0)$ with $x_n \to x_0$, we have $\lim_{n \to \infty} \frac{f(x_n) - f(x_0)}{x_n - x_0}$ exists, and in that case

$$f'_-(x_0) = \lim_{n \to \infty} \frac{f(x_n) - f(x_0)}{x_n - x_0}.$$

(ii) If $[x_0, x_0 + \delta_0) \subseteq I$ for some $\delta_0 > 0$, then $f'_+(x_0)$ exists if and only if for every sequence (x_n) in $(x_0, x_0 + \delta_0)$ with $x_n \to x_0$, we have $\lim_{n \to \infty} \frac{f(x_n) - f(x_0)}{x_n - x_0}$ exists, and in that case

$$f'_+(x_0) = \lim_{n \to \infty} \frac{f(x_n) - f(x_0)}{x_n - x_0}.$$ $\Diamond$

In view of the above discussion, we have the following:

If x_0 is an interior point of I, then $f'(x_0)$ exists if and only if $f'_+(x_0)$ and $f'_-(x_0)$ exists and $f'(x_0) = f'_+(x_0) = f'_-(x_0)$.

Example 2.3.4 Let

$$f(x) = \begin{cases} 0, & x \in [-1, 0), \\ 1, & x \in [0, 1]. \end{cases}$$

Then we have following.

1. f is differentiable at every $x_0 \in (-1, 0) \cup (0, 1)$, and $f'(x_0) = 0$,
2. f is right differentiable at -1 and $f'_+(-1) = 0$,
3. f is right differentiable at 0 and $f'_+(0) = 0$, left derivative at 0 does not exist.
4. f is left differentiable at 1, and $f'_-(1) = 0$.

Here are some details of the above statements:

(1) If $x_0 \in (-1, 0) \cup (0, 1)$, then

$$\lim_{x \to x_0} \frac{f(x) - f(x_0)}{x - x_0} = \lim_{x \to x_0} \frac{0}{x - x_0} = 0.$$

(2) If $x_0 = -1$, then

$$\lim_{x \to x_0^+} \frac{f(x) - f(x_0)}{x - x_0} = \lim_{x \to x_0^+} \frac{0 - 0}{x - x_0} = 0.$$

(3) If $x_0 = 0$, then

$$\lim_{x \to x_0^+} \frac{f(x) - f(x_0)}{x - x_0} = \lim_{x \to x_0^+} \frac{1 - 1}{x - x_0} = 0$$

and

$$\lim_{x \to x_0^-} \frac{f(x) - f(x_0)}{x - x_0} = \lim_{x \to x_0^-} \frac{0 - 1}{x} \quad \text{does not exist.}$$

(4) If $x_0 = 1$, then

$$\lim_{x \to x_0^-} \frac{f(x) - f(x_0)}{x - x_0} = \lim_{x \to x_0^-} \frac{1 - 1}{x - x_0} = 0. \qquad \diamondsuit$$

Example 2.3.5 Consider the **signum function**, $f(x) = \mathrm{sgn}(x)$, $x \in \mathbb{R}$, that is, $f : \mathbb{R} \to \mathbb{R}$ is defined by

$$f(x) = \begin{cases} x/|x| & \text{if } x \neq 0, \\ 0 & \text{if } x = 0. \end{cases}$$

Note that $f(x) = 1$ for $x > 0$, $f(x) = -1$ for $x < 0$. Hence, we obtain $f'(x) = 0$ for every $x \neq 0$. Note that

$$\frac{f(x) - f(0)}{x} = \begin{cases} 1/x, & x > 0 \\ -1/x, & x < 0. \end{cases}$$

Hence, neither $f'_+(0)$ nor $f'_-(0)$ exists. However,

$$\lim_{x \to 0^-} f'(x) = 0 = \lim_{x \to 0^+} f'(x). \qquad \diamondsuit$$

Example 2.3.6 Let $f : \mathbb{R} \to \mathbb{R}$ be defined by

$$f(x) = \begin{cases} 1 - |x| & \text{if } x \in [-1, 1], \\ 0 & \text{if } x \notin [-1, 1]. \end{cases}$$

That is,

$$f(x) = \begin{cases} 1 + x & \text{if } x \in [-1, 0), \\ 1 - x & \text{if } x \in [0, 1], \\ 0 & \text{if } x \notin [-1, 1]. \end{cases}$$

Clearly, f is differentiable at every $x_0 \notin \{-1, 0, 1\}$. Let us consider the situations at the points $-1, 0, 1$.

(i) Let $x_0 = -1$. Then

$$\lim_{x \to x_0-} \frac{f(x) - f(x_0)}{x - x_0} = \lim_{x \to x_0-} \frac{0 - 0}{x - (-1)} = 0,$$

$$\lim_{x \to x_0+} \frac{f(x) - f(x_0)}{x - x_0} = \lim_{x \to x_0+} \frac{(1 + x) - 0}{x - (-1)} = 1.$$

Hence, $f'_-(-1) = 0$ and $f'_+(-1) = 1$.

(ii) Let $x_0 = 0$. Then

$$\lim_{x \to x_0-} \frac{f(x) - f(x_0)}{x - x_0} = \lim_{x \to x_0-} \frac{(1 + x) - 1}{x - 0} = 1,$$

$$\lim_{x \to x_0+} \frac{f(x) - f(x_0)}{x - x_0} = \lim_{x \to x_0+} \frac{(1 - x) - 1}{x - 0} = -1.$$

Hence, $f'_-(0) = 1$ and $f'_+(0) = -1$.

(iii) Let $x_0 = 1$. Then

$$\lim_{x \to x_0-} \frac{f(x) - f(x_0)}{x - x_0} = \lim_{x \to x_0-} \frac{(1 - x) - 0}{x - 1} = -1,$$

$$\lim_{x \to x_0+} \frac{f(x) - f(x_0)}{x - x_0} = \lim_{x \to x_0+} \frac{0 - 0}{x - 1} = 0.$$

Hence, $f'_-(1) = -1$ and $f'_+(0) = 0$. Thus left and right derivatives of f at the points $-1, 0, 1$ exist, but they are not the same, and hence f is not differentiable at these points.   ◊

2.3.3 Some Properties of Differentiable Functions

The proof of the following theorem is easy, and hence it is left as an exercise.

Theorem 2.3.2 *Suppose f and g are differentiable at a point x_0 and $\alpha \in \mathbb{R}$. Then $f + g$ and αf are differentiable at x_0, and*

$$(f + g)'(x_0) = f'(x_0) + g'(x_0), \qquad (\alpha f)'(x_0) = \alpha f'(x_0).$$

Here is a necessary condition for differentiability.

Theorem 2.3.3 (Differentiability implies continuity) *Suppose f is differentiable at a point x_0. Then f is continuous at x_0.*

Proof Let $x_0 \in I$. Then for every $x \in I$ with $x \neq x_0$, we have

$$f(x) - f(x_0) = \left[\frac{f(x) - f(x_0)}{x - x_0} \right] (x - x_0).$$

Hence, $\lim_{x \to x_0} (f(x) - f(x_0)) = f'(x_0).0 = 0$, showing that f is continuous at x_0. ∎

Exercise 2.3.1 Prove that if f is left (respectively, right) differentiable at x_0, then it is left (respectively, right) continuous at x_0. ◁

For the following theorem, we may recall that if a function g is continuous at a point x_0 and $g(x_0) \neq 0$, then there exists an open interval I_0 containing x_0 such that $g(x) \neq 0$ for all $x \in I_0$ (cf. Theorem 2.2.5).

Theorem 2.3.4 (Product and quotient rules) *Suppose f and g are defined on an open interval I and differentiable at $x_0 \in I$. Then fg is differentiable at x_0, and*

$$(fg)'(x_0) = f'(x_0)g(x_0) + f(x_0)g'(x_0).$$

If g is nonzero in a neighbourhood of x_0, then f/g is differentiable at x_0, and

$$\left(\frac{f}{g} \right)'(x_0) = \frac{g(x_0)f'(x_0) - f(x_0)g'(x_0)}{[g(x_0)]^2}.$$

Proof Let $\varphi := fg$ and $\psi := f/g$. For any $x \in I$, we have

$$\varphi(x) - \varphi(x_0) = f(x)g(x) - f(x_0)g(x_0)$$
$$= [f(x) - f(x_0)]g(x) + f(x_0)[g(x) - g(x_0)]$$

so that, using the facts that $f'(x_0)$ and $g'(x_0)$ exist and g is continuous at x_0, we obtain

$$\frac{\varphi(x) - \varphi(x_0)}{x - x_0} = \frac{f(x) - f(x_0)}{x - x_0} g(x) + f(x_0) \frac{g(x) - g(x_0)}{x - x_0}$$
$$\rightarrow f'(x_0)g(x_0) + f(x_0)g'(x_0) \quad \text{as} \quad x \rightarrow x_0.$$

Hence, φ is differentiable at x_0, and

$$\varphi'(x_0) = f'(x_0)g(x_0) + f(x_0)g'(x_0).$$

Also, for $x \in I$ with $x \neq x_0$, we have

$$\psi(x) - \psi(x_0) = \frac{f(x)g(x_0) - f(x_0)g(x)}{g(x)g(x_0)}$$
$$= \frac{[f(x) - f(x_0)]g(x_0) - f(x_0)[g(x) - g(x_0)]}{g(x)g(x_0)},$$

so that, using differentiability of f and g at x_0 and continuity of g at x_0, we have

$$\frac{\psi(x) - \psi(x_0)}{x - x_0} = \frac{1}{g(x)g(x_0)} \left[\frac{f(x) - f(x_0)}{x - x_0} g(x_0) - f(x_0) \frac{g(x) - g(x_0)}{x - x_0} \right]$$
$$\rightarrow \frac{f'(x_0)g(x_0) - f(x_0)g'(x_0)}{[g(x_0)]^2}$$

as $x \rightarrow x_0$. Thus, ψ is differentiable at x_0, and

$$\psi'(x_0) = \frac{g(x_0)f'(x_0) - f(x_0)g'(x_0)}{[g(x_0)]^2}.$$

This completes the proof. ∎

Theorem 2.3.5 (Composition rule) *Suppose f is differentiable at x_0 and g is differentiable at $y_0 := f(x_0)$. Then $g \circ f$ is differentiable at x_0 and*

$$(g \circ f)'(x_0) = g'(y_0)f'(x_0).$$

Proof Let (x_n) be a sequence in a deleted neighbourhood of x_0 which converges to x_0. We have to prove that $\lim_{n \to \infty} \dfrac{(g \circ f)(x_n) - (g \circ f)(x_0)}{x_n - x_0}$ exists and the limit is $g'(y_0)f'(x_0)$. For this, let $y_n := f(x_n)$ for $n \in \mathbb{N}$ and $y_0 = f(x_0)$. Let us look at the formal expression

$$\frac{(g \circ f)(x_n) - (g \circ f)(x_0)}{x_n - x_0} = \frac{g(y_n) - g(y_0)}{x_n - x_0}$$
$$= \frac{g(y_n) - g(y_0)}{y_n - y_0} \times \frac{f(x_n) - f(x_0)}{x_n - x_0}.$$

Since $f'(x_0)$ exists, $\lim_{n\to\infty} \dfrac{f(x_n) - f(x_0)}{x_n - x_0} = f'(x_0)$. However, we will not be able

to write $\lim_{n\to\infty} \dfrac{g(y_n) - g(y_0)}{y_n - y_0} = g'(x_0)$, because (y_n) may not be in a deleted neigh-

bourhood of y_0, although $y_n \to y_0$, by continuity of f at x_0. To take care of this situation, for each $n \in \mathbb{N}$, we define

$$\alpha_n = \begin{cases} \frac{g(y_n)-g(y_0)}{y_n-y_0} & \text{if } y_n \neq y_0, \\ g'(y_0) & \text{if } y_n = y_0. \end{cases}$$

Then $\alpha_n \to g'(y_0)$. Hence,

$$\frac{(g \circ f)(x_n) - (g \circ f)(x_0)}{x_n - x_0} = \alpha_n \times \frac{f(x_n) - f(x_0)}{x_n - x_0} \to g'(y_0)f'(x_0)$$

showing that $(g \circ f)'(x_0) = g'(y_0)f'(x_0)$. ∎

In view of the formula in Theorem 2.3.5, the following result is not surprising.

Theorem 2.3.6 *Suppose $g \circ f$ is differentiable at x_0, g is differentiable at $y_0 := f(x_0)$ with $g'(y_0) \neq 0$, and f is continuous at x_0. Then f is differentiable at x_0 and*

$$f'(x_0) = \frac{(g \circ f)'(x_0)}{g'(y_0)}.$$

Proof Let (x_n) be a sequence in a deleted neighbourhood of x_0 which converges to x_0, $y_n := f(x_n)$ for $n \in \mathbb{N}$ and $y_0 = f(x_0)$. Let (α_n) be as in the proof of Theorem 2.3.5, that is,

$$\alpha_n = \begin{cases} \frac{g(y_n)-g(y_0)}{y_n-y_0} & \text{if } y_n \neq y_0, \\ g'(y_0) & \text{if } y_n = y_0. \end{cases}$$

Since f is continuous at x_0, $y_n \to y_0$. Hence, $\alpha_n \to g'(y_0) \neq 0$ and $\alpha_n \neq 0$ for all large enough n. Then, for all such n, we have

$$\frac{(g \circ f)(x_n) - (g \circ f)(x_0)}{x_n - x_0} = \frac{g(y_n) - g(y_0)}{x_n - x_0} = \alpha_n \frac{f(x_n) - f(x_0)}{x_n - x_0},$$

so that

$$\frac{f(x_n) - f(x_0)}{x_n - x_0} = \frac{1}{\alpha_n} \times \frac{(g \circ f)(x_n) - (g \circ f)(x_0)}{x_n - x_0}$$

$$\to \frac{(g \circ f)'(x_0)}{g'(y_0)} \quad \text{as } n \to \infty.$$

Thus $f'(x_0)$ exists and $f'(x_0) = \dfrac{(g \circ f)'(x_0)}{g'(y_0)}$ ∎

As a corollary to the above theorem we have a formula for the derivative of the inverse of a function.

Theorem 2.3.7 *Suppose* $f : I \to J$ *is a bijective function between open intervals I and J. Suppose* f *is differentiable at a point* $x_0 \in I$ *and* $f'(x_0) \neq 0$ *and* f^{-1} *is continuous at* $y_0 := f(x_0)$. *Then* f^{-1} *is differentiable at* y_0, *and*

$$(f^{-1})'(y_0) = \frac{1}{f'(x_0)}.$$

Proof Note that $(f \circ f^{-1})(y) = y$ for every $y \in J$. Hence by Theorem 2.3.6, f^{-1} is differentiable at y_0 and $(f^{-1})'(y_0) = 1/f'(x_0)$.   ∎

Remark 2.3.3 Recall that in Theorem 2.3.6 and Theorem 2.3.7 we assumed $g'(y_0) \neq 0$ and $f'(x_0) \neq 0$, respectively. Can we obtain at least differentiability without the above assumptions? Theorem 2.3.5 shows that the condition $f'(x_0) \neq 0$ is necessary in Theorem 2.3.7 for the differentiability of f^{-1} at x_0. What about the case of Theorem 2.3.6? That is, can we say that f is differentiable at x_0 and $f'(x_0)g'(y_0) = (g \circ f)'(x_0)$? The answer is not in the affirmative. For example, consider

$$f(x) = |x|, \quad g(x) = x^2, \quad x \in \mathbb{R}.$$

Then $(g \circ f)(x) = x^2$. In this case, $g \circ f$ and g are differentiable at every point in $\mathbb{R}$, but f is not differentiable at $x_0 = 0$. Note that $g'(y_0) = 0$.   ◇

Example 2.3.7 In the following we consider some applications of Theorems 2.3.2–2.3.7.

(i) For $n \in \mathbb{N}$, let $f(x) = x^n$, $x \in \mathbb{R}$. We have already seen that $f'(x) = nx^{n-1}$ for $x \in \mathbb{R}$. This can also be seen by the product formula: Let $f_k(x) := x^k$ for $k \in \mathbb{N}$. Then we have $f_k(x) = xf_{k-1}(x)$ for all $k \in \mathbb{N}$. Now, $f_1'(x) = 1$. Assuming that $f_k'(x) = kx^{k-1}$ for some $k \in \mathbb{N}$, we have

$$f_{k+1}'(x) = \frac{d}{dx}[xf_k(x)] = xf_k'(x) + f_k(x) = x(kx^{k-1}) + x^k = (k+1)x^k.$$

Thus, by induction, $\frac{d}{dx}x^n = nx^{n-1}$ for all $x \in \mathbb{R}$.

(ii) Let $f(x) = \cos x$. Since $\cos x = 1 - 2\sin^2(x/2)$, $x \in \mathbb{R}$, applying Theorems 2.3.2 and 2.3.5,

$$\frac{d}{dx}[1 - 2\sin^2(x/2)] = \frac{d}{dx}[-2\sin^2(x/2)] = -2\frac{d}{dx}\sin^2(x/2),$$

$$\frac{d}{dx}\sin^2(x/2) = 2\sin(x/2)\frac{d}{dx}\sin(x/2) = 2\sin(x/2)[(1/2)\cos(x/2)].$$

Hence,

$$f'(x) = -2\frac{d}{dx}\sin^2(x/2) = -2\sin(x/2)\cos(x/2) = -\sin x.$$

(iii) Let $f(x) = \tan x$ for $x \in D := \{x \in \mathbb{R} : \cos x \neq 0\}$. Then

$$f'(x) = \frac{d}{dx}\frac{\sin x}{\cos x} = \frac{\cos x \cos x - \sin x(-\sin x)}{\cos^2 x} = \frac{1}{\cos x} = \sec^2 x.$$

(iv) By Theorem 2.3.7, if $y = \tan x$ for $-\pi/2 < x < \pi/2$, then

$$\frac{d}{dy}\tan^{-1} y = \frac{1}{\tan' x} = \frac{1}{\sec^2 x} = \frac{1}{1 + \tan^2 x} = \frac{1}{1 + y^2}.$$

Thus,

$$\frac{d}{dy}\tan^{-1} y = 1/(1 + y^2) \quad \text{for all} \quad y \in \mathbb{R}. \qquad \diamondsuit$$

Example 2.3.8 Let $f : \mathbb{R} \to \mathbb{R}$ be defined by

$$f(x) = \begin{cases} x \sin(1/x) & \text{if } x \neq 0, \\ 0 & \text{if } x = 0. \end{cases}$$

From the composition and product rules, it can be seen that f is differentiable at every $x_0 \neq 0$. Now, let us check the differentiability at $x_0 = 0$. For x in a deleted neighbourhood of 0, we have

$$\frac{f(x) - f(0)}{x} = \frac{x \sin(1/x)}{x} = \sin(1/x).$$

Hence $f'(0)$ does not exist (cf. Example 2.1.7). $\qquad \diamondsuit$

Example 2.3.9 Let $f : \mathbb{R} \to \mathbb{R}$ be defined by

$$f(x) = \begin{cases} x^2 \sin(1/x) & \text{if } x \neq 0, \\ 0 & \text{if } x = 0. \end{cases}$$

In this case also, f is differentiable at every $x_0 \neq 0$ which follows from the composition and product rules. Now, let $x_0 = 0$ and x be in a deleted neighbourhood of 0. Then

$$\frac{f(x) - f(0)}{x} = \frac{x^2 \sin(1/x)}{x} = x \sin(1/x).$$

Since $0 \leq |x \sin(1/x)| \leq |x|$, $\lim_{x \to 0} x \sin(1/x)$ exists and it is equal to 0. Hence $f'(0)$ exists and $f'(0) = 0$. $\qquad \diamondsuit$

Example 2.3.10 Let $f(x) = x|x|$, $x \in \mathbb{R}$, that is,

$$f(x) = \begin{cases} x^2 & \text{if } x \geq 0, \\ -x^2 & \text{if } x < 0. \end{cases}$$

Note that, for $x \neq 0$, f is differentiable at x and $f'(x) = 2|x|$. Now, let us check the differentiability at 0. We have

$$\lim_{x \to 0-} \frac{f(x) - f(0)}{x} = \lim_{x \to 0-} \frac{x^2}{x} = 0,$$

$$\lim_{x \to 0+} \frac{f(x) - f(0)}{x} = \lim_{x \to 0+} \frac{-x^2}{x} = 0.$$

Thus, $f'_-(0) = 0 = f'_+(0)$ so that f is differentiable at 0 and $f'(0) = 0$. Hence, $f'(x) = 2|x|$ for every $x \in \mathbb{R}$.   ◊

Example 2.3.11 For $a > 0$, the function a^x is differentiable at every $x \in \mathbb{R}$ and

$$(a^x)' = a^x \ln a \qquad \forall x \in \mathbb{R}.$$

Indeed, by the composition rule (Theorem 2.3.5),

$$(a^x)' = (e^{x \ln a})' = e^{x \ln a} \ln a = a^x \ln a.$$ ◊

Example 2.3.12 The function $\ln x$ is differentiable for every $x > 0$, and

$$(\ln x)' = \frac{1}{x}, \qquad x > 0.$$

To see this, let $f(x) = \ln x$ and $g(x) = e^x$. Then we have $g(f(x)) = x$ for every $x > 0$. Since $g \circ f$ is differentiable, and $g'(y) = e^y \neq 0$ for every $y \in \mathbb{R}$, Theorem 2.3.6 implies that f is differentiable at every $x > 0$ and $g'(f(x))f'(x) = 1$. Thus,

$$1 = e^{\ln x}(\ln x)' = x(\ln x)'$$

so that $(\ln x)' = 1/x$.   ◊

Example 2.3.13 For $a > 0$, the function $\log_a x$ is differentiable for every $x > 0$, and

$$(\log_a x)' = \frac{1}{x \ln a}, \qquad x > 0.$$

To see this, recall that $\log_a x = \frac{\ln x}{\ln a}$. Hence, $(\log_a x)' = \frac{1}{x \ln a}$ for every $x > 0$. ◊

Example 2.3.14 For $r \in \mathbb{R}$, let $f(x) = x^r$ for $x > 0$. Then f is differentiable for every $x > 0$ and

$$f'(x) = rx^{r-1}, \qquad x > 0.$$

By the composition rule in Theorem 2.3.5,

$$f'(x) = (e^{r \ln x})' = e^{r \ln x} \frac{r}{x} = \frac{x^r r}{x} = rx^{r-1}. \qquad \Diamond$$

Exercise 2.3.2 Prove the following.

(i) The function $\ln |x|$ is differentiable for every $x \in \mathbb{R}$ with $x \neq 0$, and

$$(\ln |x|)' = \frac{1}{x}, \qquad x \neq 0.$$

(ii) For $a > 0$, the function $\log_a |x|$ is differentiable for every $x \in \mathbb{R}$ with $x \neq 0$, and

$$(\log_a |x|)' = \frac{1}{x \ln a}, \qquad x \neq 0. \qquad \triangleleft$$

2.3.4 Local Maxima and Local Minima

Recall from Theorem 2.2.8 that if $f : [a, b] \to \mathbb{R}$ is a continuous function, then there exists x_0, y_0 in $[a, b]$ such that

$$f(x_0) \leq f(x) \leq f(y_0) \qquad \forall x \in [a, b].$$

In Remark 2.2.4 we have seen that a function f defined on an interval I need not attain maximum or minimum if either I is not closed and bounded or if f is not continuous. However, maximum or minimum can attain if we restrict the function to a subinterval. To take care of these cases, we introduce the following definition.

Definition 2.3.4 Let f be a (real valued) function defined on an interval I. Then f is said to attain a

(1) **local maximum** at a point $x_1 \in I$ if $f(x) \leq f(x_1)$ for all x in a deleted neighbourhood of x_1,
(2) **local minimum** at a point $x_2 \in I$ if $f(x) \geq f(x_2)$ for all x in a deleted neighbourhood of x_2,
(3) **strict local maximum** and **strict local minimum** at x_1 and x_2, respectively, if strict inequality holds in (a) and (b), respectively.

The function f is said to attain a

(4) **local extremum** at a point $x_0 \in I$ if f attains either a local maximum or a local minimum at x_0,
(5) **strict local extremum** at a point $x_0 \in I$ if f attains either a strict local maximum or a strict local minimum at x_0.   ◊

Remark 2.3.4 Sometimes the adjective *local* in local maximum, local minimum and local extremum are omitted, if there is no confusion with the maximum and minimum on the whole interval. In order to avoid any possible confusion, the maximum and minimum on the whole interval are also called **global maximum** and **global minimum**.   ◊

Theorem 2.3.8 (A necessary condition) *Suppose f is a continuous function defined on an interval I having a local extremum at a point $x_0 \in I$. If x_0 is an interior point of I and f is differentiable at x_0, then $f'(x_0) = 0$.*

Proof Suppose f attains local maximum at x_0 which is an interior point of I. Then there exists $\delta > 0$ such that $(x_0 - \delta, x_0 + \delta) \subseteq I$ and $f(x_0) \geq f(x_0 + h)$ for all h with $|h| < \delta$. Hence, for all h with $|h| < \delta$,

$$\frac{f(x_0 + h) - f(x_0)}{h} \geq 0 \quad \text{if} \quad h < 0,$$

$$\frac{f(x_0 + h) - f(x_0)}{h} \leq 0 \quad \text{if} \quad h > 0.$$

Letting $h \to 0$, we get $f'(x_0) \geq 0$ and $f'(x_0) \leq 0$. Thus, $f'(x_0) = 0$.

By similar arguments, it can be shown that if f attains local minimum at a point $y_0 \in (a, b)$, then $f'(y_0) = 0$.   ∎

If a differentiable function f has a local extremum at a point x_0, then $f'(x_0) = 0$

Definition 2.3.5 Suppose f is defined on an interval I and x_0 is an interior point of I. If $f'(x_0)$ exists and $f'(x_0) = 0$ or if $f'(x_0)$ does not exist, then x_0 is called a **critical point** of f.   ◊

Points at which the derivative of a function vanish are also called **stationary points** of the function.

Remark 2.3.5 A function can have more than one local maximum and local minimum. For example, consider

$$f(x) = \sin(4x), \qquad [0, \pi].$$

We see that f has a local maximum value 1 at $\pi/8$ and $5\pi/8$, and has a local minimum value -1 at $3\pi/8$ and $7\pi/8$.   ◊

As a consequence of Theorem 2.3.8, we have the following result on intermediate value property of derivatives.

Theorem 2.3.9 (Darboux's theorem) *Suppose f is differentiable in an open interval I containing [a, b] and λ lies between f′(a) and f′(b). Then there exists α ∈ [a, b] such that f′(α) = λ.*

Proof Assume, without loss of generality, that $f'(a) < \lambda < f'(b)$. Let

$$g(x) = f(x) - \lambda x, \quad x \in [a, b].$$

Since g is differentiable at every point in $[a, b]$ and

$$g'(x) = f'(x) - \lambda \quad \forall x \in [a, b],$$

we have

$$g'(a) = f'(a) - \lambda < 0, \quad g'(b) = f'(b) - \lambda > 0.$$

Since g is continuous, by Theorem 2.2.8, it has maximum and minimum at some points in $[a, b]$. Since

$$\lim_{x \to a} \frac{g(x) - g(a)}{x - a} = g'(a) < 0, \quad \lim_{y \to b} \frac{g(y) - g(b)}{y - b} = g'(b) > 0,$$

there exists $\delta > 0$ such that

$$g(x) < g(a) \quad \forall x \in [a, a + \delta] \quad \text{and} \quad g(y) < g(b) \quad \forall y \in [b - \delta, b].$$

Hence the point at which g has a local minimum can neither be at a nor at b. Therefore, g has a local minimum at an interior point, say at $x = \alpha$. Therefore, by Theorem 2.3.8, $g'(\alpha) = 0$, that is, $f'(\alpha) = \lambda$. ∎

Darboux's theorem (Theorem 2.3.9) can be rephrased as follows:

> The derivative of a differentiable function has intermediate value property

Here is an illustration of the Theorem 2.3.9 .

Example 2.3.15 Let

$$f(x) = \begin{cases} -1, & x \neq 0, \\ 1, & x = 0. \end{cases}$$

Then there is no function g which is differentiable on $(-1, 1)$ such that $g'(x) = f(x)$ for all $x \in (-1, 1)$. This follows from Darboux's theorem (Theorem 2.3.9), since f does not take values between 0 and 1. ◇

Remark 2.3.6 (a) In view of Theorem 2.3.8, if a function f is differentiable at an interior point x_0 of an interval I and $f'(x_0) \neq 0$, then f cannot have local maximum or local minimum at x_0.

(b) In order to have a local maximum or local minimum at a point, the function need not be differentiable at that point. For example

$$f(x) = 1 - |x|, \qquad |x| \leq 1,$$

has a local maximum at 0 and

$$g(x) = |x|, \qquad |x| \leq 1,$$

has a local minimum at 0. Both f and g are not differentiable at 0.

(c) Also, if a function is differentiable at a point x_0 and $f'(x_0) = 0$, then it is not necessary that it has local maximum or local minimum at x_0. For example, consider

$$f(x) = x^3, \qquad |x| < 1.$$

In this example, we have $f'(0) = 0$. Note that f has neither local maximum nor local minimum at 0. $\Diamond$

In Sects. 2.3.6 and 2.3.8, we shall give some sufficient conditions for existence of local extrema of functions. Now, let us derive some important consequences of Theorem 2.3.8.

2.3.5 Rolle's Theorem and Mean Value Theorems

Suppose we have a curve

$$y = f(x)$$

defined on a closed interval $[a, b]$. If the values of f at both the end points a and b are the same, and if the curve has unique tangents at every point x with $a < x < b$, can we say that there is a point at which the tangent is parallel to the x-axis? The answer is in the affirmative. That is the *Rolle's theorem* (Fig. 2.14).

Theorem 2.3.10 (Rolle's theorem) *Suppose f is a continuous function defined on a closed and bounded interval $[a, b]$ such that it is differentiable at every $x \in (a, b)$. If $f(a) = f(b)$, then there exists $c \in (a, b)$ such that $f'(c) = 0$.*

Proof Let $g(x) = f(x) - f(a)$. Then we have

$$g(a) = 0 = g(b) \quad \text{and} \quad g'(x) = f'(x) \quad \forall x \in (a, b). \tag{$*$}$$

Since g is continuous on $[a, b]$, it attains the (global) maximum and (global) minimum at some points x_1 and x_2, respectively, in $[a, b]$, i.e., there exists x_1, x_2 in $[a, b]$ such that

Fig. 2.14 Illustration of rolle's theorem

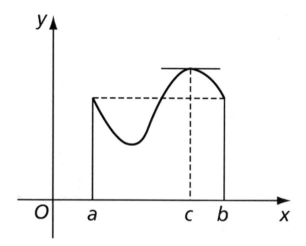

$$g(x_2) \le g(x) \le g(x_1) \qquad \forall x \in [a, b].$$

If $g(x_1) = g(x_2)$, then g is a constant function and hence $g'(x) = 0$ for all $x \in [a, b]$. Hence, assume that $g(x_2) < g(x_1)$. Then, either $g(x_1) \ne 0$ or $g(x_2) \ne 0$. If $g(x_1) \ne 0$, then by $(*)$, $x_1 \notin \{a, b\}$, i.e., $x_1 \in (a, b)$ so that by Theorem 2.3.8, $g'(x_1) = 0$ and hence, $f'(x_1) = 0$.

Similarly, if $g(x_2) \ne 0$, then we shall arrive at $f'(x_2) = 0$. ∎

Remark 2.3.7 Rolle's theorem guarantees the existence of at least one point c such that $f'(c) = 0$. There can be more than one point as illustrated in the following example. ◊

Example 2.3.16 Let $f(x) = x^3 - x + 1$. Note that

$$f(-1) = 1 \quad \text{and} \quad f(1) = 1.$$

So, Rolle's theorem ensures that there exists a point $c \in (-1, 1)$ such that $f'(c) = 0$. In fact,

$$f'(x) = 0 \iff 3x^2 - 1 = 0 \iff x = \pm 1/\sqrt{3}.$$

Thus, $c = 1/\sqrt{3}$ or $c = -1/\sqrt{3}$. ◊

Remark 2.3.8 In Rolle's theorem, the stated conditions on f are not necessary conditions. That is, in the absence of some of the conditions also, there can be a point in (a, b) at which f is differentiable and its derivative vanishes. For example, consider the function $f(x) = x^3$ for $x \in [-1, 1]$. In this case, $f(x) \ne f(y)$ for every $x \ne y$. But, $f'(0) = 0$.

Also, if we drop any of the conditions in Rolle's theorem, then the conclusion need not hold. To see this, consider $f(x) = |x|$ for $x \in [-1, 1]$. Note that, f is continuous

Fig. 2.15 Illustration of
Lagrange's mean value
theorem

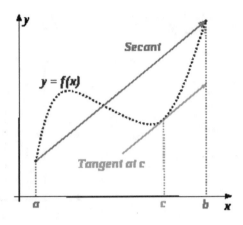

and $f(-1) = f(1) = 1$, but there is no point $c \in (-1, 1)$ such that $f'(c) = 0$. In this
case, the continuous function f is not differentiable at 0. Thus, in Rolle's theorem, the
assumption on differentiability of f in the whole open interval cannot be dropped.$\lozenge$

In Rolle's theorem, we imposed the condition that $f(a) = f(b)$ and obtained a
point in the open interval (a, b) at which f' vanishes.

What can we say if $f(a) \neq f(b)$? The geometry suggests that there must be a
point in the open interval (a, b) at which the tangent to the curve $y = f(x)$ is parallel
to line joining the points $A(a, f(a))$ and $B(b, f(b))$ (Fig. 2.15).

Theorem 2.3.11 (Lagrange's mean value theorem (MVT)) *Let f be a function which
is continuous on $[a, b]$ and differentiable on (a, b). Then there exists $c \in (a, b)$ such
that*

$$f'(c) = \frac{f(b) - f(a)}{b - a}.$$

Proof Let

$$\varphi(x) := f(x) - \frac{f(b) - f(a)}{b - a}(x - a), \qquad x \in [a, b].$$

Then φ is continuous on $[a, b]$, differentiable in (a, b), $\varphi(a) = \varphi(b)$, and

$$\varphi'(x) := f'(x) - \frac{f(b) - f(a)}{b - a}, \qquad x \in (a, b).$$

By Rolle's theorem (Theorem 2.3.10), there exists $c \in (a, b)$ such that $\varphi'(c) = 0$.
Thus, $f(b) - f(a) = f'(c)(b - a)$. ∎

Remark 2.3.9 Rolle's theorem is a special case of MVT, in the sense that, if we
have $f(a) = f(b)$, then we have $f'(c) = 0$. $\lozenge$

Example 2.3.17 Let $f(x) = 2x^3 - 3x + 1$. Note that

$$f(0) = 1, \quad f(1) = 0 \quad \text{and} \quad \frac{f(1) - f(0)}{1 - 0} = -1.$$

So, MVT ensures that there exists a point $c \in (0, 1)$ such that $f'(c) = -1$. In fact,

$$f'(x) = -1 \iff 6x^2 - 3 = -1 \iff 3x^2 - 1 = 0 \iff x = \pm 1/\sqrt{3}.$$

Thus, in this case, $c = 1/\sqrt{3}$.   ◊

Example 2.3.18 Let f be continuous on $[a, b]$ and differentiable at every point in (a, b). Suppose there exists $c \in \mathbb{R}$ such that

$$f'(x) = c \quad x \in (a, b).$$

Then there exists $b \in \mathbb{R}$ such that

$$f(x) = cx + b \quad \forall x \in [a, b].$$

In particular, if $f'(x) = 0$ for all $x \in (a, b)$, then f is a constant function.

To see this consider $x_0 \in (a, b)$. Then, for any $x \in [a, b]$, there exists ξ_x between x_0 and x such that

$$f(x) - f(x_0) = f'(\xi_x)(x - x_0) = c(x - x_0).$$

Hence, $f(x) = cx + b$ with $b = f(x_0) - c x_0$.   ◊

Convention: If f is defined in a closed interval $[a, b]$, then by $\lim_{x \to a} f(x)$ and $\lim_{x \to b} f(x)$ we mean $\lim_{x \to a+} f(x)$ and $\lim_{x \to b-} f(x)$, respectively.

Theorem 2.3.12 *Let f be continuous on $[a, b]$ and differentiable on (a, b). Then the following hold.*

(i) f is increasing iff $f'(x) \geq 0$ for all $x \in (a, b)$.
(ii) f is decreasing iff $f'(x) \leq 0$ for all $x \in (a, b)$.
(iii) f is strictly increasing if $f'(x) > 0$ for all $x \in (a, b)$.
(iv) f is strictly decreasing if $f'(x) < 0$ for all $x \in (a, b)$.

Proof (i) Suppose f is increasing and $x \in (a, b)$. Then

$$\frac{f(x + h) - f(x)}{h} \geq 0$$

for all h such that $x + h \in (a, b)$. Hence $f'(x) \geq 0$.

Conversely, suppose $f'(x) \geq 0$ for all $x \in (a, b)$. Let $x_1, x_2 \in [a, b]$ with $x_1 < x_2$. Then, by the Lagrange's MVT, there exists $\xi \in (x_1, x_2)$ such that

$$f(x_2) - f(x_1) = f'(\xi)(x_2 - x_1).$$

Since $f'(\xi) \geq 0$, the above equation shows that $f(x_1) \leq f(x_2)$.

(ii) As in the proof of (i) by reversing the inequalities.

(iii) Follows from the converse part of the proof of (i) using $f'(\xi) > 0$.

(iv) As in the converse part of the proof of (i) by using $f'(\xi) < 0$. ∎

Example 2.3.19 Consider the function $f(x) = x^4$ for $x \in \mathbb{R}$. Then we have $f'(x) = 4x^3$ for all $x \in \mathbb{R}$. Note that

$$f'(x) > 0 \quad \forall x > 0 \quad \text{and} \quad f'(x) < 0 \quad \forall x < 0.$$

Hence, f is strictly increasing on $(0, \infty)$, and f is strictly decreasing on $(-\infty, 0)$. ◊

Suppose f and g are continuous functions on $[a, b]$ which are differentiable on (a, b). Suppose further that $g'(x) \neq 0$ for all $x \in (a, b)$. Then, by the mean value theorem, there exist c_1, c_2 in (a, b) such that

$$\frac{f(b) - f(a)}{g(b) - g(a)} = \frac{f'(c_1)}{g'(c_2)}.$$

One may ask whether there exists a single point $c \in (a, b)$ such that

$$\frac{f(b) - f(a)}{g(b) - g(a)} = \frac{f'(c)}{g'(c)}.$$

Answer is in the affirmative as the following theorem shows.

Theorem 2.3.13 (Cauchy's generalized mean value theorem) *Suppose f and g are continuous functions on $[a, b]$ which are differentiable at every point in (a, b). Suppose further that $g'(x) \neq 0$ for all $x \in (a, b)$. Then, $g(a) \neq g(b)$ and there exists $c \in (a, b)$ such that*

$$\frac{f(b) - f(a)}{g(b) - g(a)} = \frac{f'(c)}{g'(c)}.$$

Proof First note that, since $g'(x) \neq 0$ for all $x \in (a, b)$, by Lagrange's MVT, $g(b) \neq g(a)$. Now, let

$$\varphi(x) := f(x) - \frac{f(b) - f(a)}{g(b) - g(a)}[g(x) - g(a)], \qquad x \in [a, b].$$

Note that φ is continuous on $[a, b]$, differentiable in (a, b), $\varphi(a) = \varphi(b)$, and

$$\varphi'(x) := f'(x) - \frac{f(b) - f(a)}{g(b) - g(a)}g'(x), \qquad x \in (a, b).$$

By Rolle's theorem (Theorem 2.3.10), there exists $c \in (a, b)$ such that $\varphi'(c) = 0$. This completes the proof. ∎

Exercise 2.3.3 Let $0 < a < b$. Using Cauchy's generalized mean value theorem, show that

$$a < \frac{n[b^{n+1} - a^{n+1}]}{(n+1)[b^n - a^n]} < b \quad \forall n \in \mathbb{N}.$$

[*Hint:* take $f(x) = x^{n+1}$ and $g(x) = x^n$.] ◁

2.3.6 A Sufficient Condition for a Local Extremum Point

Theorem 2.3.14 *Suppose f is continuous on an interval I and x_0 is an interior point of I. Further suppose that f is differentiable in a deleted neighbourhood of x_0.*

(i) *If there exists an open interval $I_0 \subseteq I$ containing x_0 such that*

$$f'(x) > 0 \quad \forall x \in I_0, \ x < x_0 \quad and \quad f'(x) < 0 \quad \forall x \in I_0, \ x > x_0,$$

then f has a local maximum at x_0.

(ii) *If there exists an open interval $I_0 \subseteq I$ containing x_0 such that*

$$f'(x) < 0 \quad \forall x \in I_0, \ x < x_0 \quad and \quad f'(x) > 0 \quad \forall x \in I_0, \ x > x_0,$$

then f has a local minimum at x_0.

Proof (i) Let $x \in I_0$. Then, by mean value theorem, there exists ξ_x between x_0 and x such that

$$f(x) - f(x_0) = f'(\xi_x)(x - x_0).$$

By assumption,

$$x < x_0 \Rightarrow f'(\xi_x) > 0 \quad \text{and} \quad x > x_0 \Rightarrow f'(\xi_x) < 0.$$

Hence, in both the cases, we have $f(x) < f(x_0)$ so that f has local maximum at x_0. Thus, (i) is proved.

Similar arguments will lead to the proof of (ii). ∎

Example 2.3.20 Consider

$$f(x) = x^4, \qquad g(x) = 1 - x^4, \quad |x| < 1.$$

Then $f'(x) = 4x^3$ is negative for $x < 0$ and positive for $x > 0$. Hence, by Theorem 2.3.14, f has local minimum at 0. Also, $g'(x) = -4x^3$ is positive for $x < 0$ and negative for $x > 0$. Hence, by Theorem 2.3.14, g has local maximum at 0. ◇

Remark 2.3.10 The conditions given in Theorem 2.3.14 cannot be dropped. For example, consider $f(x) = x^3$, $x \in \mathbb{R}$. Then $f'(x) = 3x^2 > 0$ for all $x \neq 0$. Note that f does not have extremum at 0.   ◇

2.3.7 L'Hospital's Rules

Recall from Theorem 2.1.8 that if $\lim_{x \to a} f(x) = b$ and $\lim_{x \to a} g(x) = c$, and if $c \neq 0$, then $g(x) \neq 0$ in a deleted neighbourhood of a and

$$\lim_{x \to a} \frac{f(x)}{g(x)} = \frac{b}{c}.$$

Clearly, if $c = 0$, then the above result is meaningless. However, the limit $\lim_{x \to a} f(x)/g(x)$ can exist whenever $g(x) \neq 0$ for x in a deleted neighbourhood of x_0. For example, we have seen that

$$\lim_{x \to a} \frac{x^2 - a^2}{x - a} = 2a \quad \text{and} \quad \lim_{x \to 0} \frac{\sin x}{x} = 1.$$

Note that, in these examples, the numerator of the quotient $f(x)/g(x)$ also vanish at the point where the limit is taken. So, one may tend to hypothesize that $\lim_{x \to a} f(x)/g(x)$ may exist if both $f(x)$ and $g(x)$ vanish at a. Again, not always. Look at the quotient

$$\frac{x^2 - a^2}{(x - a)^2}$$

for x in a deleted neighbourhood of a. In this case, taking $x_n = a + 1/n$, we have $x_n \to a$, but

$$\frac{x_n^2 - a^2}{(x_n - a)^2} = \frac{(a + 1/n)^2 - a^2}{1/n^2} = n^2 \left(\frac{2a}{n} + \frac{1}{n^2} \right) = 2an + 1 \to \infty.$$

So, under what additional condition, can one assert the existence of $\lim_{x \to a} f(x)/g(x)$? Some answers are provided in the following theorems, which are known as L'Hospital's rules.[2] The first one is the simplest among them.

Theorem 2.3.15 (L'Hospital's rule-I) *Suppose f and g are differentiable at a point x_0 and $f(x_0) = 0 = g(x_0)$. If $g'(x_0) \neq 0$, then $\lim_{x \to x_0} f(x)/g(x)$ exists and*

---

[2] *L'Hospital* (pronounced as *Lopital*) rule is named after the 17th-century French mathematician Guillaume de l'Hospital, who published the rule in his book *Analyse des Infiniment Petits pour l'Intelligence des Lignes Courbes* (i.e., Analysis of the Infinitely Small to Understand Curved Lines) (1696), the first textbook on differential calculus - curtsey Wikipedia.

$$\lim_{x \to x_0} \frac{f(x)}{g(x)} = \frac{f'(x_0)}{g'(x_0)}.$$

Proof We note that, for $x \neq x_0$,

$$\frac{f(x)}{g(x)} = \frac{f(x) - f(x_0)}{g(x) - g(x_0)} = \frac{(f(x) - f(x_0))/(x - x_0)}{(g(x) - g(x_0))/(x - x_0)}.$$

Hence, if $g'(x_0) \neq 0$, then using the differentiability of f and g at x_0, we obtain

$$\frac{f(x)}{g(x)} \to \frac{f'(x_0)}{g'(x_0)} \quad \text{as} \quad x \to x_0.$$

This completes the proof. ∎

Example 2.3.21 We have already recalled that

$$\lim_{x \to a} \frac{x^2 - a^2}{x - a} = 2a \quad \text{and} \quad \lim_{x \to 0} \frac{\sin x}{x} = 1.$$

These results follow from Theorem 2.3.15:

(i) Taking $f(x) = x^2 - a^2$ and $g(x) = x - a$ we have $f'(a) = 2a$ and $g'(a) = 1$ so that, by Theorem 2.3.15,

$$\lim_{x \to a} \frac{x^2 - a^2}{x - a} = \frac{f'(a)}{g'(a)} = 2a.$$

(ii) Taking $f(x) = \sin x$ and $g(x) = x$ we have $f'(0) = 1$ and $g'(0) = 1$ so that, by Theorem 2.3.15,

$$\lim_{x \to 0} \frac{\sin x}{x} = \frac{f'(0)}{g'(0)} = 1. \qquad \Diamond$$

Example 2.3.22 For $a \in \mathbb{R}$ with $a \neq 0$, let us find

$$\lim_{x \to 0} \frac{\log(1 + ax)}{x}$$

by using Theorem 2.3.15. Note that $\frac{\log(1+ax)}{x}$ is of the form $f(x)/g(x)$ with $f(x) = \log(1 + ax)$ and $g(x) = x$. Clearly, f is well defined for x in a neighbourhood of 0, say for $|x| < 1/2|a|$ and g is defined for all $x \in \mathbb{R}$. Further, $f(0) = 0 = g(0)$. Also, we have

$$f'(x) = \frac{a}{1 + ax}, \quad g'(x) = 1 \quad \text{for} \quad |x| < \frac{1}{2|a|}.$$

In particular,

$$f'(0) = a, \quad g'(0) = 1.$$

Hence, by Theorem 2.3.15, we obtain

$$\lim_{x \to 0} \frac{\log(1 + ax)}{x} = \lim_{x \to 0} \frac{f'(x)}{g'(x)} = \lim_{x \to 0} \frac{a}{1 + ax} = a. \qquad \Diamond$$

Example 2.3.23 Using Theorem 2.3.15, we show that for $a \in \mathbb{R}$,

$$\lim_{n \to \infty} \left(1 + \frac{a}{n}\right)^n = \exp(a).$$

This is true for $a = 0$. So, assume that $a \neq 0$. We have shown in Example 2.3.28 that for $\lim_{x \to 0} \frac{\log(1+ax)}{x} = a$. Since the function $x \mapsto \exp(x)$ is continuous on $\mathbb{R}$, we have

$$\lim_{x \to 0} \exp\left(\frac{\log(1 + ax)}{x}\right) = \exp\left(\lim_{x \to 0} \frac{\log(1 + ax)}{x}\right) = \exp(a).$$

From this, since $1/n \to 0$ as $n \to \infty$, we have

$$\lim_{n \to \infty} \exp\left(\frac{\log(1 + a/n)}{1/n}\right) = \exp(a).$$

But, for all $n \in \mathbb{N}$ such that $1 + a/n > 0$,

$$\exp\left(\frac{\log(1 + a/n)}{1/n}\right) = \exp\left(\log(1 + a/n)^n\right) = \left(1 + \frac{a}{n}\right)^n.$$

Thus, we have shown that

$$\lim_{n \to \infty} \left(1 + \frac{a}{n}\right)^n = \exp(a). \qquad \Diamond$$

Example 2.3.24 Let us find $\lim_{x \to \pi/4} \frac{\sin x - \cos x}{x - \pi/4}$. Note that $\frac{\sin x - \cos x}{x - \pi/4}$ is of the form $f(x)/g(x)$ with $f(x) = \sin x - \cos x$ and $g(x) = x - \pi/4$, and $f(\pi/4) = 0 = g(\pi/4)$. Since $f'(\pi/4) = \sqrt{2}$ and $g'(\pi/4) = 1$, by Theorem 2.3.15, we obtain

$$\lim_{x \to \pi/4} \frac{\sin x - \cos x}{x - \pi/4} = \sqrt{2}. \qquad \Diamond$$

Example 2.3.25 Let us find $\lim_{x \to 0} \frac{\sin x}{x^2 - x}$. Applying Theorem 2.3.15, where $f(x) = \sin x$, $g(x) = x^2 - x$ which satisfy $f(0) = 0 = g(0)$ and $f'(0) = 1$, $g'(0) = -1$, we obtain

$$\lim_{x \to 0} \frac{\sin x}{x^2 - x} = \lim_{x \to 0} \frac{\cos x}{2x - 1} = \frac{1}{-1} = -1. \qquad \Diamond$$

While applying Theorem 2.3.15, we have to ensure that the conditions in the theorem are satisfied. For example,

$$\lim_{x \to 2} \frac{x-2}{x^2-2} = \lim_{x \to 2} \frac{1}{2x} = \frac{1}{4}$$

is not correct. The correct answer is

$$\lim_{x \to 2} \frac{x-2}{x^2-2} = \frac{\lim_{x \to 2}(x-2)}{\lim_{x \to 2}(x^2-2)} = \frac{0}{2} = 0.$$

Note that, writing $(x-2)/(x^2-2)$ as $f(x)/g(x)$ with $f(x) = x-2$ and $g(x) = x^2-2$, we have $f(2) = 0$, but $g(2) \neq 0$.

We know that Theorem 2.3.15 cannot be applied if $g'(x_0) = 0$. However, the condition $g'(x_0) \neq 0$ is not a necessary condition to obtain $\lim_{x \to x_0} f(x)/g(x)$. For example, consider question of finding

$$\lim_{x \to 0} \frac{\sin x^2}{x^2}.$$

Note that $\frac{\sin x^2}{x^2}$ is of the form $f(x)/g(x)$ with $g'(0) = 0$. So, we cannot apply Theorem 2.3.15. However, the limit does exist. This can be seen as follows: Let $\varphi(x) = x^2$ and $\psi(x) = \frac{\sin x}{x}$. Then

$$\frac{\sin x^2}{x^2} = \psi(\varphi(x)) \quad \text{with} \quad \lim_{x \to 0} \varphi(x) = 0, \quad \lim_{y \to 0} \psi(y) = 1.$$

Hence, by composition rule for limits (Theorem 2.1.11),

$$\lim_{x \to 0} \frac{\sin x^2}{x^2} = \lim_{x \to 0} \psi(\varphi(x)) = 1.$$

So, if $g'(x_0) = 0$, then under what additional condition, can we assert the existence of $\lim_{x \to a} \frac{f(x)}{g(x)}$? Here is a another L'Hospital's rule to deal with such situations.

Theorem 2.3.16 (L'Hospital's rule-II) *Suppose f and g are differentiable in a deleted neighbourhood of x_0. Suppose*

$$f(x_0) = 0, \quad g(x_0) = 0 \quad and \quad \lim_{x \to x_0} \frac{f'(x)}{g'(x)} \quad exists.$$

Then $\lim_{x \to x_0} f(x)/g(x)$ exists and

$$\lim_{x \to x_0} \frac{f(x)}{g(x)} = \lim_{x \to x_0} \frac{f'(x)}{g'(x)}.$$

Proof Since $\lim_{x \to x_0} f'(x)/g'(x)$ exists, there exists a deleted neighbourhood D_0 of x_0 in the domain of definition of g such that $g'(x) \neq 0$ for all $x \in D_0$. By Cauchy's generalized mean value theorem (Theorem 2.3.13), for every $x \in D_0$, there exists ξ_x between x and x_0 such that

$$\frac{f(x)}{g(x)} = \frac{f(x) - f(x_0)}{g(x) - g(x_0)} = \frac{f'(\xi_x)}{g'(\xi_x)}.$$

Since $|\xi_x - x_0| < |x - x_0|$ and $\lim_{x \to x_0} f'(x)/g'(x)$ exists, by using the limits of composition of functions, $\lim_{x \to x_0} f'(\xi_x)/g'(\xi_x)$ exists and it is equal to $\lim_{x \to x_0} f'(x)/g'(x)$. Thus, $\lim_{x \to x_0} f(x)/g(x)$ exists and

$$\lim_{x \to x_0} \frac{f(x)}{g(x)} = \lim_{x \to x_0} \frac{f'(x)}{g'(x)}.$$

This completes the proof.   ∎

Example 2.3.26 By Theorem 2.3.16,

$$\lim_{x \to 0} \frac{\sin x^2}{x^2} = \lim_{x \to 0} \frac{2x \cos x^2}{2x} = \lim_{x \to 0} \cos x^2 = 1. \qquad \Diamond$$

Example 2.3.27 By Theorem 2.3.16,

$$\lim_{x \to 0} \frac{1 - \cos x}{x^2} = \lim_{x \to 0} \frac{\sin x}{2x} = \frac{1}{2} \lim_{x \to 0} \frac{\sin x}{x} = \frac{1}{2}. \qquad \Diamond$$

We know that if a function f is defined in a neighbourhood of a point x_0 and if $\lim_{x \to x_0} f(x)$ exists, then the above limit need not be $f(x_0)$. For example, if

$$f(x) = \begin{cases} 0, & x \neq 0, \\ 1, & x = 0, \end{cases}$$

then $\lim_{x \to 0} f(x) = 0$, but $f(0) = 1$. This kind of situation will not occur for the derivative of a function, as the following theorem shows.

Theorem 2.3.17 *Suppose f is differentiable in a deleted neighbourhood of x_0 and $\lim_{x \to x_0} f'(x)$ exists. Then*

$$f'(x_0) = \lim_{x \to x_0} f'(x).$$

Proof Applying L'Hospital's rule-II (Theorem 2.3.16) to the quotient $g(x)/h(x)$, where $g(x) = f(x) - f(x_0)$ and $h(x) = x - x_0$, we have

$$\lim_{x \to x_0} \frac{f(x) - f(x_0)}{x - x_0} = \lim_{x \to x_0} \frac{g'(x)}{h'(x)} = \lim_{x \to x_0} \frac{f'(x)}{1} = \lim_{x \to x_0} f'(x),$$

so that $f'(x_0)$ exists and $f'(x_0) = \lim_{x \to x_0} f'(x)$. ∎

In the following L'Hospital's rule, we further relax the conditions on f and g, by replacing the condition that f and g vanish at x_0 by requiring the limits of f and g to be 0 as $x \to x_0$.

Theorem 2.3.18 (L'Hospital's rule-III) *Suppose f and g are differentiable in a deleted neighbourhood of x_0. Suppose*

$$\lim_{x \to x_0} f(x) = 0, \quad \lim_{x \to x_0} g(x) = 0 \quad and \quad \lim_{x \to x_0} \frac{f'(x)}{g'(x)} \text{ exists.}$$

Then $\lim_{x \to x_0} f(x)/g(x)$ exists and

$$\lim_{x \to x_0} \frac{f(x)}{g(x)} = \lim_{x \to x_0} \frac{f'(x)}{g'(x)}.$$

Proof Let

$$\tilde{f}(x) = \begin{cases} f(x) & \text{if } x \neq x_0 \\ 0 & \text{if } x = x_0 \end{cases}, \quad \tilde{g}(x) = \begin{cases} g(x) & \text{if } x \neq x_0 \\ 0 & \text{if } x = x_0. \end{cases}$$

Then, the result is obtained from Theorem 2.3.16 by taking $\tilde{f}$ and $\tilde{g}$ in place of f and g, respectively. ∎

Next we discus case of finding the limit of $f(x)/g(x)$ as $x \to \infty$.

Theorem 2.3.19 (L'Hospital's rule-IV) *Suppose f and g are differentiable at every point in (a, ∞) for some $a > 0$. Suppose*

$$\lim_{x \to \infty} f(x) = 0, \quad \lim_{x \to \infty} g(x) = 0 \quad and \quad \lim_{x \to \infty} \frac{f'(x)}{g'(x)} \text{ exists.}$$

Then $\lim_{x \to \infty} f(x)/g(x)$ exists and

$$\lim_{x \to \infty} \frac{f(x)}{g(x)} = \lim_{x \to \infty} \frac{f'(x)}{g'(x)}.$$

Proof Let $\tilde{f}(y) = f(1/y)$ and $\tilde{g}(y) = g(1/y)$ for $0 < y < 1/a$. We note that

$$\lim_{x \to \infty} f(x) = 0 = \lim_{x \to \infty} g(x) \iff \lim_{y \to 0} \tilde{f}(y) = 0 = \lim_{y \to 0} \tilde{g}(y).$$

Also, since

$$\tilde{f}'(y) = [f(1/y)]' = f'(1/y)(-1/y^2), \qquad \tilde{g}'(y) = [g(1/y)]' = g'(1/y)(-1/y^2),$$

we have

$$\lim_{x \to \infty} \frac{f'(x)}{g'(x)} \quad \text{exists} \iff \lim_{y \to 0} \frac{\tilde{f}'(y)}{\tilde{g}'(y)} \quad \text{exists}.$$

Hence, applying Theorem 2.3.16 to $\tilde{f}$, $\tilde{g}$ instead of f, g, we obtain the result. ∎

Theorem 2.3.20 (L'Hospital's rule-V) *Let $x_0 \in [a, b]$. Suppose f and g are differentiable in a deleted neighbourhood of a point x_0. Suppose*

$$\lim_{x \to x_0} f(x) = \infty, \quad \lim_{x \to x_0} g(x) = \infty \quad \text{and} \quad \lim_{x \to x_0} \frac{f'(x)}{g'(x)} \quad \text{exists}.$$

Then $\lim_{x \to x_0} f(x)/g(x)$ exists and

$$\lim_{x \to x_0} \frac{f(x)}{g(x)} = \lim_{x \to x_0} \frac{f'(x)}{g'(x)}.$$

Proof Let $\beta := \lim_{x \to x_0} \dfrac{f'(x)}{g'(x)}$. First we consider the case of $\beta \neq 0$. In this case, since

$$\lim_{x \to x_0} f(x) = \infty = \lim_{x \to \infty} g(x) \iff \lim_{x \to x_0}(1/f(x)) = 0 = \lim_{x \to x_0}(1/g(x)),$$

the result follows from Theorem 2.3.18 by interchanging the roles of the functions f and g.

To consider the general case when β is not necessarily non-zero, let x, y be distinct points in a deleted neighbourhood of x_0. Since $g'(x) \neq 0$ for x sufficiently close to x_0, in view of Lagrange's MVT, we can assume that $g(x) \neq g(y)$. Note that,

$$\frac{f(x) - f(y)}{g(x) - g(y)} = \frac{f(x)}{g(x)} \frac{\left[1 - \frac{f(y)}{f(x)}\right]}{\left[1 - \frac{g(y)}{g(x)}\right]} \tag{2.1}$$

Since $f(x) \to \infty$ and $g(x) \to \infty$ as $x \to x_0$, the above expression is meaningful for each fixed y and x close enough to x_0, and

$$\lim_{x \to x_0} \left[1 - \frac{f(y)}{f(x)} \right] = 1 = \lim_{x \to x_0} \left[1 - \frac{g(y)}{g(x)} \right]. \tag{2.2}$$

Also, by the Cauchy's generalized MVT, there exists $\xi_{x,y}$ lying between x and y such that

$$\frac{f(x) - f(y)}{g(x) - g(y)} = \frac{f'(\xi_{x,y})}{g'(\xi_{x,y})}. \tag{2.3}$$

From (2.1) and (2.3) above we have

$$\frac{f(x)}{g(x)} = \frac{f'(\xi_{x,y})}{g'(\xi_{x,y})} \frac{\left[1 - \frac{g(y)}{g(x)} \right]}{\left[1 - \frac{f(y)}{f(x)} \right]}. \tag{2.4}$$

We observe that

$$|\xi_{x,y} - x_0| \le |\xi_{x,y} - y| + |y - x_0| \le |x - y| + |y - x_0|.$$

Hence, $\xi_{x,y} \to x_0$ as $x \to x_0$ and $y \to y_0$. Hence, by using the limits of composition of functions, we obtain

$$\lim_{y \to x_0} \frac{f'(\xi_{x,y})}{g'(\xi_{x,y})} = \lim_{x \to x_0} \frac{f'(x)}{g'(x)}. \tag{2.5}$$

Therefore, (2.2), (2.4), (2.5) imply that $\lim_{x \to x_0} \dfrac{f(x)}{g(x)}$ exists and

$$\lim_{x \to x_0} \frac{f(x)}{g(x)} = \lim_{y \to x_0} \frac{f'(\xi_{x,\alpha})}{g'(\xi_{x,\alpha})} \frac{\left[1 - \frac{f(y)}{f(x)} \right]}{\left[1 - \frac{g(y)}{g(x)} \right]} = \lim_{x \to x_0} \frac{f'(x)}{g'(x)}.$$

This completes the proof. ∎

Remark 2.3.11 The cases

$$\lim_{x \to -\infty} f(x) = 0 = \lim_{x \to -\infty} g(x), \quad \lim_{x \to x_0} f(x) = -\infty = \lim_{x \to x_0} g(x)$$

can be treated analogously to the cases already discussed in the above theorems. ◊

We shall make use of Theorem 2.3.20 to prove the following theorem.

Theorem 2.3.21 (L'Hospital's rule-VI) *Suppose f and g are differentiable on $(0, \infty)$. Suppose*

$$\lim_{x \to \infty} f(x) = \infty = \lim_{x \to \infty} g(x) \quad and \quad \lim_{x \to \infty} \frac{f'(x)}{g'(x)} \ exists.$$

Then $\lim\limits_{x\to\infty} f(x)/g(x)$ *exists and*

$$\lim_{x\to\infty} \frac{f(x)}{g(x)} = \lim_{x\to\infty} \frac{f'(x)}{g'(x)}.$$

Proof We observe that

$$\lim_{x\to\infty} \frac{f(x)}{g(x)} \text{ exists } \iff \lim_{x\to 0^+} \frac{f(1/x)}{g(1/x)} \text{ exists },\qquad (2.1)$$

and in that case they are same, and

$$\lim_{x\to\infty} \frac{f'(x)}{g'(x)} \text{ exists } \iff \lim_{x\to 0^+} \frac{f'(1/x)}{g'(1/x)} \text{ exists },\qquad (2.2)$$

and in that case they are same. Therefore, it is enough to show that

$$\lim_{x\to 0^+} \frac{f(1/x)}{g(1/x)} = \lim_{x\to 0^+} \frac{f'(1/x)}{g'(1/x)}.\qquad (2.3)$$

To show this, let φ and ψ be defined by

$$\varphi(x) = f(1/x) \quad \text{and} \quad \psi(x) = g(1/x)$$

for $x > 0$. Then we have

$$\varphi'(x) = -\frac{f'(1/x)}{x^2}, \quad \psi'(x) = -\frac{g'(1/x)}{x^2}$$

so that

$$\frac{\varphi(x)}{\psi(x)} = \frac{f(1/x)}{g(1/x)} \quad \text{and} \quad \frac{\varphi'(x)}{\psi'(x)} = \frac{f'(1/x)}{g'(1/x)}\qquad (2.4)$$

for all $x > 0$. By assumption, $\lim_{x\to\infty} \frac{f'(x)}{g'(x)}$ exists. Hence, by (2.2) and (2.4), $\lim_{x\to\infty} \frac{\varphi'(x)}{\psi'(x)}$ exists. Therefore, by Theorem 2.3.20, $\lim_{x\to\infty} \frac{\varphi(x)}{\psi(x)}$ exists and are equal, that is

$$\lim_{x\to\infty} \frac{\varphi(x)}{\psi(x)} = \lim_{x\to\infty} \frac{\varphi'(x)}{\psi'(x)}$$

In view of (2.4), the above equality is same as (2.3), completing the proof. ∎

Example 2.3.28 We have seen in Example 2.2.17 that $\lim_{n\to\infty} \frac{\log(1+n)}{n} = 0$. Now, using Theorem 2.3.21, we show

$$\lim_{n\to\infty} \frac{\log(1+n)}{n^p} = 0$$

for any $p > 0$. In fact, we show

$$\lim_{x \to \infty} \frac{\log(1 + x)}{x^p} = 0$$

for any $p > 0$. For this, let $p > 0$ and let

$$f(x) = \log(1 + x), \quad g(x) = x^p \quad \text{for} \quad x > 0.$$

Then we have $\lim_{x \to \infty} f(x) = \infty = \lim_{x \to \infty} g(x)$. Note that, for $x > 0$,

$$\frac{f'(x)}{g'(x)} = \frac{1/(1 + x)}{px^{p-1}} = \left(\frac{x}{1 + x}\right)\frac{1}{px^p}.$$

Hence, by Theorem 2.3.21,

$$\lim_{x \to \infty} \frac{\log(1 + x)}{x^p} = \lim_{x \to \infty} \frac{f(x)}{g(x)} = \lim_{x \to \infty} \frac{f'(x)}{g'(x)} = \lim_{x \to \infty} \left(\frac{x}{1 + x}\right)\frac{1}{px^p} = 0.$$

From this we have $\lim_{n \to \infty} \frac{1}{n^p} \log(1 + n) = 0.$ ◊

2.3.8 Higher Derivatives and Taylor's Formula

Suppose f is defined on an open interval I and $x_0 \in I$. If f is differentiable in a neighbourhood of x_0, then the derivative f' is a function defined in that neighbourhood, and in that case we can talk about the existence of derivative of f' at x_0, and similarly, we can talk about *higher derivatives* of f at x_0.

Definition 2.3.6 Suppose f is differentiable in a neighbourhood of x_0. If

$$\lim_{x \to x_0} \frac{f'(x) - f'(x_0)}{x - x_0}$$

exists, then f is said to be **twice differentiable** at x_0, and in that case the limit is called the **second derivative** of f at x_0 and it is denoted by

$$f''(x_0).$$ ◊

Definition 2.3.7 Let f be defined in an open interval I. Then, for $k \in \mathbb{N}$ with $k \geq 2$, f is said to be k **times differentiable** at $x_0 \in I$ if $f^{(1)}(x)$, $f^{(2)}(x), \ldots, f^{(k-1)}(x)$ are defined iteratively as

$$f^{(j)}(x) := [f^{(j-1)}]'(x), \quad j = 1, \ldots, k - 1$$

for x in a neighbourhood of x_0 with $f^{(0)}(x) = f(x)$ and

$$f^{(k)}(x_0) := [f^{(k-1)}]'(x_0)$$

exists. Then $f^{(k)}(x_0)$ is called the kth-**derivative** of f at x_0. $\Diamond$

Note that $f^{(2)}(x_0)$ is the second derivative of f at x_0. The kth-derivative of f is also denoted by

$$\frac{d^k f}{dx^k}.$$

It is also customary to use the notation Df for f' and $D^k f$ for $f^{(k)}$, so that it is $D^k f = D(D^{k-1} f)$.

Definition 2.3.8 The function f is said to be **infinitely differentiable** at a point $x_0 \in I$ if f has k^{th}-derivative at x_0 for every $k \in \mathbb{N}$. $\Diamond$

We may observe the following:

If f is infinitely differentiable at a point $x_0 \in I$, then for every $k \in \mathbb{N}$, f has k^{th}-derivative not only at x_0 but also at every point in some neighbourhood of x_0.

Example 2.3.29 For $n \in \mathbb{N}$, let $f(x) = x^n$, $x \in \mathbb{R}$. Then we know that $f^{(1)}(x) = f'(x) = nx^{n-1}$. Hence, for $k \le n$, we have

$$f^{(k)}(x) = n(n-1)\cdots(n-k+1)x^{n-k}$$

and $f^{(k)}(x) = 0$ for $k > n$. Thus, f is infinitely differentiable in $\mathbb{R}$. More generally, if f is a polynomial, then f is infinitely differentiable in $\mathbb{R}$. $\Diamond$

Example 2.3.30 Let $f(x) = \sin x$, $x \in \mathbb{R}$. Then we have

$$f^{(1)}(x) = \cos x, \quad f^{(2)}(x) = -\sin x, \quad f^{(3)}(x) = -\cos x, \quad f^{(4)}(x) = \sin x,$$

and more generally for any $k \in \mathbb{N}$,

$$f^{(2k-1)}(x) = (-1)^{k+1}\cos x, \quad f^{(2k)}(x) = (-1)^k \sin x.$$

Thus, f is infinitely differentiable in $\mathbb{R}$. $\Diamond$

Example 2.3.31 Let $f(x) = e^x$, $x \in \mathbb{R}$. We know that $f'(x) = e^x$. Hence, it follows that $f^{(k)}(x) = e^x$ for every $k \in \mathbb{N}$ so that f is infinitely differentiable in $\mathbb{R}$. $\Diamond$

Example 2.3.32 Let $f(x) = x|x|$, $x \in \mathbb{R}$. We have seen in Example 2.3.10 that f is differentiable at every point in $\mathbb{R}$ and $f'(x) = 2|x|$. Thus, f is infinitely differentiable at every $x \ne 0$, but differentiable only once at 0.

For $k \in \mathbb{N}$, if $f(x) = x^k|x|$, $x \in \mathbb{R}$, then it can be verified that f is infinitely differentiable at every $x \ne 0$ and $f^{(k)}(0)$ exists, but $f^{(k+1)}(0)$ does not exist. $\Diamond$

L'Hospital's Rule Revisited

We know, by L'Hospital's rule (Theorem 2.3.18) using first derivatives, that if

$$\lim_{x \to x_0} f(x) = 0, \quad \lim_{x \to x_0} g(x) = 0 \quad \text{and} \quad \lim_{x \to x_0} \frac{f'(x)}{g'(x)} \quad \text{exists,}$$

then $\lim_{x \to x_0} \frac{f(x)}{g(x)}$ exists and it is equal to $\lim_{x \to x_0} \frac{f'(x)}{g'(x)}$.

Suppose it happens that

$$\lim_{x \to x_0} f'(x) = 0, \quad \lim_{x \to x_0} g'(x) = 0 \quad \text{but} \quad \lim_{x \to x_0} \frac{f''(x)}{g''(x)} \quad \text{exists.}$$

Then, Theorem 2.3.18 can be applied to f', g' instead of f, g so that $\lim_{x \to x_0} \frac{f'(x)}{g'(x)}$ exists and $\lim_{x \to x_0} \frac{f'(x)}{g'(x)} = \lim_{x \to x_0} \frac{f''(x)}{g''(x)}$. Hence, applying again Theorem 2.3.18 to f, g, we have

$$\lim_{x \to x_0} \frac{f(x)}{g(x)} = \lim_{x \to x_0} \frac{f'(x)}{g'(x)} = \lim_{x \to x_0} \frac{f''(x)}{g''(x)}.$$

More generally we have the following result.

Theorem 2.3.22 (L'Hospital's rule) *Suppose the functions f and g are k times differentiable in a deleted neighbourhood of x_0,*

$$\lim_{x \to x_0} f^{(j-1)}(x) = 0, \quad \lim_{x \to x_0} g^{(j-1)}(x) = 0$$

for $j = 1, \ldots, k$ and $\lim_{x \to x_0} \frac{f^{(k)}(x)}{g^{(k)}(x)}$ exists. Then $\lim_{x \to x_0} f(x)/g(x)$ exists and

$$\lim_{x \to x_0} \frac{f(x)}{g(x)} = \lim_{x \to x_0} \frac{f^{(k)}(x)}{g^{(k)}(x)}.$$

Remark 2.3.12 Other forms of L'Hospital's rules in the general setting as in Theorem 2.3.22 are also valid. ◊

Exercise 2.3.4 Supply details of the proof of Theorem 2.3.22. ◁

Example 2.3.33 Let us check whether the $\lim_{x \to \infty} \frac{x^3}{e^x}$ exists. Let $f(x) = x^3$ and $g(x) = e^x$. Then, we have the following:

$$\lim_{x \to \infty} f(x) = \lim_{x \to \infty} x^3 = \infty, \quad \lim_{x \to \infty} g(x) = \lim_{x \to \infty} e^x = \infty,$$

$$\lim_{x \to \infty} f'(x) = \lim_{x \to \infty} 3x^2 = \infty, \quad \lim_{x \to \infty} g'(x) = \lim_{x \to \infty} e^x = \infty,$$

$$\lim_{x\to\infty} f''(x) = \lim_{x\to\infty} 6x = \infty, \quad \lim_{x\to\infty} g''(x) = \lim_{x\to\infty} e^x = \infty,$$

$$\lim_{x\to\infty} f'''(x) = 6, \quad \lim_{x\to\infty} g'''(x) = \lim_{x\to\infty} e^x = \infty.$$

Hence, $\lim_{x\to x_0} \dfrac{x^3}{e^x} = \lim_{x\to x_0} \dfrac{6}{e^x} = 0$. More generally, for any $k \in \mathbb{N}$, it can be shown (*verify*) that $\lim_{x\to\infty} \frac{x^k}{e^x} = 0$. $\qquad\qquad\qquad\qquad\qquad\Diamond$

Taylor's Formula

Recall from Lagrange's MVT that if f is differentiable in an open interval I and $x_0 \in I$, then for every $x \in I$, there exists ξ_x between x_0 and x such that

$$f(x) = f(x_0) + f'(\xi_x)(x - x_0).$$

Can we say something more if f has higher-order derivatives in a neighbourhood of x_0?

Suppose $f(x)$ is a polynomial of degree $n \in \mathbb{N}$ and $x_0 \in \mathbb{R}$. Since $f(x) - f(x_0)$ vanishes at $x = x_0$, we can write

$$f(x) = f(x_0) + (x - x_0)f_1(x),$$

where $f_1(x)$ is a polynomial of degree $n - 1$. By the same argument, if $n > 1$, then f_1 can be written as

$$f_1(x) = f_1(x_0) + (x - x_0)f_2(x),$$

where $f_2(x)$ is a polynomial of degree $n - 2$. Thus,

$$f(x) = f(x_0) + f_1(x_0)(x - x_0) + (x - x_0)^2 f_2(x).$$

Continuing this, there are polynomials $f_1(x), f_2(x), \ldots, f_{n-2}(x), f_{n-1}(x), f_n(x)$ of degree $n - 1, n - 2, \ldots, 2, 1, 0$, respectively, such that

$$f(x) = f(x_0) + f_1(x_0)(x - x_0) + f_2(x_0)(x - x_0)^2 + \cdots + f_n(x_0)(x - x_0)^n.$$

Note that

$$f^{(1)}(x_0) = f_1(x_0), \quad f^{(2)}(x_0) = 2! f_2(x_0), \ldots, f^{(n)}(x_0) = n! f_n(x_0),$$

so that

$$f(x) = f(x_0) + \frac{f^{(1)}(x_0)}{1!}(x - x_0) + \frac{f^{(2)}(x_0)}{2!}(x - x_0)^2 + \cdots + \frac{f^{(n)}(x_0)}{n!}(x - x_0)^n.$$

Now, suppose that f is a function which is $n + 1$ times differentiable in a neighbourhood of x_0 for some $n \in \mathbb{N}$. If we write,

$$P(x) = f(x_0) + \sum_{k=1}^{n} \frac{f^{(k)}(x_0)}{k!}(x - x_0)^k,$$

then we can write

$$f(x) = P(x) + R(x)$$

where $R(x) := f(x) - P(x)$ is $n + 1$ times differentiable and $R(x_0) = 0$. We may also observe that

$$R^{(k)}(x_0) = 0 \quad \text{for} \quad k = 1, \dots, n.$$

Taylor's formula in the following theorem gives a specific expression for $R(x)$ in terms of the $(n + 1)^{\text{th}}$ derivative of f at a point ξ lying between x_0 and x.

Theorem 2.3.23 (Taylor's formula) *Suppose f is defined and has derivatives $f^{(1)}(x), f^{(2)}(x), \dots, f^{(n+1)}(x)$ for x in an open interval I and $x_0 \in I$. Then, for every $x \in I$, there exists ξ_x between x and x_0 such that*

$$f(x) = f(x_0) + \sum_{j=1}^{n} \frac{f^{(j)}(x_0)}{j!}(x - x_0)^j + \frac{f^{(n+1)}(\xi_x)}{(n+1)!}(x - x_0)^{n+1}.$$

Proof Let $x \in I$ with $x \neq x_0$, and let

$$P_n(y) = f(x_0) + \sum_{j=1}^{n} \frac{f^{(j)}(x_0)}{j!}(y - x_0)^j, \quad y \in I.$$

Then $P_n(y)$ is a polynomial of degree n, $P_n(x_0) = f(x_0)$ and

$$P_n^{(j)}(x_0) = f^{(j)}(x_0), \quad j = 1, \dots, n.$$

Now, let

$$g(y) = f(y) - P_n(y) - \varphi(x)(y - x_0)^{n+1}, \quad y \in I,$$

where

$$\varphi(x) := \frac{f(x) - P_n(x)}{(x - x_0)^{n+1}}.$$

Note that, by this choice of $\varphi(x)$, we have $g(x_0) = 0$ and $g(x) = 0$. Also, we have

$$g^{(1)}(x_0) = 0, \quad g^{(2)}(x_0) = 0, \quad \dots, \quad g^{(n)}(x_0) = 0.$$

Since $g(x_0) = 0 = g(x)$, by Rolle's theorem, there exists x_1 between x_0 and x such that $g'(x_1) = 0$. Since $g'(x_0) = 0 = g'(x_1)$, again by Rolle's theorem, there exists x_2 between x_0 and x_1 such that $g''(x_2) = 0$. Continuing this, there exists $\xi_x := x_{n+1}$ between x_0 and x_n such that $g^{(n+1)}(\xi_x) = 0$. But,

$$g^{(n+1)}(y) = f^{(n+1)}(y) - P_n^{(n+1)}(y) - \varphi(x)(n+1)!$$
$$= f^{(n+1)}(y) - \varphi(x)(n+1)!.$$

Thus, using the fact that $g^{(n+1)}(\xi_x) = 0$, we have

$$\varphi(x) = \frac{f^{(n+1)}(\xi_x)}{(n+1)!}.$$

Thus,

$$f(x) = P_n(x) + \frac{f^{(n+1)}(\xi_x)}{(n+1)!}(x - x_0)^{n+1},$$

and the proof is complete.   ∎

Definition 2.3.9 With the notations as in Theorem 2.3.23, the polynomial

$$P_n(x) = f(x_0) + \sum_{j=1}^{n} \frac{f^{(j)}(x_0)}{j!}(x - x_0)^j$$

is called the **Taylor's polynomial** of f of degree n around x_0, and the term

$$R_n(x) := \frac{f^{(n+1)}(\xi_x)}{(n+1)!}(x - x_0)^{n+1}$$

is called the **remainder term** in the formula corresponding to the n-th degree Taylor's polynomial.   ◊

Remark 2.3.13 The Taylor's formula given in Theorem 2.3.23 is usually known as *Lagrange form of the Taylor's formula*. The proof given above is adapted from [6]. In the next chapter, we shall give another form of the Taylor's formula, called, the *Cauchy form of the Taylor's formula*, and derive the formula given in Theorem 2.3.23.   ◊

We observe that if f is infinitely differentiable and if

$$|R_n(x)| \to 0 \quad \text{as} \quad n \to \infty$$

for every $x \in I$, then

$$f(x) = f(x_0) + \sum_{n=1}^{\infty} \frac{f^{(n)}(x_0)}{n!}(x - x_0)^n, \quad x \in I.$$

Definition 2.3.10 If f is infinitely differentiable in a neighbourhood of x_0 and if it can be represented as a series

$$f(x) = f(x_0) + \sum_{n=1}^{\infty} \frac{f^{(n)}(x_0)}{n!}(x - x_0)^n, \qquad x \in I$$

for all x in a neighbourhood of x_0, then such a series is called the **Taylor series** of f around the point x_0. The Taylor series with $x_0 = 0$ is called the **Maclaurin series** of f.   ◊

Observe that if $f^{(n+1)}$ is bounded in a neighbourhood of x_0, that is, there exists $M_n > 0$ such that $|f^{(n+1)}(x)| \leq M_n$ for all x in that neighbourhood, then

$$|f(x) - P_n(x)| \leq \frac{M_n |x - x_0|^{n+1}}{(n+1)!}.$$

In particular, if f is infinitely differentiable, and if there exists $M > 0$, independent of n such that $|f^{(n+1)}(x)| \leq M$ for all x in a neighbourhood I_0 of x_0, then

$$|f(x) - P_n(x)| \leq \frac{M |x - x_0|^{n+1}}{(n+1)!} \to 0 \quad \text{as} \quad n \to \infty.$$

Thus, we have proved the following theorem.

Theorem 2.3.24 *Suppose f is infinitely differentiable in an open interval I and $x_0 \in I$. Further, suppose that there exists $M > 0$ such that*

$$|f^{(k)}(x)| \leq M \qquad \forall x \in I, \quad \forall k \in \mathbb{N} \cup \{0\}.$$

Then f has the Taylor series expansion

$$f(x) = f(x_0) + \sum_{n=1}^{\infty} \frac{f^{(n)}(x_0)}{n!}(x - x_0)^n$$

for all $x \in I_0$.

Remark 2.3.14 A natural question that one may ask is:

Does every infinitely differentiable function in a neighbourhood of x_0 has a Taylor's series expansion?

Unfortunately, the answer is in the negative. For example, if we define

$$f(x) = \begin{cases} e^{-1/x^2}, & x \neq 0, \\ 0, & x = 0, \end{cases}$$

then it can be seen that $f(0) = 0$ and $f^{(k)}(0) = 0$ for all $k \in \mathbb{N}$. Thus, f does not have the Taylor's series expansion around the point 0.   ◊

Example 2.3.34 Let $f(x) = e^x$ for $x \in \mathbb{R}$. Then $f^{(k)}(x) = e^x$ so that for any $x_0, x \in \mathbb{R}$,

$$R_n(x) := \frac{f^{(n+1)}(\xi_x)}{(n+1)!}(x - x_0)^{n+1} = \frac{e^{\xi_x}}{(n+1)!}(x - x_0)^{n+1}.$$

Since $\dfrac{(x - x_0)^{n+1}}{(n+1)!} \to 0$ as $n \to \infty$, we have

$$R_n(x) \to 0 \quad \text{as} \quad n \to \infty.$$

Hence, f has the Taylor series expansion

$$e^x = e^{x_0}\left[1 + \sum_{n=1}^{\infty} \frac{(x - x_0)^n}{n!}\right].$$

for every $x, x_0 \in \mathbb{R}$. Observe that the function which represents the series within the bracket is nothing but e^{x-x_0}. In particular, taking $x_0 = 0$, we have $e^x = 1 + \sum_{n=1}^{\infty} \frac{x^n}{n!}$.   ◊

Example 2.3.35 Using the Taylor's formula, we shall show that

$$\sin x = \sum_{n=0}^{\infty} \frac{(-1)^n x^{2n+1}}{(2n+1)!} \qquad \forall x \in \mathbb{R}.$$

For this, let $f(x) = \sin x$ and $x_0 = 0$. Since f is infinitely differentiable, and

$$f^{(2j)}(0) = 0, \qquad f^{(2j-1)}(0) = (-1)^j \qquad \forall j \in \mathbb{N},$$

we have

$$f(x) = f(x_0) + \sum_{j=1}^{2n+1} \frac{f^{(j)}(0)}{j!}x^j + \frac{f^{(2n+2)}(\xi_x)}{(2n+2)!}x^{2n+2}$$

$$= f(x_0) + \sum_{j=0}^{n} \frac{f^{(2j+1)}(0)}{(2j+1)!}x^{2j+1} + \frac{f^{(2n+2)}(\xi_x)}{(2n+2)!}x^{2n+2}$$

$$= f(x_0) + \sum_{j=0}^{n} \frac{(-1)^j}{(2j+1)!}x^{2j+1} + \frac{f^{(2n+2)}(\xi_x)}{(2n+2)!}x^{2n+2}$$

Also, since $|\sin x| \leq 1$, we have

$$\left|\frac{f^{(2n+2)}(\xi_x)x^{2n+2}}{(2n+2)!}\right| \leq \frac{|x|^{2n+2}}{(2n+2)!} \to 0 \quad \text{as} \quad n \to \infty.$$

Therefore,

$$\left| f(x) - \left[f(x_0) + \sum_{j=0}^{n} \frac{(-1)^j}{(2j+1)!} x^{2j+1} \right] \right| \to 0 \quad \text{as} \quad n \to \infty$$

and hence, $\sin x = \sum_{n=0}^{\infty} \frac{(-1)^n x^{2n+1}}{(2n+1)!} \quad \forall x \in \mathbb{R}.$ ◊

Exercise 2.3.5 Using the Taylor's formula, prove the following:

(i) $\cos x = \sum_{n=0}^{\infty} \frac{(-1)^n x^{2n}}{(2n)!}$ for all $x \in \mathbb{R}$.

(ii) $\dfrac{1}{1-x} = \sum_{n=0}^{\infty} x^n$ for all x with $|x| < 1$.

(iii) $\tan^{-1} x = \sum_{n=0}^{\infty} \frac{(-1)^n x^{2n+1}}{2n+1}$ for all $x \in \mathbb{R}$.

(iv) From (iii), deduce $\dfrac{\pi}{4} = \sum_{n=0}^{\infty} \frac{(-1)^n}{2n+1}$, the *Madhava-Nilakantha* series for $\pi/4$. ◁

   Taylor's formula will help us to find approximate values for a complicated function by the evaluation of a polynomial. Let us look at a few examples.

Example 2.3.36 Let $f(x) = \sin x$, $x \in \mathbb{R}$. Let us find approximate value for $\sin 1$: By the Taylor's formula, we have $\sin x = P_n(x) + R_n(x)$, where

$$P_n(x) := \sum_{k=0}^{n} \frac{(-1)^k x^{2k+1}}{(2k+1)!}, \quad R_n(x) = \frac{f^{(2n+2)}(\xi_x) x^{2n+2}}{(2n+2)!}.$$

Thus, we obtain

$$|R_n(x)| = \left| \frac{f^{(2n+2)}(\xi_x) x^{2n+2}}{(2n+2)!} \right| \le \frac{|x|^{2n+2}}{(2n+2)!} = \varepsilon_n(x).$$

Here are the values $f_n(1)$ and the corresponding error $\varepsilon_n(1)$ for some values of n.

n	0	1	2	3
$P_n(1)$	1	$1 - \dfrac{1}{3!}$	$1 - \dfrac{1}{3!} + \dfrac{1}{5!}$	$1 - \dfrac{1}{3!} + \dfrac{1}{5!} - \dfrac{1}{7!}$
$\varepsilon_n(1)$	0.5	$\dfrac{1}{4!} \simeq 0.4166$	$\dfrac{1}{6!} \simeq 0.001388$	$\dfrac{1}{8!} \simeq 0.0001736$

In particular,

$$\sin 1 \simeq 1 - 0.16666 + 0.00833 - 0.0001984 \simeq 0.3415316$$

and
$$| \sin 1 - 0.3415316| \simeq 0.0001736. \qquad \Diamond$$

Sufficient Condition for Extremum Points Revisited

Using the second derivative of a function and the Taylor's formula, now we give another sufficient condition for a critical point to be an extreme point.

Theorem 2.3.25 *Suppose f is defined on an interval I and x_0 is an interior point of I. Suppose that x_0 is a critical point of f and f has continuous second derivative in a neighbourhood of x_0. Then the following hold.*

(i) If $f''(x_0) < 0$, then f has local maximum at x_0.

(ii) If $f''(x_0) > 0$, then f has local minimum at x_0.

Proof By Taylor's theorem, there exists an open interval I_0 containing x_0 such that for every $x \in I_0$, there exists ξ_x between x_0 and x such that

$$f(x) - f(x_0) = f'(x_0)(x - x_0) + \frac{f''(\xi_x)}{2}(x - x_0)^2.$$

Since x_0 is a critical point of f and f is differentiable at x_0, $f'(x_0) = 0$. Hence,

$$f(x) - f(x_0) = \frac{f''(\xi_x)}{2}(x - x_0)^2. \qquad (*)$$

(i) Suppose $f''(x_0) < 0$. Since f'' is continuous in a neighbourhood of x_0, there exists an open interval I_1 containing x_0 such that for all $x \in I_1$,

$$f''(x) \leq \frac{f''(x_0)}{2}.$$

In particular, from $(*)$, we obtain

$$f(x) - f(x_0) = \frac{f''(\xi_x)}{2}(x - x_0)^2 < 0 \qquad \forall x \in I_1.$$

Thus, f has a local maximum at x_0. (ii) Suppose $f''(x_0) > 0$. In this case, we obtain reverse of the inequalities in the proof of (i), and arrive at the conclusion that f has a local minimum at x_0. ∎

Remark 2.3.15 The conditions given in Theorem 2.3.25 are only sufficient conditions. There are functions f for which none of the conditions (i) and (ii) of Theorem 2.3.25 are satisfied at a point x_0, still f can have local extremum at x_0. For example, consider
$$f(x) = x^4, \qquad g(x) = 1 - x^4, \quad |x| < 1.$$

Then $f'(0) = 0 = g'(0)$, f has local minimum at 0 and g has local maximum at 0. But, $f''(0) = 0 = g''(0)$. $\qquad \Diamond$

Remark 2.3.16 How to identify critical points and extreme points of a function?

1. Suppose f is defined on an open interval I.

   (a) Find those points at which either f is not differentiable or f' vanish. These points are the critical points of f.
   (b) Suppose f is differentiable in a neighbourhood of x_0 and $f'(x_0) = 0$. If $f'(x)$ has the same sign for x on both side of x_0, then f does not have an extremum at x_0; otherwise, use the test for maximum or minimum given in Theorem 2.3.14.

2. Suppose f is continuous on $[a, b]$ and differentiable on (a, b).

   (a) f can have maximum or minimum only at the end points of $[a, b]$ or at those points in (a, b) at which f' vanishes.
   (b) Use the tests as in Theorem 2.3.14 or Theorem 2.3.25.   ◊

2.3.9 Determination of Shapes of a Curves

We shall use conditions on derivatives of a function to find out certain nature of the curve determined by a function. First we spell out what is meant by a curve determined by a function.

Definition 2.3.11 Let f be a continuous function defined on an interval I. Then the graph of f, i.e.,

$$G_f := \{(x, f(x)) : x \in I\},$$

is called the **curve determined by** f.   ◊

A curve determined by a function $f : I \to \mathbb{R}$ is often written as an equation

$$y = f(x), \quad x \in I.$$

Definition 2.3.12 Let f be a continuous function defined on an interval I. Then the curve determined by f is said to be

(1) **convex upwards** or **concave downwards** if f is differentiable at all interior points of I and the tangent line at each point $x \in I$ lies above the curve,
(2) **convex downwards** or **concave upwards** if f is differentiable at all interior points of I and the tangent line at each point $x \in I$ lies below the curve.   ◊

Thus, if f is defined on an interval I and differentiable at all interior points of I, then the curve determined by f is

Fig. 2.16 Concave down
and concave up on
subintervals

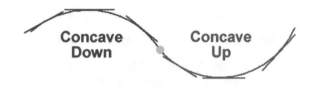

(1) convex upwards if and only if for any interior point x_0 of I,

$$x \in I \setminus \{x_0\}, \ y = f(x_0) + f'(x_0)(x - x_0) \quad \Rightarrow \quad f(x) < y,$$

(2) convex downwards if and only if for any interior point x_0 of I,

$$x \in I \setminus \{x_0\}, \ y = f(x_0) + f'(x_0)(x - x_0) \quad \Rightarrow \quad f(x) > y.$$

It is also conventional to define a function to be convex or concave in the following sense.

Definition 2.3.13 Let I be an interval. Then a function $f : I \to \mathbb{R}$ is said to be a **convex function** if for every $x, y \in I$ and $0 < \lambda < 1$,

$$f((1 - \lambda)x + \lambda y) \leq (1 - \lambda)f(x) + \lambda f(y).$$

The function $f : I \to \mathbb{R}$ is said to be a **concave function** if for every $x, y \in I$ and $0 < \lambda < 1$,

$$f((1 - \lambda)x + \lambda y) \geq (1 - \lambda)f(x) + \lambda f(y). \qquad \Diamond$$

Thus, if $f : I \to \mathbb{R}$ is differentiable at every interior points of I, then

(1) f is convex if and only if f is convex downwards if and only if f is concave upwards, and
(2) f is concave if and only if f is concave downwards if and only if f is convex upwards. $\qquad \Diamond$

Exercise 2.3.6 Give an example of a function f defined on some interval I which is

(i) convex, but not concave downwards,
(ii) concave, but not convex upwards. $\qquad \triangleleft$

Theorem 2.3.26 *Let f be a continuous function defined on an interval I. Suppose f has second derivative at all interior points of I. Then the curve determined by f is*

(i) *convex upwards if $f''(x) < 0$ for all interior points x in I;*
(ii) *convex downwards if $f''(x) > 0$ for all interior points x in I.*

Fig. 2.17 Concave down and concave up on subintervals

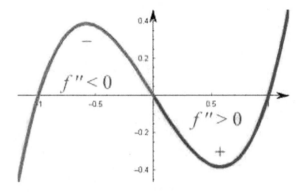

Proof Suppose $f''(x) < 0$ for all interior points x in I. Let x_0 be any point in the interior of I. We have to show that

$$x \in I, \; y = f(x_0) + f'(x_0)(x - x_0) \quad \Rightarrow \quad f(x) < y.$$

So let $x \in I$ and $y = f(x_0) + f'(x_0)(x - x_0)$. By Taylor's theorem, there exists ξ_x between x and x_0 such that

$$f(x) = f(x_0) + f'(x_0)(x - x_0) + \frac{f''(\xi_x)}{2}(x - x_0)^2$$

so that, using the fact that $f''(\xi_x) < 0$,

$$f(x) = y + \frac{f''(\xi_x)}{2}(x - x_0)^2 < y.$$

Hence, G_f is convex upwards, proving (i). Proof of (ii) follows analogously (Fig. 2.17). ∎

Exercise 2.3.7 Prove that the converse statements in (i) and (ii) in Theorem 2.3.26 are also true. ◁

Example 2.3.37 Let us consider a few examples.

(i) Let
$$f(x) = x^2 \quad \text{and} \quad g(x) = 1 - x^2 \quad \text{for} \quad x \in \mathbb{R}.$$

Then G_f is convex downwards and G_g is convex upwards.

(ii) Let $f(x) = e^x$, $x \in \mathbb{R}$. Note that $f''(x) > 0$ for all $x \in \mathbb{R}$. Hence, by the Theorem 2.3.26, $y = e^x$ is convex downwards on $\mathbb{R}$.

(iii) Let $f(x) = x^3$, $x \in \mathbb{R}$. Note that $f''(x) = 6x$ so that, by the Theorem 2.3.26, the curve $y = x^3$ is convex upwards for $x < 0$ and convex downwards for $x > 0$. ◊

Fig. 2.18 Point of inflexion

Definition 2.3.14 A point (x_0, y_0) on the curve determined by a function f is said to be a **point of inflexion** of the curve if in a neighbourhood of x_0, the curve is convex upwards on one side of x_0 and convex downwards on other side of x_0 (Fig. 2.18). ◊

Example 2.3.38 In view of the conclusions in Example 2.3.37(iii), the point $(0, 0)$ on the curve $y = x^3$ is a point of inflexion. ◊

Theorem 2.3.27 *Suppose f has second derivative in a deleted neighbourhood of a point x_0. Then the point $(x_0, f(x_0))$ is a point of inflexion of the curve G_f if f'' has constant but different signs on each side of x_0, and at the point x_0, either $f''(x_0)$ does not exist or $f''(x_0) = 0$.*

Proof This is a consequence of Theorem 2.3.26. ∎

Theorem 2.3.28 *Suppose f has second derivative in a neighbourhood I_0 of a point x_0. If $(x_0, f(x_0))$ is a point of inflexion of the curve G_f and if f'' is continuous at x_0, then $f''(x_0) = 0$.*

Proof Suppose $(x_0, f(x_0))$ is a point of inflexion of the curve G_f and f'' is continuous at x_0. Without loss of generality, we may assume that G_f is convex upwards for $x \in I_0$, $x < x_0$ and it is convex downwards for $x \in I_0$, $x > x_0$. Thus,

$$x \in I_0, \; x < x_0 \quad \Rightarrow \quad f(x) < f(x_0) + f'(x_0)(x - x_0), \qquad (2.1)$$

$$x \in I_0, \; x > x_0 \quad \Rightarrow \quad f(x) > f(x_0) + f'(x_0)(x - x_0). \qquad (2.2)$$

So, let $x \in I_0$. By Taylor's theorem, there exists ξ_x between x and x_0 such that

$$f(x) = f(x_0) + f'(x_0)(x - x_0) + \frac{f''(\xi_x)}{2}(x - x_0)^2.$$

Now, (2.1) implies that $f''(\xi_x) < 0$ so that by letting $x \to x_0$, we have $f''(x_0) \le 0$. Also, (2.2) implies that $f''(\xi_x) > 0$ so that by letting $x \to x_0$, we have $f''(x_0) \ge 0$. Thus, $f''(x_0) = 0$.   ∎

2.4 Additional Exercises

2.4.1 Limit

1. Using the definition of limit, show that $\lim_{x \to 3} \dfrac{x}{4x-9} = 1$.

2. Show that the function f defined by $f(x) = \begin{cases} x, & \text{if } x < 1, \\ 1+x, & \text{if } x \ge 1 \end{cases}$ does not have the limit as $x \to 1$.

3. Let f be defined by $f(x) = \begin{cases} 3-x, & \text{if } x > 1, \\ 1, & \text{if } x = 1, \\ 2x, & \text{if } x < 1. \end{cases}$ Find $\lim_{x \to 1} f(x)$. Is this limit equal to $f(1)$?

4. In the following cases, find the left limit and right limit of f at 1 and check whether they are equal and equal to $f(1)$:

   (i) $f(x) = \begin{cases} x, & \text{if } x \le 1, \\ 2, & \text{if } x > 1. \end{cases}$ (ii) $f(x) = \begin{cases} x, & \text{if } x \le 1, \\ 1, & \text{if } x > 1. \end{cases}$

5. Let f be defined on a deleted neighbourhood D_0 of a point x_0 and $\lim_{x \to x_0} f(x) = b$. If $b \ne 0$, then show that there exists $\delta > 0$ such that $f(x) \ne 0$ for every $x \in (x_0 - \delta, x_0 + \delta) \cap D_0$.

6. Let f be defined by $f(x) = \begin{cases} 1, & \text{if } x \in \mathbb{Q}, \\ 0, & \text{if } x \notin \mathbb{Q}. \end{cases}$ Show that

   (a) $\lim_{x \to a} f(x)$ does not exist for any $a \in \mathbb{R}$, and
   (b) $\lim_{x \to a} (x-a) f(x) = 0$ for every $a \in \mathbb{R}$.

7. Suppose $f, g : \mathbb{R} \to \mathbb{R}$ and $b \in \mathbb{R}$. Show that, if $\lim_{x \to \infty} f(x) = \infty$ and $\lim_{x \to \infty} g(x) = b$, then $\lim_{x \to \infty} (g \circ f)(x) = b$.

8. Give an example in which $\lim_{x \to a} f(x) = b$ and $\lim_{y \to b} g(y) = c$, but $\lim_{x \to a} (g \circ f)(x) \ne c$.

9. Let $f : (0, \infty) \to \mathbb{R}$. Show that $\lim_{x \to 0} f(x) = b$ if and only if $\lim_{x \to \infty} f(1/x) = b$.

10. Suppose $\lim_{x \to \infty} f(x) = b$ and $\lim_{x \to \infty} g(x) = c$. Verify the following.

    (a) $\lim_{x \to \infty} [f(x) + g(x)] = b + c$ and $\lim_{x \to \infty} f(x)g(x) = bc$.
    (b) If $c \ne 0$, then there exists $M_0 > 0$ such that $g(x) \ne 0$ for all $x > M_0$ and $\lim_{x \to \infty} \dfrac{f(x)}{g(x)} = \dfrac{b}{c}$.

11. State and prove sequential characterization for the following:

$$\lim_{x \to a} f(x) = \infty, \quad \lim_{x \to a} f(x) = -\infty, \quad \lim_{x \to +\infty} f(x) = \infty,$$
$$\lim_{x \to +\infty} f(x) = -\infty, \quad \lim_{x \to -\infty} f(x) = \infty, \quad \lim_{x \to -\infty} f(x) = -\infty.$$

12. Let $f : \mathbb{R} \to \mathbb{R}$ be such that $f(x + y) = f(x) + f(y)$. Suppose $\lim_{x \to 0} f(x)$ exists. Prove that $\lim_{x \to 0} f(x) = 0$ and $\lim_{x \to c} f(x) = f(c)$ for any $c \in \mathbb{R}$.

2.4.2 Continuity

1. Suppose $f : [a, b] \to \mathbb{R}$ is continuous. If $c \in (a, b)$ is such that $f(c) > 0$, and if $0 < \beta < f(c)$, then show that there exists $\delta > 0$ such that $f(x) > \beta$ for all $x \in (c - \delta, c + \delta) \cap [a, b]$.
2. Let $f : \mathbb{R} \to \mathbb{R}$ satisfy the relation $f(x + y) = f(x) + f(y)$ for every $x, y \in \mathbb{R}$. If f is continuous at 0, then show that f is continuous at every $x \in \mathbb{R}$, and in that case $f(x) = xf(1)$ for every $x \in \mathbb{R}$.
3. There does not exist a continuous function f from $[0, 1]$ onto $\mathbb{R}$. Why?
4. Find a continuous function f from $(0, 1)$ onto $\mathbb{R}$.
5. Prove that if $f : [a, b] \to \mathbb{R}$ is continuous on the intervals $[a, c]$ and $[c, b]$ for some $c \in (a, b)$, then f is continuous on $[a, b]$.
6. Give an example of a discontinuous function $f : [0, 1] \to \mathbb{R}$ such that f is continuous on the intervals $[0, 1/2]$ and $(1/2, 1]$.
7. Suppose $f : [a, b] \to [a, b]$ is continuous. Show that there exists $c \in [a, b]$ such that $f(c) = c$.
8. Suppose $f : [a, b] \to [a, b]$ is such that there exists r satisfying $0 < r < 1$ and $|f(x) - f(y)| \le r|x - y|$ for all $x, y \in [a, b]$. Let $x_1 \in [a, b]$ and for $n \in \mathbb{N}$, let $x_{n+1} := f(x_n)$. Prove that $x_n \to c$ for some c and $f(c) = c$.
9. There exists $x \in \mathbb{R}$ such that $17x^{19} - 19x^{17} - 1 = 0$. Why?
10. Let $f : \mathbb{R} \to \mathbb{R}$ be defined by $f(x) = 1 + x + x^2$. Without solving a quadratic equation, can you assert that there is some x_0 such that $f(x_0) = 2$?
11. If $p(x)$ is a polynomial of odd degree, then there exists at least one $x_0 \in \mathbb{R}$ such that $p(x_0) = 0$. Why?
12. Suppose $f : \mathbb{R} \to \mathbb{R}$ is continuous such that $f(x) \to 0$ as $|x| \to \infty$. Prove that f attains either a maximum or a minimum.
13. Suppose $f : [a, b] \to \mathbb{R}$ is continuous such that for every $x \in [a, b]$, there exists a $y \in [a, b]$ such that $|f(y)| \le |f(x)|/2$. Show that there exists $x_0 \in [a, b]$ such that $f(x_0) = 0$.
14. Let $f : [a, b] \to [a, b]$ be such that $|f(x) - f(y)| \le |x - y|/2$ for all $x, y \in [a, b]$. Show that there exists $x_0 \in [a, b]$ such that $f(x_0) = x_0$.
15. Write details of the proof of Corollary 2.2.10
16. Prove the following.

   (a) Let $f : (a, b) \to \mathbb{R}$ be a continuous function. If $\lim_{x \to a} f(x) = c$ and $\lim_{x \to b} f(x) = d$, where $c < d$, then for every $y \in (c, d)$, there exists $x \in (a, b)$ such that $f(x) = y$.

   (b) Let $f : \mathbb{R} \to \mathbb{R}$ be a continuous function. If $\lim_{x \to -\infty} f(x) = c$ and $\lim_{x \to \infty} f(x) = d$, where $c < d$, then for every $y \in (c, d)$, there exists $x \in \mathbb{R}$ such that $f(x) = y$.

   (c) Let $f : \mathbb{R} \to \mathbb{R}$ be a continuous function. If $\lim_{x \to -\infty} f(x) \to c$ and $\lim_{x \to \infty} f(x) = \infty$, where $c < d$, then for every $y \in (c, \infty)$, there exists $x \in \mathbb{R}$ such that $f(x) = y$.

   (d) From (c) above, deduce that for every $y \in (0, \infty)$, there exists $x \in \mathbb{R}$ such that $e^x = y$.

17. Prove that if f is strictly monotonic on an interval I, then f is injective on I.

18. Let $n \in \mathbb{N}$. Prove that for every $y \geq 0$, there exists a unique $x \geq 0$ such that $x^n = y$.

19. Suppose f is a continuous function defined on an interval I and x_0 is an interior point of I. Prove the following.

   (a) If f is increasing on $(x_0 - h, x_0)$ and decreasing on $(x_0, x_0 + h)$ for some $h > 0$, then f attains local maximum at x_0.

   (b) If "increasing" and "decreasing" in (a) above are interchanged, then in the conclusion "maximum" can be replaced by "minimum".

   (c) If "increasing" and "decreasing" in (a) are replaced by "strictly increasing" and "strictly decreasing", respectively, then we obtain "strict local maximum".

20. Let f be a continuous function defined on an interval I. Show that if f is injective, then it is strictly monotonic on I [Hint: Use Intermediate Value Theorem].

2.4.3 Differentiation

1. Prove that the function $f(x) = |x|, x \in \mathbb{R}$ is not differentiable at 0.

2. Let x_0 be an interior point of an interval I. Prove that $f : I \to \mathbb{R}$ is differentiable at x_0 if and only if $f'_+(x_0)$ and $f'_-(x_0)$ exist and $f'_-(x_0) = f'_+(x_0)$, and in that case $f'(x_0) = f'_-(x_0) = f'_+(x_0)$.

3. Consider a polynomial $p(x) = a_0 + a_1 x^2 + \ldots + a_n x^n$ with real coefficients $a_0, a_1, \ldots, a_n$ such that $a_0 + \dfrac{a_1}{2} + \dfrac{a_2}{3} + \ldots + \dfrac{a_n}{n+1} = 0$. Show that there exists $x_0 \in \mathbb{R}$ such that $p(x_0) = 0$. [Note that the conclusion need not hold if the condition imposed on the coefficients is dropped. To see this, consider $p(x) = 1 + x^2$.]

4. Let I and J be open intervals and $f : I \to J$ be bijective and differentiable at every $x_0 \in I$. If $f'(x_0) \neq 0$, then show that the inverse function $f^{-1} : J \to I$ is also differentiable at x_0 and $(f^{-1})'(x_0) = 1/f'(x_0)$.

5. Let $f : \mathbb{R} \to \mathbb{R}$ be defined by

$$f(x) = \begin{cases} x^2, & x \text{ rational,} \\ 0, & x \text{ irrational.} \end{cases}$$

Show that f is differentiable at 0 and $f'(0) = 0$.

6. Let $f : \mathbb{R} \to \mathbb{R}$ be defined by

$$f(x) = \begin{cases} x^2 \sin(1/x), & x \neq 0, \\ 0, & x = 0. \end{cases}$$

Show that f is differentiable at every $x \in \mathbb{R}$. Is f' continuous?

7. Let $f : \mathbb{R} \to \mathbb{R}$ be defined by

$$f(x) = \begin{cases} x^2, & x \geq 0, \\ 0, & x < 0. \end{cases}$$

Show that f is differentiable at every $x \in \mathbb{R}$. Is f' continuous?

8. The function $f : \mathbb{R} \to \mathbb{R}$ defined by $f(x) = 1 + x + x^3$ has neither maximum nor minimum at any point. Why?

9. Find the points at which the function $f : \mathbb{R} \to \mathbb{R}$ defined by $f(x) = 1 = x - x^3$ attain local maxima and local minima.

10. Using Taylor's theorem, show that

$$(1 + x)^n = 1 + nx + \frac{n(n-1)}{2!}x^2 + \frac{n(n-1)(n-2)}{3!}x^3 + \ldots + x^n.$$

11. Show that there does not exist a function $f : [0, 1] \to \mathbb{R}$ which is differentiable on $(0, 1)$ such that

$$f'(x) = \begin{cases} 0, & \text{if } 0 < x < 1/2, \\ 1, & \text{if } 1/2 \leq x < 1. \end{cases}$$

[Hint: Use Example 2.3.18 in the interval $[0, 1/2]$ and $[1/2, 1]$ taking $x_0 = 1/2$, and show that the resulting function f is not differentiable at $x_0 = 1/2$.]

12. Suppose f is differentiable on $(0, \infty)$ and $\lim_{x \to \infty} f'(x) = 0$. Prove that $\lim_{x \to \infty} [f(x + 1) - f(x)] = 0$.

13. Let $f : \mathbb{R} \to \mathbb{R}$ be defined by $f(x) = 1 + x + x^3$. Prove that f is a bijection. [*Hint:* Use Intermediate value theorem and mean value theorem.]

14. Let $f : [a, b] \to \mathbb{R}$ be a continuous function which is differentiable on (a, b). Prove the following.

   (a) If $f'(x) \geq 0$ (respectively, $f'(x) > 0$) for all $x \in (a, b)$, then f is monotonically increasing (respectively, strictly increasing) on $[a, b]$.

   (b) If $\lim_{x \to a} f'(x)$ exists, then $f'(a)$ exists and $f'(a) = \lim_{x \to a} f'(x)$. [*Hint:* Use mean value theorem.]

15. Let $f : [a, b] \to \mathbb{R}$ be a continuous function which is differentiable on (a, b). Prove that if f' is strictly positive or strictly negative on (a, b), then it is a one-one function.

16. Let $f : [a, b] \to \mathbb{R}$ be a continuous function which is differentiable on (a, b). Prove that if $f'(x) \neq 0$ for all $x \in (a, b)$, then f is either strictly increasing or strictly decreasing on $[a, b]$.

17. Let $f : [0, 1] \to \mathbb{R}$ be a continuous function which is differentiable on $(0, 1)$. Prove that, if $f(0) = 0$ and $f(1) = 1$, then there exists $c \in (0, 1)$ such that $f'(c) = 1$.

18. Using MVT, prove the following:

   (a) $e^x \geq 1 + x$ for all $x \in \mathbb{R}$.

   (b) $|\sin x - \sin y| \leq |x - y|$ for all $x \in \mathbb{R}$.

   (c) $\dfrac{x - 1}{x} \leq \ln x \leq x - 1$ for all $x \in \mathbb{R}$.

19. Let $f(x) = x^2 + 1$ and $g(x) = x + 2$ for all $x \in \mathbb{R}$. What is wrong with the following statement?

$$\text{Since } g'(0) \neq 0, \lim_{x \to 0} \frac{f(x)}{g(x)} = \frac{f'(0)}{g'(0)} = 0.$$

20. Prove that $\lim_{x \to \infty} x \ln(1 + 1/x)$ exists and it is equal to 1. [*Hint*: L'Hospital's rule]

21. Find the following limits.

   (a) $\lim_{x \to \pi/2} \dfrac{\cos x}{x - \pi/2}$

   (b) $\lim_{x \to 0} \dfrac{1}{x(\ln x)^2}$

   (c) $\lim_{x \to \infty} \dfrac{\ln x}{x^2}$

   (d) $\lim_{x \to \infty} \dfrac{\ln x}{\sqrt{x}}$

22. Let f be a function defined on an interval I. Prove the following.

   (a) f is convex (i.e., convex downwards) if and only if the line segment joining any tow points on the graph of f lies in the region $\{(x, y) : f(x) \leq y, x \in I\}$.

   (b) f is concave (i.e., concave upwards) if and only if the line segment joining any tow points on the graph of f lies in the region $\{(x, y) : y \leq f(x), x \in I\}$.

23. Suppose f is convex on an open interval I. Prove the following.

   (a) $f'(x-)$ and $f'(x+)$ exist for every $x \in I$.

   (b) f is continuous on I.

24. Give an example to show that a convex function defined on a closed interval need not be continuous.

25. Find approximate values of $\sin x$, $\cos x$, $\tan^{-1} x$ at the point $x = 1/2$ using Taylor's formula and find the estimates for the errors for different values of n.

Chapter 3
Definite Integral

Integration is as ubiquitous in calculus as differentiation is. As the word suggests, it represents, in some sense, aggregate or sums of certain quantities associated with the values of certain functions. The integration that we deal with is the *definite integral* which is different from the so called *indefinite integral* or *antiderivative* of a function. Definite integral also can be defined using indefinite integral. But, the class of functions that has indefinite integral is much smaller than the class of functions which can be integrated using the concept of definite integral. Moreover, the indefinite integral becomes a special case of definite integral, thanks to an important theorem, called the *fundamental theorem of integration*.

3.1 Integrability and Integral

3.1.1 Introduction

In school one comes across the definition of the integral of a real valued function defined on a closed and bounded interval $[a, b]$ between the limits a and b, that is,

$$\int_a^b f(x)\mathrm{d}x$$

as the number $g(b) - g(a)$, where g is a function whose derivative is f at all points in the open interval (a, b). One immediate question that one would like to ask is the following:

Given any function $f : [a, b] \to \mathbb{R}$, is it possible to find a a function $g : [a, b] \to \mathbb{R}$ which is differentiable on (a, b) and $g'(x) = f(x)$ for all $x \in (a, b)$?

The answer is in the negative, as following example shows.

© The Author(s), under exclusive license to Springer Nature Switzerland AG 2021
M. T. Nair, *Calculus of One Variable*,
https://doi.org/10.1007/978-3-030-88637-0_3

Fig. 3.1 Area under the
curve $y = f(x)$

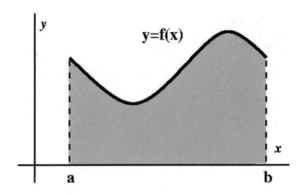

Example 3.1.1 Consider the function

$$f(x) = \begin{cases} -1, & -1 \le x \le 0, \\ 1, & 0 < x \le 1. \end{cases}$$

In this case, there does not exist a function $g : [-1, 1] \to \mathbb{R}$ which is differentiable
on $(-1, 1)$ and $g'(x) = f(x)$ for all $x \in (-1, 1)$ (cf. Theorem 2.3.9).   ◊

Another point one recalls from school mathematics is that if g is a differentiable
function defined on an open interval, then $g'(x)$ has a geometric meaning, namely,
it represents the slope of the tangent to the graph of g at the point x.

Do we have a geometric meaning to the integral $\int_a^b f(x)dx$?

We answer to the above question in the affirmative for certain class of functions as
we shall see.

Suppose $f : [a, b] \to \mathbb{R}$ is a bounded function. Our attempt is to associate a
number γ to such a function such that, in case $f(x) \ge 0$ for $x \in [a, b]$, then γ is the
area of the region under the graph of f, i.e., the region R bounded by the graph of f,
x-axis, and the ordinates at a and b. We may not succeed to do this for *all* bounded
functions f (Fig. 3.1).

Suppose, for a moment, that $f(x) \ge 0$ for all $x \in [a, b]$. Let us agree that we
have some idea about the area A of the region R under the curve $y = f(x)$. For
example, we require that the area A lies between two quantities A_1 and A_2, where
A_1 is the area of the rectangle that is inscribed within the region R and A_2 is the area
of the region which subscribe R, with their sides parallel to the coordinate axes. In
particular, $A_1 = m(b - a)$, $A_2 = M(b - a)$ and

$$m(b - a) \le A \le M(b - a), \tag{3.1}$$

where

$$m = \inf_{x \in [a,b]} f(x) \quad \text{and} \quad M = \sup_{x \in [a,b]} f(x).$$

Thus, we get an upper and lower bound for A. To get better estimates, let us consider a point c such that $a < c < b$. Then we have

$$m_1 \le f(x) \le M_1 \quad \forall x \in [a, c]; \quad m_2 \le f(x) \le M_2 \quad \forall x \in [c, b],$$

where

$$m_1 = \inf_{x \in [a,c]} f(x), \qquad m_2 = \inf_{x \in [c,b]} f(x),$$

$$M_1 = \sup_{x \in [a,c]} f(x), \qquad M_2 = \sup_{x \in [c,b]} f(x).$$

Then, we must have

$$m_1(c - a) + m_2(b - c) \le A \le M_1(c - a) + M_2(b - c). \tag{3.2}$$

Since $m \le \min\{m_1, m_2\}$ and $\max\{M_1, M_2\} \le M$, we have

$$m(b - a) = m(c - a) + m(b - c) \le m_1(c - a) + m_2(b - c),$$

$$M(b - a) = M(c - a) + M(b - c) \ge M_1(c - a) + M_2(b - c).$$

Thus, we can infer that the estimates in (3.2) are better than those in (3.1). We may be able to improve these bounds by taking more and more points in $[a, b]$. This is the basic idea of *Riemann integration* that we describe below.

3.1.2 Lower and Upper Sums

Let $f : [a, b] \to \mathbb{R}$ be a bounded function and let P be a *partition* of $[a, b]$, i.e., a finite set $P = \{x_0, x_1, x_2, \ldots, x_k\}$ of points in $[a, b]$ such that

$$a = x_0 < x_1 < x_2 < \ldots < x_k = b.$$

We shall denote such partition also by $P = \{x_i\}_{i=0}^k$. Corresponding to the partition $P = \{x_i\}_{i=0}^k$ and the function f, we associate two numbers:

$$L(P, f) := \sum_{i=1}^k m_i(x_i - x_{i-1}), \quad U(P, f) := \sum_{i=1}^k M_i(x_i - x_{i-1}),$$

where

$$m_i = \inf_{x \in [x_{i-1}, x_i]} f(x), \quad M_i = \sup_{x \in [x_{i-1}, x_i]} f(x)$$

for $i = 1, \ldots, k$.

Note that for the definition of $L(P, f)$ and $U(P, f)$ above, we used the fact that f is a bounded function (How?).

Definition 3.1.1 The quantities $L(P, f)$ and $U(P, f)$ are called the **lower sum** and the **upper sum**, respectively, of the function f associated with the partition P. ◊

Note that if $f(x) \geq 0$ for all $x \in [a, b]$, then $L(P, f)$ is the *total area* of the rectangles with lengths m_i and widths $x_i - x_{i-1}$, and $U(P, f)$ is the *total area* of the rectangle with lengths M_i and widths $x_i - x_{i-1}$ for $i = 1, \ldots, k$. Thus, it is intuitively clear that the required area A must satisfy the relation

$$L(P, f) \leq A \leq U(P, f)$$

for all partitions P of $[a, b]$ (Figs. 3.2 and 3.3).

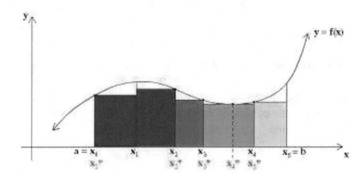

Fig. 3.2 Lower sum

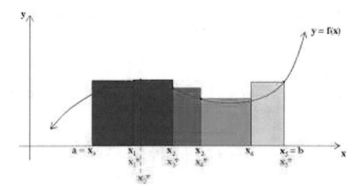

Fig. 3.3 Upper sum

Remark 3.1.1 If f is continuous, then $L(P, f)$ and $U(P, f)$ can be represented as

$$L(P, f) := \sum_{i=1}^{k} f(c_i)(x_i - x_{i-1}), \quad U(P, f) := \sum_{i=1}^{k} f(d_i)(x_i - x_{i-1}),$$

respectively, for some c_i, d_i, $i \in \{1, \ldots, k\}$ (Why?). ◊

3.1.3 The Integral and Its Characterizations

Throughout this chapter, functions defined on $[a, b]$ are considered to be bounded functions, and we use the notation $\mathcal{P}$ to denote the set of all partitions of $[a, b]$. Thus we have

$$L(P, f) \leq U(P, f) \quad \forall P \in \mathcal{P}.$$

Definition 3.1.2 Let $f : [a, b] \to \mathbb{R}$ be a bounded function. Then f is said to be **Riemann integrable** on $[a, b]$, if there exists a unique γ such that for every $P \in \mathcal{P}$,

$$L(P, f) \leq \gamma \leq U(P, f),$$

and in that case, γ is called the **Riemann integral**[1] of f, and it is denoted by

$$\int_a^b f(x)dx$$ ◊

CONVENTION: In the due course, Riemann integral will be simply referred to as the *integral*.

Remark 3.1.2 In higher mathematics one will come across another form of integral called *Lebesgue integral*, which is more general than Riemann integral. It is dealt in a course on Measure and Integration (see, e.g., Nair [9]). ◊

For a bounded function $f : [a, b] \to \mathbb{R}$, though $L(P, f) \leq U(P, f)$ for all $P \in \mathcal{P}$, it is not obvious that

$$\sup_{P \in \mathcal{P}} L(P, f) \leq \inf_{P \in \mathcal{P}} U(P, f).$$

In fact, it is true. Before showing this, let us introduce a definition.

---

[1] Named after the German mathematician Georg Friedrich Bernhard Riemann (September 17, 1826 July 20, 1866) who placed the concept of integration in a rigorous footing.

Definition 3.1.3 Let P and Q be a partitions of $[a, b]$. Then Q is called a **refinement** of P if $P \subseteq Q$. $\diamond$

If Q is a refinement of P, then it can be shown easily that

$$L(P, f) \le L(Q, f) \quad \text{and} \quad U(Q, f) \le U(P, f).$$

Theorem 3.1.1 *Let $f : [a, b] \to \mathbb{R}$ be a bounded function. Then*

(i) $L(P, f) \le U(Q, f) \quad \forall P, Q \in \mathcal{P}$,

(ii) $\sup\limits_{P \in \mathcal{P}} L(P, f) \le \inf\limits_{P \in \mathcal{P}} U(P, f).$

Proof Let P and Q be partitions of $[a, b]$, and let $\tilde{P}$ be the partition obtained by taking points in P and Q without repetitions. Then we have

$$L(P, f) \le L(\tilde{P}, f) \le U(\tilde{P}, f) \le U(Q, f).$$

Thus, (i) is proved. By (i),

$$\sup\limits_{P \in \mathcal{P}} L(P, f) \le U(Q, f) \quad \forall Q \in \mathcal{P}.$$

Hence,

$$\sup\limits_{P \in \mathcal{P}} L(P, f) \le \inf\limits_{Q \in \mathcal{P}} U(Q, f).$$

Thus, (ii) is also proved. ∎

Notation 3.1.1 Let P and Q be partitions of $[a, b]$. Then we denote by $P \cup Q$, the partition obtained by taking points in P and Q without repetitions. $\diamond$

We have already observed in Theorem 3.1.1 that, if P and Q are partitions of $[a, b]$, then $P \cup Q$ is a refinement of both P and Q.

Suppose f is Riemann integrable over $[a, b]$. Then we know that

$$L(P, f) \le \int_a^b f(x)\, dx \le U(P, f) \quad \forall P \in \mathcal{P}. \tag{1}$$

Hence,

$$\sup\limits_{P \in \mathcal{P}} L(P, f) \le \int_a^b f(x)\, dx \le \inf\limits_{P \in \mathcal{P}} U(P, f).$$

Since $\gamma := \int_a^b f(x)\,dx$ is the only number satisfying (1), we can assert that

$$\sup_{P \in \mathcal{P}} L(P, f) = \inf_{P \in \mathcal{P}} U(P, f). \tag{2}$$

Conversely, suppose (2) holds, and let

$$\gamma := \sup_{P \in \mathcal{P}} L(P, f) = \inf_{P \in \mathcal{P}} U(P, f).$$

Then it is clear that $L(P, f) \le \gamma \le U(P, f)$ for all $P \in \mathcal{P}$. If $\tilde{\gamma} \in \mathbb{R}$ is such that $L(P, f) \le \tilde{\gamma} \le U(P, f)$ for all $P \in \mathcal{P}$, then we obtain

$$\sup_P L(P, f) \le \tilde{\gamma} \le \inf_P U(P, f)$$

so that by (2), $\tilde{\gamma} = \gamma$.

Thus, we have proved the following theorem.

Theorem 3.1.2 *A bounded function $f : [a, b] \to \mathbb{R}$ is Riemann integrable if and only if*

$$L(f) := \sup_{P \in \mathcal{P}} L(P, f) \quad and \quad U(f) := \inf_{P \in \mathcal{P}} U(P, f)$$

are equal, and in that case, the common value is the integral of f.

Notation 3.1.2 In view of Theorem 3.1.2, the quantities

$$L(f) := \sup_{P \in \mathcal{P}} L(P, f) \quad and \quad U(f) := \inf_{P \in \mathcal{P}} U(P, f)$$

are known as *lower integral* and *upper integral*, respectively, and they are also denoted by

$$\underline{\int_a^b} f(x)dx \quad and \quad \overline{\int_a^b} f(x)dx,$$

respectively. ◇

Definition 3.1.4 Suppose $f : [a, b] \to \mathbb{R}$ is integrable. Then we define

$$\int_b^a f(x)\,dx = -\int_a^b f(x)\,dx,$$

and for any $\tau \in [a, b]$, we define $\int_\tau^\tau f(x)\,dx = 0$. ◇

Example 3.1.2 Let $f(x) = c$ for all $x \in [a, b]$, for some $c \in \mathbb{R}$. Then we have

$$L(P, f) = c(b - a), \quad U(P, f) = c(b - a)$$

Hence

$$c(b - a) = L(P, f) \le L(f) \le U(f) \le U(P, f) = c(b - a).$$

Thus, f is integrable and $\int_a^b f(x)\mathrm{d}x = c(b - a)$.   ◊

Example 3.1.3 Let $f(x) = x$ for all $x \in [a, b]$. Let us consider an arbitrary partition $P : a = x_0 < x_1 < \cdots < x_k = b$. Then

$$L(P, f) = \sum_{i=1}^{k} x_{i-1}(x_i - x_{i-1}), \quad U(P, f) = \sum_{i=1}^{k} x_i(x_i - x_{i-1}).$$

Since

$$x_{i-1}(x_i - x_{i-1}) \le \frac{1}{2}(x_i + x_{i-1})(x_i - x_{i-1}) \le x_i(x_i - x_{i-1})$$

we have

$$L(P, f) \le \frac{1}{2} \sum_{i=1}^{k} (x_i^2 - x_{i-1}^2) \le U(P, f).$$

Thus,

$$L(P, f) \le \frac{b^2 - a^2}{2} \le U(P, f)$$

for all partitions P. Also, we have

$$U(P, f) - L(P, f) = \sum_{i=1}^{k} (x_i - x_{i-1})^2.$$

Thus, taking $x_i = a + i\frac{(b-a)}{k}$ for $i = 0, 1, \ldots, k$, we have

$$U(P, f) - L(P, f) = k\left(\frac{b - a}{k}\right)^2 = \frac{(b - a)^2}{k}.$$

Since $L(P, f) \le L(f) \le U(f) \le U(P, f)$, we obtain,

$$0 \le U(f) - L(f) \le \frac{(b - a)^2}{k} \quad \forall k \in \mathbb{N}.$$

Thus, $L(f) = U(f)$ so that f is integrable, and the value of $\int_a^b f(x)\mathrm{d}x$ is $(b^2 - a^2)/2$. $\diamond$

Remark 3.1.3 Not all functions are integrable! For example, consider $f : [a, b] \to \mathbb{R}$ defined by

$$f(x) = \begin{cases} 0, & x \in \mathbb{Q}, \\ 1, & x \notin \mathbb{Q}. \end{cases}$$

For this function we have $L(P, f) = 0$ and $U(P, f) = b - a$ for any partition P of $[a, b]$. Thus, in this case $L(f) = 0$, $U(f) = b - a$, and hence, f is not integrable. The above function f is called the **Dirichlet function**. $\diamond$

Another sum corresponding to a partition P of $[a, b]$ is the so-called *Riemann sum*.

Definition 3.1.5 Let $P : a = x_0 < x_1 < \cdots < x_k = b$ of $[a, b]$. Then any set $T := \{\xi_1, \ldots, \ldots \xi_k\}$ of points in $[a, b]$ with $\xi_i \in [x_{i-1}, x_i]$ for $i = 1, \ldots, k$ is called a set of **tags** on P, and the sum

$$S(P, f, T) := \sum_{i=1}^{k} f(\xi_i)(x_i - x_{i-1})$$

is called the **Riemann sum** of f corresponding to (P, T). $\diamond$

We may observe that for any partition P of $[a, b]$,

$$L(P, f) \le S(P, f, T) \le U(P, f)$$

for every tag T on P.

Theorem 3.1.3 *A bounded function $f : [a, b] \to \mathbb{R}$ is Riemann integrable if and only if for every $\varepsilon > 0$, there exists a partition P of $[a, b]$ such that*

$$U(P, f) - L(P, f) < \varepsilon.$$

Proof Suppose f is integrable and γ is its integral. Then γ is the unique number such that for every partition P of $[a, b]$,

$$L(P, f) \le \gamma \le U(P, f). \tag{$*$}$$

Hence, if there exists an $\varepsilon_0 > 0$ satisfying $U(P, f) - L(P, f) \ge \varepsilon_0$ for all partitions P of $[a, b]$, then we arrive at a contradiction to the fact that γ is the unique number satisfying $(*)$. This proves the necessary part.

Conversely, suppose that for every $\varepsilon > 0$, there exists a partition P_ε of $[a, b]$ such that $U(P_\varepsilon, f) - L(P_\varepsilon, f) < \varepsilon$. Since

$$L(P_\varepsilon, f) \leq L(f) \leq U(f) \leq U(P_\varepsilon, f)$$

we obtain that $U(f) - L(f) < \varepsilon$ for every $\varepsilon > 0$. This shows that $L(f) = U(f)$, so that by Theorem 3.1.2, f is integrable. ∎

The proof of the following corollary is immediate from Theorem 3.1.3.

Corollary 3.1.4 (Archimedes–Riemann theorem)[2] *A bounded function $f : [a, b] \to \mathbb{R}$ is Riemann integrable if and only if there exists a sequence (P_n) of partitions of $[a, b]$ such that*

$$U(P_n, f) - L(P_n, f) \to 0 \quad as \quad n \to \infty$$

and in that case

$$L(P_n, f) \to \int_a^b f(x)\, dx, \quad U(P_n, f) \to \int_a^b f(x)\, dx$$

and

$$S(P_n, f, T_n) \to \int_a^b f(x)\, dx,$$

where T_n is any set of tags on P_n, $n \in \mathbb{N}$. as $n \to \infty$.

Exercise 3.1.1 Supply details of the proof of Corollary 3.1.4. ◁

Here is another application of Theorem 3.1.3.

Theorem 3.1.5 *Every monotonic function $f : [a, b] \to \mathbb{R}$ is Riemann integrable.*

Proof Assume first that f is monotonically increasing, that is, for every $x, y \in [a, b], x \leq y$ implies $f(x) \leq f(y)$. This assumption, in particular implies that f is a bounded function, as $f(a) \leq f(x) \leq f(b)$ for every $x \in [a, b]$. We obtain the proof by using Theorem 3.1.3. For this, let $\varepsilon > 0$ be given. Let P be a partition of $[a, b]$ with *equidistant points*, that is,

$$x_i = a + \frac{i(b - a)}{n}, \quad i = 0, 1, \ldots, n.$$

Then, we have

$$U(P, f) - L(P, f) = \sum_{i=1}^{n} (M_i - m_i)(x_i - x_{i-1}),$$

---

[2] Attributed to the Greek mathematician of the yore, Archimedes, as he devised procedures to compute areas of non-polygonal regions using the idea of approximation using polygonal regions, more than 2000 years before Riemann considered the procedure in a general framework in the year 1845.

where

$$M_i = f(x_i), \quad m_i = f(x_{i-1}), \quad x_i - x_{i-1} = \frac{b-a}{n}.$$

Thus,

$$U(P, f) - L(P, f) = \frac{b-a}{n} \sum_{i=1}^{n} [f(x_i) - f(x_{i-1})] = \frac{b-a}{n}[f(b) - f(a)]$$

Taking n large enough such that $\dfrac{b-a}{n}[f(b) - f(a)] < \varepsilon$, we obtain $U(P, f) -$ $L(P, f) < \varepsilon$ so that, by Theorem 3.1.3, f is integrable. ∎

In the appendix (Sect. 3.6), we shall also give a characterization of Riemann integrability in terms of Riemann sums (see Theorem 3.6.1).

Remark 3.1.4 Suppose for each $n = 2, 3, \ldots$, we have a partition P_n of k_n sub-intervals. It is to be observed that, as n varies, not only that k_n may vary, but also the partition points in P_n also may vary. Thus, we may write P_n as

$$P_n := x_0^{(n)} < x_1^{(n)} < \cdots < x_{k_n}^{(n)} = b.$$

For example, let P_n be the partition obtained by dividing $[a, b]$ into *equidistant points* with n partition points. Then we have $k_n = n$ and

$$x_i^{(n)} = a + i \frac{b-a}{n}, \quad i = 0, 1, \ldots, n.$$

Note that P_{n+1} is not a refinement of P_n. However, if we take Q_n to be the partition with 2^n equidistant points, that is,

$$y_i^{(n)} = a + i \frac{b-a}{2^n}, \quad i = 0, 1, \ldots, 2^n,$$

then Q_{n+1} is a refinement of Q_n. In this case $k_n = 2^n$. ◊

3.1.4 Some Basic Properties of Integral

Theorem 3.1.6 *Suppose f is integrable on $[a, b]$, and m, M are such that $m \leq f(x) \leq M$ for all $x \in [a, b]$. Then*

$$m(b - a) \leq \int_a^b f(x) \, dx \leq M(b - a).$$

Proof We know that for any partition P on $[a, b]$,

$$m(b-a) \le L(P, f) \le \int_a^b f(x)\, dx \le U(P, f) \le M(b-a).$$

Hence the result. ∎

Observation: For a bounded function φ defined on a closed interval I, let

$$m_\varphi := \inf_{x \in I} \varphi(x), \quad M_\varphi := \sup_{x \in I} \varphi(x).$$

Then, we see that, for any two bounded functions f and g on I and for every $y \in I$,

$$\inf_{x \in I} f(x) + \inf_{x \in I} g(x) \le f(y) + g(y) \le \sup_{x \in I} f(x) + \sup_{x \in I} g(x).$$

Hence, we obtain

$$m_f + m_g \le m_{f+g} \le M_{f+g} \le M_f + M_g.$$

Theorem 3.1.7 *Let f and g be integrable over $[a, b]$. Then $f + g$ is integrable and*

$$\int_a^b [f(x) + g(x)]dx = \int_a^b f(x)\, dx + \int_a^b g(x)\, dx,$$

Proof From the observation that we made before the statement of the theorem, we obtain

$$L(P, f) + L(P, g) \le L(P, f + g) \le U(P, f + g) \le U(P, f) + U(P, g)$$

for any partition P of $[a, b]$ (Verify). Now, let P_1 and P_2 be any two partitions of $[a, b]$ and $P = P_1 \cup P_2$. Since

$$L(P_1, f) + L(P_2, g) \le L(P, f) + L(P, g) \le L(P, f + g),$$

$$U(P, f + g) \le U(P, f) + U(P, g) \le U(P_1, f) + U(P_2, g),$$

and since

$$L(P, f + g) \le L(f + g) \le U(f + g) \le U(P, f + g),$$

we obtain

$$L(P_1, f) + L(P_2, g) \le L(f + g) \le U(f + g) \le U(P_1, f) + U(P_2, g).$$

This is true for any two arbitrary partitions P_1, P_2 of $[a, b]$. Therefore (How?),

$$L(f) + L(g) \le L(f + g) \le U(f + g) \le U(f) + U(g).$$

But,

$$L(f) + L(g) = \int_a^b f(x)dx + \int_a^b g(x)dx = U(f) + U(g).$$

Hence, $L(f + g) = U(f + g)$ and it is equal to $\int_a^b f(x)dx + \int_a^b g(x)dx$ so that $f + g$ is integrable and

$$\int_a^b [f(x) + g(x)]dx = \int_a^b f(x)dx + \int_a^b g(x)dx.$$

Thus, the proof is complete.   ∎

Exercise 3.1.2 Prove Theorem 3.1.7 using Corollary 3.1.4.   ◁

Observation: For the next theorem, we shall be using the following fact: For any bounded set $S \subseteq \mathbb{R}$,

$$\inf\{-s : s \in S\} = -\sup\{s : s \in S\},$$

$$\sup\{-s : s \in S\} = -\inf\{s : s \in S\}.$$

Theorem 3.1.8 *If f is integrable on $[a, b]$ and $c \in \mathbb{R}$, then cf is integrable on $[a, b]$, and*

$$\int_a^b cf(x)dx = c \int_a^b f(x)dx.$$

Proof Let $c \ge 0$. Then for any given partition P, we have

$$L(P, cf) = cL(P, f), \quad U(P, cf) = cU(P, f)$$

so that

$$\sup_P L(P, cf) = c \int_a^b f(x)dx, \quad \inf_P U(P, cf) = c \int_a^b f(x)dx.$$

Hence,

$$\sup_P L(P, cf) = c \int_a^b f(x)dx = \inf_P U(P, cf),$$

showing that $\int_a^b cf(x)dx = c \int_a^b f(x)dx$.

Next, let us assume that $c < 0$. Then we have

$$\int_a^b cf(x)dx = \int_a^b (-c)(-f)(x)dx = (-c) \int_a^b (-f)(x)dx.$$

Thus, once we prove

$$\int_a^b [-f(x)]dx = - \int_a^b f(x)dx, \tag{1}$$

we obtain

$$\int_a^b cf(x)dx = (-c)(-1) \int_a^b f(x)dx = c \int_a^b f(x)dx. \tag{2}$$

To prove (1), let P be any partition of $[a, b]$. Now, using the observation made before the statement of this theorem, we obtain (Verify)

$$L(P, -f) = -U(P, f), \quad -L(P, f) = U(P, -f)$$

and

$$\sup L(P, -f) = - \inf U(P, f), \quad \inf U(P, -f) = - \sup L(P, f).$$

Thus,

$$\sup L(P, -f) = - \int_a^b f(x)dx = \inf U(P, -f)$$

Thus, we have proved (1), and hence (2).   ∎

Theorem 3.1.9 *The following results hold.*

(i) *If f and g are integrable on $[a, b]$, then fg is integrable on $[a, b]$.*

(ii) *If f is integrable on $[a, b]$, $f(x) \neq 0$ for all $x \in [a, b]$ and $1/f$ is bounded, then $1/f$ is integrable on $[a, b]$.*

Proof First we show that if f is integrable on $[a, b]$, then f^2 is integrable on $[a, b]$. Since

$$f^2(x) - f^2(y) = (|f(x)| + |f(y)|)(|f(x)| - |f(y)|)$$
$$\leq 2M(|f(x)| - |f(y)|)$$

for all $x, y \in [a, b]$, where $M > 0$ is a constant such that $|f(x)| \leq M$ for all $x \in [a, b]$. From this, it can be shown that, for any partition P of $[a, b]$,

$$U(P, f^2) - L(P, f^2) \leq 2M[U(P, f) - L(P, f)].$$

Hence, by Theorem 3.1.3, f^2 is integrable.

(i) Let f and g be integrable on $[a, b]$. We observe that

$$fg = \frac{1}{4}[(f + g)^2 - (f - g)^2].$$

Hence, the proof of (ii) follows by applying (i) above, Theorem 3.1.7 and Theorem 3.1.8.

(ii) Suppose $f(x) \neq 0$ for all $x \in [a, b]$ and $1/f$ is bounded. Let $K > 0$ be such that $1/|f(x)| \leq K$ for all $x \in [a, b]$. Then, for every $x, y \in [a, b]$, we have

$$\frac{1}{f(x)} - \frac{1}{f(y)} = \frac{f(y) - f(x)}{f(x)f(y)} \leq K^2[f(y) - f(x)].$$

Hence, we obtain

$$U(P, 1/f) - L(P, 1/f) \leq K^2[U(P, f) - L(P, f)],$$

so that by Theorem 3.1.3, $1/f$ is integrable. ∎

Example 3.1.4 For $a_0, a_1, \ldots, a_n \in \mathbb{R}$, the function $f : \mathbb{R} \to \mathbb{R}$ defined by

$$f(x) = a_0 + a_1 x + \cdots + a_n x^n, x \in \mathbb{R},$$

is integrable on every closed interval $[a, b]$. Indeed, we have seen in Example 3.1.3 that the function $g(x) = x$ is integrable on $[a, b]$, so that by Theorems 3.1.7, 3.1.8, 3.1.9, we obtain that f is integrable on every closed interval $[a, b]$. ◇

In the above example, we saw that every polynomial function f defined by $f(x) = a_0 + a_1 x + \cdots + a_n x^n$ is integrable on every closed and bounded interval $[a, b]$. However, we did not obtain the value of the integral $\int_a^b f(x) dx$. Theorems 3.1.7 and 3.1.8 do give the relation

$$\int\limits_a^b f(x)\,dx = a_0 \int_a^b dx + a_1 \int_a^b x\,dx + \cdots + a_n \int_a^b x^n\,dx.$$

So, we get the value of $\int_a^b f(x)\,dx$, once we know the value of $\int_a^b x^k\,dx$ for each $k \in \mathbb{N}$. This can also be computed using the procedure adopted in Example 3.1.3. However, there are easier ways. In fact, we have already indicated in the beginning of this chapter that if there is a differentiable function g such that $g'(x) = f(x)$ on (a, b), then

$$\int\limits_a^b f(x)\,dx = g(b) - g(a)$$

so that

$$\int\limits_a^b x^k dx = \int\limits_a^b \frac{d}{dx}\left(\frac{x^{k+1}}{k+1}\right) dx = \frac{b^{k+1} - a^{k+1}}{k+1}.$$

We shall prove this result soon.

Theorem 3.1.10 *Suppose f is integrable on $[a, c]$ and $[c, b]$. Then f is integrable on $[a, b]$, and*

$$\int\limits_a^b f(x)\,dx = \int\limits_a^c f(x)\,dx + \int\limits_c^b f(x)\,dx.$$

Proof Let $f_1 = f|_{[a,c]}$, $f_2 = f|_{[c,b]}$. Let $\varepsilon > 0$ be given. Since f_1 and f_2 are integrable, there exist partitions P_1 and P_2 of $[a, c]$ and $[c, b]$ respectively such that

$$U(P_1, f_1) - L(P_1, f_1) < \frac{\varepsilon}{2}, \quad U(P_2, f_2) - L(P_2, f_2) < \frac{\varepsilon}{2}.$$

Suppose $P = P_1 \cup P_2$, the partition obtained by taking points in P_1 and P_2. . Then, it can be seen that

$$L(P, f) = L(P_1, f_1) + L(P_2, f_2), \quad U(P, f) = U(P_1, f_1) + U(P_2, f_2).$$

Hence,

$$U(P, f) - L(P, f) = [U(P_1, f_1) - L(P_1, f_1)] + [U(P_2, f_2) - L(P_2, f_2)]$$
$$< \varepsilon.$$

Thus f is integrable. Since

$$L(P, f) \leq \int_a^b f(x)dx \leq U(P, f),$$

$$L(P_1, f_1) + L(P_2, f_2) \leq \int_a^c f(x)dx + \int_c^b f(x)dx \leq U(P_1, f_1) + U(P_2, f_2),$$

it follows that

$$\left| \int_a^c f(x)\,dx + \int_c^b f(x)\,dx - \int_a^b f(x)\,dx \right| < \varepsilon.$$

This is true for all $\varepsilon > 0$. Hence the final result. ∎

We shall make use of the following lemma.

Lemma 3.1.11 *Let f, g, h be bounded functions on $[a, b]$ such that*

$$g(x) \leq f(x) \leq h(x) \quad \forall x \in [a, b].$$

Then

$$\int_a^b g(x)\,dx \leq \underline{\int_a^b} f(x)\,dx \leq \overline{\int_a^b} f(x)dx \leq \overline{\int_a^b} h(x)dx. \qquad (*)$$

Proof Let P and Q be any two partitions of $[a, b]$. Then, from the assumption that $g(x) \leq f(x) \leq h(x)$ for all $x \in [a, b]$, we have

$$L(P, g) \leq L(P, f) \leq U(Q, f) \leq U(Q, h).$$

From this we have (*how?*)

$$\sup_{P \in \mathcal{P}} L(P, g) \leq \sup_{P \in \mathcal{P}} L(P, f) \leq \inf_{P \in \mathcal{P}} U(Q, f) \leq \inf_{P \in \mathcal{P}} U(Q, h).$$

This is exactly the relations in $(*)$. ∎

The following theorem is a simple consequence of the above lemma.

Theorem 3.1.12 *Let f be integrable on $[a, b]$. Then*

$$f(x) \geq 0 \quad \forall x \in [a, b] \quad \Rightarrow \quad \int_a^b f(x)\,dx \geq 0.$$

More generally, if f and g are integrable on [a, b], then

$$f(x) \le g(x) \quad \forall x \in [a, b] \quad \Rightarrow \quad \int_a^b f(x)\, dx \le \int_a^b g(x)\, dx.$$

Proof Suppose $f(x) \ge 0$ forall $x \in [a, b]$. Taking $g = 0$ and $h = f$ in Lemma 3.1.11, we obtain $\int_a^b f(x)\, dx \ge 0$. Next, let f and g be integrable on $[a, b]$ such that $f(x) \le g(x)$ forall $x \in [a, b]$. Applying the above result for the function $g - f$, we obtain $\int_a^b f(x)\, dx \le \int_a^b g(x)\, dx$.   ∎

Corollary 3.1.13 *Let f be integrable on [a, b] such that $f(x) \ge 0$ for all $x \in [a, b]$. Then*

$$[c, d] \subseteq [a, b] \quad \Rightarrow \quad \int_c^d f(x)\, dx \le \int_a^b f(x)\, dx.$$

Proof Let $[c, d] \subseteq [a, b]$. By Theorem 3.1.10, we have

$$\int_a^b f(x)\, dx = \int_a^c f(x)\, dx + \int_c^d f(x)\, dx + \int_d^b f(x)\, dx$$

and by Theorem 3.1.12, w have

$$\int_a^c f(x)\, dx + \int_d^b f(x) \ge 0.$$

Hence, we obtain the required result.   ∎

Theorem 3.1.14 *Let f be a bounded function on [a, b], and for $n \in \mathbb{N}$, let g_n and h_n be integrable functions on [a, b] such that*

$$g_n(x) \le f(x) \le h_n(x) \quad \forall x \in [a, b], \forall n \in \mathbb{N},$$

$$\int_a^b [h_n(x) - g_n(x)]dx \to 0.$$

Then f is integrable and

$$\lim_{n \to \infty} \int_a^b g_n(x)\, dx = \int_a^b f(x)\, dx = \lim_{n \to \infty} \int_a^b h_n(x)dx.$$

Proof Since $g_n(x) \le f(x) \le h_n(x)$ for all $x \in [a, b]$ and g_n and h_n are integrable for all $n \in \mathbb{N}$, it follows that f is a bounded function and by Lemma 3.1.11, we have,

$$\int_a^b g_n(x)\, dx \le \underline{\int_a^b} f(x)\, dx \le \overline{\int_a^b} f(x)dx \le \int_a^b h_n(x)dx \tag{*}$$

for each $n \in \mathbb{N}$. From this, since $\int_a^b [h_n(x) - g_n(x)]dx \to 0$, we have $\underline{\int_a^b} f(x)\, dx = \overline{\int_a^b} f(x)dx$, so that f is integrable, and hence by (*), we have $\lim_{n \to \infty} \int_a^b g_n(x)\, dx = \int_a^b f(x)\, dx = \lim_{n \to \infty} \int_a^b h_n(x)dx$. ∎

Theorem 3.1.15 *Suppose f is integrable on $[a, b]$. Then $|f|$ is integrable and*

$$\left| \int_a^b f(x)dx \right| \le \int_a^b |f(x)|dx.$$

Proof First we show that $|f|$ is integrable. For this, let $\varepsilon > 0$ be given and let $P := \{x_i\}_{i=0}^n$ be a partition of $[a, b]$ such that

$$U(P, f) - L(P, f) < \varepsilon.$$

For $i = 1, 2, \ldots, n$, let

$$M_i = \sup_{x \in [x_{i-1}, x_i]} f(x), \quad m_i = \inf_{x \in [x_{i-1}, x_i]} f(x).$$

Then, for $i \in \{1, \ldots, n\}$ and for any $x, y \in [x_{i-1}, x_i]$, we have

$$|f(x)| - |f(y)| \le |f(x) - f(y)|$$
$$= \max\{f(x) - f(y), \ f(y) - f(x)\}$$
$$\le M_i - m_i.$$

From this, it follows that

$$U(P, |f|) - L(P, |f|) \le U(P, f) - L(P, f) < \varepsilon.$$

Hence, by Theorem 3.1.3, $|f|$ is integrable.

Next, since $f(x) \le |f(x)|$ and $-f(x) \le |f(x)|$ for all $x \in [a, b]$, by Theorem 3.1.12, we have

$$\int_a^b f(x)dx \le \int_a^b |f(x)|dx \quad \text{and} \quad -\int_a^b f(x)dx \le \int_a^b |f(x)|dx.$$

Hence,

$$\left| \int_a^b f(x)\mathrm{d}x \right| = \max \left\{ \int_a^b f(x)\mathrm{d}x, \ -\int_a^b f(x)\mathrm{d}x \right\} \le \int_a^b |f(x)|\mathrm{d}x.$$

This completes the proof.   ∎

3.1.5 Integral of Continuous Functions

We have shown that every monotone function on $[a, b]$ is Riemann integrable (Theorem 3.1.5). Our aim in this section is to show that every continuous function defined on $[a, b]$ is Riemann integrable. We shall also deduce some important properties of integrals of continuous functions.

Definition 3.1.6 Given a partition $P = \{x_i : i = 0, 1, \ldots, k\}$ of $[a, b]$, the quantity

$$\mu(P) := \max\{x_i - x_{i-1} : i = 1, \ldots, k\}$$

is called the **mesh** or **norm** of the partition P.   ◊

Theorem 3.1.16 *Suppose $f : [a, b] \to \mathbb{R}$ is a continuous function. Then f is integrable.*

Further, if (P_n) is a sequence of partitions of $[a, b]$ such that $\mu(P_n) \to 0$ as $n \to \infty$, and for each $n \in \mathbb{N}$, if T_n is a set of tags on P_n, then the sequences $\{U(P_n, f)\}$, $\{L(P_n, f)\}$ $\{S(P_n, f, T_n)\}$ converge to the same limit $\int_a^b f(x)\,dx$.

For its proof, we shall make use of a property of continuous functions defined on a closed and bounded intervals.

Theorem 3.1.17 *Let f be a real valued continuous function defined on a closed and bounded interval $[a, b]$. Then for every $\varepsilon > 0$, there exists $\delta > 0$ such that*

$$x, y \in [a, b], \quad |x - y| < \delta \quad \Rightarrow \quad |f(x) - f(y)| < \varepsilon.$$

Proof Suppose $f : [a, b] \to \mathbb{R}$ is continuous, and let $\varepsilon > 0$ be given. Suppose that the conclusion in the theorem is not true. Then there exists $\varepsilon_0 > 0$ such that for every $n \in \mathbb{N}$, there exist $x_n, y_n \in [a, b]$ such that

$$|x_n - y_n| < 1/n \quad \text{but} \quad |f(x_n) - f(y_n)| \ge \varepsilon_0. \tag{$*$}$$

Since (x_n) is a bounded sequence, it has a convergent subsequence, say $x_{k_n} \to c$ for some $c \in \mathbb{R}$. Since $|x_{k_n} - y_{k_n}| \to 0$, we also have the convergence $y_{k_n} \to c$. Now, by the continuity of f, we have $f(x_{k_n}) \to f(c)$ and $f(y_{k_n}) \to f(c)$. Thus, $|f(x_{k_n}) - f(y_{k_n})| \to 0$. This is a contradiction to $(*)$ above.   ∎

The property described in the conclusion in Theorem 3.1.17 is called *uniform continuity*.

Definition 3.1.7 A real valued function f defined on an interval I is said to be **uniformly continuous** on I if for every $\varepsilon > 0$, there exists $\delta > 0$ (depending on ε) such that

$$x, y \in I, \quad |x - y| < \delta \Rightarrow |f(x) - f(y)| < \epsilon. \qquad \Diamond$$

Clearly, every uniformly continuous function is continuous. But, the converse is not true, as the following examples show.

Example 3.1.5 Consider the function

$$f(x) = \frac{1}{x}, \quad 0 < x \leq 1.$$

Note that f is continuous on $I := (0, 1]$. But, it is not uniformly continuous. To see this, consider $x_n = 1/n$ and $y_n = 1/(n + 1)$. Then we know that $|x_n - y_n| \to 0$ as $n \to \infty$ and $|f(x_n) - f(y_n)| = 1$ for all $n \in \mathbb{N}$. Hence, the condition in the definition of uniform continuity is not satisfied if we take $\varepsilon < 1$. $\qquad \Diamond$

Example 3.1.6 Consider the function

$$f(x) = \sin(1/x), \quad 0 < x \leq 1.$$

Here, f is a bounded continuous on $I := (0, 1]$. But, it is not uniformly continuous. To see this, consider $x_n = 2/[(2n + 1)\pi]$ for $n \in \mathbb{N}$. Then taking $y_n = x_{n+1}$, we have $|x_n - y_n| \to 0$ as $n \to \infty$, but

$$|f(x_n) - f(x_{n+1})| = |\sin(2n + 1)\frac{\pi}{2}) - \sin((2n + 5)\frac{\pi}{2})| = 2 \quad \forall\, n \in \mathbb{N}.$$

Thus, the condition in the definition of uniform continuity is not satisfied if we take $\varepsilon < 2$. $\qquad \Diamond$

Remark 3.1.5 The above two examples show that each of those functions cannot be extended to a continuous function on $[0, 1]$ by assigning any value at 0. $\qquad \Diamond$

Exercise 3.1.3 Suppose f is a real valued function defined on an interval I such that there exists $K > 0$ satisfying

$$|f(x) - f(y)| \leq K|x - y| \quad \forall x, y \in I.$$

Show that f is uniformly continuous. (Functions satisfying the condition above are called *Lipschitz continuous functions*, and the constant K is called a *Lipschitz constant* for f.) $\qquad \triangleleft$

Proof (**Proofs of Theorem** 3.1.16) Let $f : [a, b] \to \mathbb{R}$ be a continuous function. First we prove that for every $\varepsilon > 0$ there exists a partition P of $[a, b]$ such that

$$U(P, f) - L(P, f) < \varepsilon$$

so that by Theorem 3.1.3, f is integrable.

Let $P : a = x_0 < x_1 < x_2 \ldots < x_k = b$ be any partition of $[a, b]$. Then

$$U(P, f) - L(P, f) = \sum_{i=1}^{k} (M_i - m_i)(x_i - x_{i-1}).$$

Since f is continuous on each closed interval $[x_{i-1}, x_i]$, there exists ξ_i, η_i in $[x_{i-1}, x_i]$ such that $M_i = f(\xi_i), m_i = f(\eta_i)$ for $i = 1, \ldots, k$. Hence,

$$U(P, f) - L(P, f) = \sum_{i=1}^{k} [f(\xi_i) - f(\eta_i)](x_i - x_{i-1}).$$

Again, since f is uniformly continuous on $[a, b]$, there exists $\delta > 0$ such that

$$|f(t) - f(s)| < \varepsilon/(b - a) \quad \text{whenever} \quad |t - s| < \delta. \tag{1}$$

Hence, if we take P such that $\mu(P) < \delta$, then we have

$$U(P, f) - L(P, f) = \sum_{i=1}^{k} [f(\xi_i) - f(\eta_i)](x_i - x_{i-1}) < \varepsilon. \tag{2}$$

Thus, we have proved that f is integrable.

Now, let (P_n) be a sequence of partitions such that $\mu(P_n) \to 0$ as $n \to \infty$. Let $\delta > 0$ be as in (1), and let $N \in \mathbb{N}$ be such that $\mu(P_n) < \delta$ for all $n \geq N$. Then, taking P_n in place of P in (2), we obtain

$$U(P_n, f) - L(P_n, f) < \varepsilon \quad \forall n \geq N.$$

Thus, $U(P_n, f) - L(P_n, f) \to 0$ as $n \to \infty$. Now, the conclusions follow from the observations

$$L(P_n, f) \leq \int_a^b f(x)\mathrm{d}x \leq U(P_n, f),$$

$$L(P_n, f) \leq S(P_n, f, T_n) \leq U(P_n, f)$$

for all $n \in \mathbb{N}$.   ∎

Example 3.1.7 Let $f(x) = e^x$ for all $x \in [a, b]$. Then, f is continuous. Let $x_0 = a$,

$$h_n = \frac{(b - a)}{n}, \quad x_i = a + ih_n, \quad t_i = x_{i-1}, \quad i = 1, \ldots, n.$$

Then with $P_n = \{x_i\}_{i=1}^n$ and $T = \{t_i\}_{i=1}^n$, we have $\mu(P_n) \to 0$, and

$$S(P_n, f, T_n) = h_n \sum_{i=1}^n e^{a+(i-1)h_n} = h_n e^a \sum_{i=1}^n \alpha_n^{(i-1)} = h_n e^a \frac{\alpha_n^n - 1}{\alpha_n - 1},$$

where $\alpha_n = e^{h_n}$. Since $\alpha_n^n = e^{b-a}$, we have

$$S(P_n, f, T_n) = h_n e^a \frac{\alpha_n^n - 1}{\alpha_n - 1} = e^a [e^{b-a} - 1] \frac{h_n}{e^{h_n} - 1} = [e^b - e^a] \frac{h_n}{e^{h_n} - 1}.$$

Since $\displaystyle\lim_{n\to\infty} \frac{e^{h_n} - 1}{h_n} = 1$, we have

$$S(P_n, f, T_n) \to e^b - e^a.$$

Hence, by Theorem 3.1.16, $\displaystyle\int_a^b e^x dx = e^b = e^a$. $\Diamond$

Remark 3.1.6 It can be shown that if a bounded function $f : [a, b] \to \mathbb{R}$ is *piecewise continuous*,, i.e., there are at most a finite number of points in $[a, b]$ at which f is discontinuous, then f is integrable. For a proof of this, see Theorem 3.6.4 in the appendix (Sect. 3.6). $\Diamond$

In view of the above remark, we can assert the following:

Every bounded function $f : [a, b] \to \mathbb{R}$ which is discontinuous at most at a finite number of points in $[a, b]$ is integrable.

Example 3.1.8 For example, the function $f : [-1, 1] \to \mathbb{R}$ defined by

$$f(x) = \begin{cases} -1, & -1 \le x \le 0, \\ 1, & 0 < x \le 1 \end{cases}$$

is bounded, and it is discontinuous only at the point 0. Hence, f is integrable, but the function $g : [-1, 1] \to \mathbb{R}$ defined by

$$g(x) = \begin{cases} 1/x, & x \ne 0, \\ 1, & x = 0 \end{cases}$$

is not integrable, although g is discontinuous only at the point 0, because g is not a bounded function.   ◊

Theorem 3.1.18 *Suppose $f : [a, b] \to \mathbb{R}$ and $g : [c, d] \to \mathbb{R}$ are continuous functions such that $f(x) \in [c, d]$ for all $x \in [a, b]$. Then $g \circ f : [a, b] \to \mathbb{R}$ is integrable.*

Proof The proof follows using the fact that composition of two continuous functions is continuous.   ∎

As a particular case of Theorem 3.1.18, it follows that, if $f : [a, b] \to \mathbb{R}$ is a continuous function, then for any $n \in \mathbb{N}$, the function

$$f^n(x) := [f(x)]^n, \quad x \in [a, b],$$

is integrable, so also the functions

$$\sin f(x), \quad \cos f(x), \quad \exp[f(x)], \quad x \in [a, b].$$

Remark 3.1.7 Theorem 3.1.18 is true without using the continuity of f, but using only its integrability. Interested reader can find the proof in [6] or [4]. Thus, if f is integrable on $[a, b]$, then $|f|$ is integrable. However, integrability of $|f|$ does not imply integrability of f. For example, consider the function $f : [0, 1] \to \mathbb{R}$ defined by

$$f(x) = \begin{cases} -1, & x \text{ rational,} \\ 1, & x \text{ irrational.} \end{cases}$$

Note that $L(P, f) = -1$ and $U(P, f) = 1$ for all partitions P of $[a, b]$, so that f is not integrable. However, $|f|$ is integrable on $[0, 1]$.   ◊

Theorem 3.1.19 *Suppose $f : [a, b] \to \mathbb{R}$ is continuous, $f(x) \geq 0$ for all $x \in [a, b]$ and $\int_a^b f(x)dx = 0$. Then $f(x) = 0$ for all $x \in [a, b]$.*

Proof Suppose the conclusion is not true. Then there exists $x_0 \in [a, b]$ such that $f(x_0) \neq 0$. Without loss of generality, assume that $f(x_0) > 0$. Then, by continuity of f at x_0, given any α with $0 < \alpha < 1$, there exists a closed interval $[c, d] \subseteq [a, b]$ containing x_0 such that $f(x) \geq \alpha f(x_0)$ for all $x \in [c, d]$. Therefore,

$$0 = \int_a^b f(x)dx \geq \int_c^d f(x)dx \geq \alpha f(x_0) \int_a^b dx = \alpha f(x_0)(d - c) \neq 0,$$

which is a contradiction.   ∎

We close this section by a test for convergence of series using integrals.

Theorem 3.1.20 (**INTEGRAL TEST**) *Suppose $f(x)$ is a continuous, non-negative and decreasing function for $x \in [1, \infty)$. For each $n \in \mathbb{N}$, let $a_n := \int_1^n f(x)dx$. Then*

$$\sum_{n=1}^{\infty} f(n) \quad \text{converges} \quad \Longleftrightarrow \quad (a_n) \quad \text{converges}.$$

Proof First we observe that $a_n = \int_1^n f(x)\,dx = \sum_{k=2}^n \int_{k-1}^k f(x)\,dx$. Now, since $f(x)$ is a decreasing function for $x \in [1, \infty)$, we have for each $k \in \mathbb{N}$,

$$k - 1 \le x \le k \Rightarrow f(k) \le f(x) \le f(k-1).$$

Hence, for $k = 2, 3, \ldots,$

$$f(k) \le \int_{k-1}^k f(x)\,dx \le f(k-1)$$

so that

$$\sum_{k=2}^n f(k) \le \sum_{k=2}^n \int_{k-1}^k f(x)\,dx \le \sum_{k=2}^n f(k-1).$$

Thus,

$$\sum_{k=2}^n f(k) \le \int_1^n f(x)\,dx \le \sum_{k=1}^{n-1} f(k).$$

Now, let $s_n := \sum_{k=1}^n f(k)$ for $n \in \mathbb{N}$. Then from the above inequalities, together with the fact that (s_n) is a monotonically increasing sequence, it follows that (a_n) converges if and only if (s_n) converges. ∎

Remark 3.1.8 Suppose $f : [a, \infty) \to \mathbb{R}$ is a continuous function. If $\lim_{n \to \infty} \int_a^n f(x)dx$ exists, then it is denoted by $\int_a^{\infty} f(x)dx$. Thus, by Theorem 3.1.20, if $f : [1, \infty) \to \mathbb{R}$ is continuous and decreasing, then

$\sum_{n=1}^{\infty} f(n)$ converges if and only if $\int_1^{\infty} f(x)dx$ exists.

More on this type of integrals will be discussed in the next chapter. ◇

3.2   Mean Value Theorems

Theorem 3.2.1 (Mean value theorem) *Suppose f is a continuous function on $[a, b]$. Then there exists $c \in [a, b]$ such that*

$$\frac{1}{b-a} \int_a^b f(x) \, dx = f(c).$$

Proof Since f is continuous, we know that there exist $u, v \in [a, b]$ such that $f(u) = m := \min f(x)$ and $f(v) = M := \max f(x)$. Hence, by Theorem 3.1.6,

$$f(u) \leq \frac{1}{b-a} \int_a^b f(x) \, dx \leq f(v).$$

Hence, by intermediate value theorem, there exists $c \in [a, b]$ such that

$$\frac{1}{b-a} \int_a^b f(x) \, dx = f(c).$$

Hence the result.   ■

Theorem 3.2.2 (Generalized mean value theorem) *Suppose f and g are continuous on $[a, b]$ with $g(x) \geq 0$ for all $x \in [a, b]$. Then there exists $c \in [a, b]$ such that*

$$\int_a^b f(x)g(x) \, dx = f(c) \int_a^b g(x) dx.$$

Proof Let $m := \inf\limits_{a \leq x \leq b} f(x)$ and $M = \sup\limits_{a \leq x \leq b} f(x)$. Then, since $g(x) \geq 0$ for every $x \in [a, b]$, we have

$$m \int_a^b g(x) dx \leq \int_a^b f(x)g(x) \, dx \leq M \int_a^b g(x) dx.$$

If $g(x) = 0$ for all $x \in [a, b]$, then the conclusion in the theorem holds trivially for any $c \in [a, b]$. So, assume that $g(x_0) \neq 0$ for some $x_0 \in [a, b]$. Now, since $g(x) \geq 0$ for all $x \in [a, b]$, it follows (how?) that $g(x_0) > 0$ and hence $\int_a^b g(x) dx > 0$. Hence,

$$m \leq \frac{\int_a^b f(x)g(x)\,dx}{\int_a^b g(x)dx} \leq M.$$

Therefore, by the intermediate value theorem, there exists $c \in [a, b]$ such that

$$f(c) = \frac{\int_a^b f(x)g(x)\,dx}{\int_a^b g(x)dx}.$$

Thus, $\int_a^b f(x)g(x)\,dx = f(c)\int_a^b g(x)dx.$ ∎

Exercise 3.2.1 The conclusion of Theorem 3.2.2 holds if $g(x) \geq 0$ is replaced by $g(x) \leq 0$. How? ◁

3.3 Fundamental Theorems

The first theorem that we prove in this section justifies what we do in school for the calculation of integrals.

Theorem 3.3.1 (Fundamental theorem-I) *Let f be a Riemann integrable function on $[a, b]$. Suppose there exists a continuous function g on $[a, b]$ such that it is differentiable in (a, b) and $g'(x) = f(x)$ for all $x \in (a, b)$. Then*

$$\int_a^b f(x)dx = g(b) - g(a).$$

Proof Let $P : a = x_0 < x_1 < \ldots < x_n = b$ be any partition of $[a, b]$. Then by Lagrange's mean value theorem, there exists $c_i \in (x_{i-1}, x_i)$ such that

$$g(x_i) - g(x_{i-1}) = g'(c_i)(x_i - x_{i-1}) = f(c_i)(x_i - x_{i-1}).$$

Hence,

$$g(b) - g(a) = \sum_{i=1}^n [g(x_i) - g(x_{i-1})] = \sum_{i=1}^n f(c_i)(x_i - x_{i-1}).$$

Thus, we have

$$L(P, f) \leq g(b) - g(a) \leq U(P, f)$$

for all partition P of $[a, b]$. Since $\int_a^b f(x)\mathrm{d}x$ is the unique number which lies between $L(P, f)$ and $U(P, f)$ for all partition P of $[a, b]$, we obtain

$$\int_a^b f(x)\mathrm{d}x = g(b) - g(a).$$

This completes the proof.   ∎

The concluding formula in Theorem 3.3.1, namely,

$$\int_a^b g'(x)\mathrm{d}x = g(b) - g(a)$$

is known as **Newton–Leibnitz formula**. The difference $g(b) - g(a)$ is usually written as $\left[g(x)\right]_a^b$, i.e.,

$$\left[g(x)\right]_a^b := g(b) - g(a).$$

Definition 3.3.1 Let $f : [a, b] \to \mathbb{R}$ be a Riemann integrable function. Then a function $g : [a, b] \to \mathbb{R}$ is called an **antiderivative** or **indefinite integral** or **primitive** of f if g is differentiable in (a, b) and $g'(x) = f(x)$ for all $x \in (a, b)$.   ◇

Difference of any two antiderivatives of a given function is a constant.

Exercise 3.3.1 Justify the above statement.   ◁

Notation 3.3.1 Antiderivative of an integrable function f, if exists, is usually denoted by

$$\int f(x)\mathrm{d}x.$$ ◇

A function can be integrable, but it need not have an antiderivative. To see this, recall the function f in Example 3.1.1, that is,

$$f(x) = \begin{cases} -1, & -1 \le x \le 0, \\ 1, & 0 < x \le 1. \end{cases}$$

We know that f is integrable on $[-1, 1]$, but there does not exist a function $g : [-1, 1] \to \mathbb{R}$ which is differentiable on $(-1, 1)$ such that $g'(x) = f(x)$ for all $x \in (-1, 1)$ (see Example 2.3.15).

However, we shall soon show that every continuous function has an antiderivative.

Remark 3.3.1 See carefully the statement of Theorem 3.3.1. It is different from the
following statement:

If $g : [a, b] \to \mathbb{R}$ is such that it is differentiable on (a, b), then $\int_a^b g'(x)dx = g(b) - g(a)$.

Because, there are functions which are derivatives of some functions, but they need
not be integrable. For example, consider the function

$$g(x) = \begin{cases} \sin(1/x), & 0 < x \le 1, \\ 0, & x = 0. \end{cases}$$

Then g is differentiable at every point in $(0, 1)$, and its derivative is

$$g'(x) = -(1/x^2)\cos(1/x), \quad x \in (0, 1).$$

But, we cannot write $\int_0^1 g'(x)dx = g(1) - g(0)$. The reason is that there is no integrable function f on $[0, 1]$ such that $f(x) = g'(x)$ for every $x \in (0, 1)$.   ◇

We shall use the concept of *extension of a function*.

Definition 3.3.2 Let f be a function defined on an interval I, and let J be another
interval such that $I \subseteq J$. Then a function $\varphi : J \to \mathbb{R}$ is said to be an **extension** of f
to J if $\varphi(x) = f(x)$ for all $x \in I$.   ◇

Indeed, every function f defined on an interval I can be extended to a bigger interval
$J \supseteq I$ by defining $\varphi : J \to \mathbb{R}$ by

$$\varphi(x) = \begin{cases} f(x), & x \in I, \\ c, & x \in J \setminus I \end{cases}$$

for any specified $c \in \mathbb{R}$. However, in applications we may require the extended function to have certain properties. Let us illustrate these by a few examples.

Example 3.3.1 Let $f : (0, 1] \to \mathbb{R}$ be defined by

$$f(x) = \frac{\sin(x)}{x}, \quad 0 < x \le 1.$$

We know that f is continuous. Then, for $c \in \mathbb{R}$, the extended function $\varphi : [0, 1] \to \mathbb{R}$
defined by

$$\varphi(x) = \begin{cases} f(x), & 0 < x \le 1, \\ c, & x = 0 \end{cases}$$

is continuous if and only if $c = 1$.   ◇

Example 3.3.2 Let $f : (0, 1] \to \mathbb{R}$ be defined by

$$f(x) = \sin(1/x), \quad 0 < x \le 1.$$

Then the function $\tilde{f} : [0, 1] \to \mathbb{R}$ defined by

$$\varphi(x) = \begin{cases} \sin(1/x), & 0 < x \le 1, \\ 0, & x = 0. \end{cases}$$

is an extension of f to $[0, 1]$. Note that f is continuous, whereas φ is not continuous. In fact, f does not have any continuous extension to all of $[0, 1]$. Also, f does not have any integrable extension to $[0, 1]$ (Why?) $\Diamond$

Example 3.3.3 Consider the function $f : [0, 1] \to \mathbb{R}$ defined by

$$f(x) = \begin{cases} x^2 \sin(1/x), & 0 < x \le 1, \\ 0, & x = 0. \end{cases}$$

It is clear that $f'(x)$ exists for every $x \in (0, 1)$ and for $x \ne 0$,

$$f'(x) = -\cos(1/x) + 2x \sin(1/x).$$

Since $\lim_{x \to 0} f'(x)$ does not exist, f' does not have a continuous extension to $[0, 1]$. However, since f' is continuous and $|f'(x)| \le 3$ on $(0, 1)$, f' has integrable extension to $[0, 1]$ (Exercise). $\Diamond$

Recall that a bounded function defined on $[a, b]$ which is continuous except possibly at a finite number of points in $[a, b]$ is integrable. Therefore, as a special case of Theorem 3.3.1 we have the following:

Let $f : [a, b] \to \mathbb{R}$ be differentiable in (a, b) and f' is continuous and bounded on (a, b). Then $\int_a^b f'(x)dx = f(b) - f(a)$.

The following examples have been worked out by knowing the antiderivatives of the functions involved.

Example 3.3.4 Recall that for $k \ge 0$, $\dfrac{d}{dx} x^{k+1} = (k+1)x^k$ so that by Theorem 3.3.1,

$$\int_a^b x^k \, dx = \left[\frac{x^{k+1}}{k+1} \right]_a^b = \frac{b^{k+1} - a^{k+1}}{k+1}. \qquad \Diamond$$

Example 3.3.5 Recall that for any $c \ne 0$, $\dfrac{d}{dx}(e^{cx}) = ce^{cx}$ so that by Theorem 3.3.1,

$$\int_a^b e^{cx}\, dx = \left[\frac{e^{cx}}{c}\right]_a^b = \frac{e^{cb} - e^{ca}}{c}. \qquad \Diamond$$

Example 3.3.6 Recall that for any $c \neq 0$,

$$\frac{d}{dx}\sin(cx) = c\cos(cx) \quad \text{and} \quad \frac{d}{dx}\cos(cx) = -c\sin(cx),$$

so that by Theorem 3.3.1,

(i) $\displaystyle\int_a^b \cos cx\, dx = \left[\frac{\sin cx}{c}\right]_a^b = \frac{1}{c}(\sin cb - \sin ca).$

(ii) $\displaystyle\int_a^b \sin cx\, dx = \left[-\frac{\cos cx}{c}\right]_a^b = \frac{1}{c}(\cos ca - \cos cb). \qquad \Diamond$

Example 3.3.7 Recall that $\dfrac{d}{dx}\log x = \dfrac{1}{x}$ for $x > 0$. Hence, by Theorem 3.3.1, we have

$$\int_1^x \frac{1}{x}dx = \log x - \log 1 = \log x. \qquad \Diamond$$

Remark 3.3.2 In some books, $\log x$ is defined as the integral $\int_1^x \frac{1}{x}dx$, whereas we defined $\log x$ in the last chapter as the inverse of the exponential function. $\qquad \Diamond$

Theorem 3.1.16 together with Theorem 3.3.1 can be used to compute the limit of certain sequences. For example, see the following.

Example 3.3.8 For $p \geq 0$, we show that[3]

$$\lim_{n\to\infty} \frac{1}{n^{p+1}}\sum_{k=0}^n k^p = \frac{1}{p+1}.$$

We note that

$$\frac{1}{n^{p+1}}\sum_{k=0}^n k^p = \sum_{k=0}^n \frac{(k/n)^p}{n} = \sum_{k=1}^n f(\xi_k^{(n)})(x_k^{(n)} - x_{k-1}^{(n)}),$$

---

[3] It is to be remarked that, for $p \in \mathbb{N}$, this result was first found in the Sanskrit text *Yuktibhasha* of *Jyeshthadeva* (AD:1500-1575) of Kerala School of Mathematics, more than hundred years before it was considered by European mathematicians Fermat, Pascal and others in the seventeenth century (cf. [7], Chap. 10).

where

$$f(x) = x^p, \quad 0 \le x \le 1,$$

$x_k^{(n)} := \frac{k}{n}$ for $k=0, 1, \ldots, n$ and $\xi_k^{(n)} := \frac{k-1}{n}$ for $k = 1, \ldots, n$. Thus, $\frac{1}{n^{p+1}} \sum_{k=0}^{n} k^p$ is a Riemann sum of the continuous function f corresponding to the partition $P_n : 0 = x_0^{(n)} < x_1^{(n)} \cdots < x_n^{(n)} = 1$ with $\mu(P_n) \to 0$ as $n \to \infty$. Hence, by Theorem 3.1.16 and Example 3.3.4,

$$\frac{1}{n^{p+1}} \sum_{k=0}^{n} k^p \sum_{k=1}^{n} f(\xi_k^{(n)})(x_k^{(n)} - x_{k-1}^{(n)}) \to \int_0^1 x^p \, dx = \frac{1}{p+1}.$$

Thus, $\lim_{n \to \infty} \frac{1}{n^{p+1}} \sum_{k=0}^{n} k^p = 1/(p+1)$.   ◊

Example 3.3.9 We show that

$$\lim_{n \to \infty} \sum_{k=0}^{n} \frac{n^2}{(n+k)^3} = \frac{3}{8}.$$

Note that

$$\sum_{k=0}^{n} \frac{n^2}{(n+k)^3} = \sum_{k=0}^{n} \frac{1}{n} \frac{1}{(1+k/n)^3} = \sum_{k=1}^{n} f(\xi_k^{(n)})(x_k^{(n)} - x_{k-1}^{(n)}),$$

where

$$f(x) = \frac{1}{(1+x)^3}, \quad 0 \le x \le 1,$$

$x_k^{(n)} := \frac{k}{n}$ for $k = 0, 1, \ldots, n$ and $\xi_k^{(n)} := \frac{k-1}{n}$ for $k = 1, \ldots, n$. Thus, $\sum_{k=0}^{n} \frac{n^2}{(n+k)^3}$ is a Riemann sum of the continuous function f corresponding to the partition $P_n : 0 = x_0^{(n)} < x_1^{(n)} \cdots < x_n^{(n)} = 1$ with $\mu(P_n) \to 0$ as $n \to \infty$. Hence, by Theorem 3.1.16,

$$\sum_{k=0}^{n} \frac{n^2}{(n+k)^3} = \sum_{k=1}^{n} f(\xi_k^{(n)})(x_k^{(n)} - x_{k-1}^{(n)}) \to \int_0^1 \frac{dx}{(1+x)^3}.$$

We know that $\frac{d}{dx}\left[\frac{1}{(1+x)^2}\right] = \frac{-2}{(1+x)^3}$. Hence, by Theorem 3.3.1,

$$\int_0^1 \frac{dx}{(1+x)^3} = \left[-\frac{1}{2}\frac{1}{(1+x)^2}\right]_0^1 = \frac{3}{8}.$$

Thus, $\displaystyle\lim_{n\to\infty}\sum_{k=0}^{n}\frac{n^2}{(n+k)^3} = 3/8.$ ◊

Now, we address the question: Does every integrable function have an antideriva-tive? As stated earlier, we answer this question affirmatively when the function is continuous.

Theorem 3.3.2 (Fundamental theorem-II) *Suppose f is Riemann integrable on $[a, b]$, and*

$$g(x) = \int_a^x f(t)dt, \quad x \in [a, b].$$

Then g is continuous on $[a, b]$. If, in addition, f is continuous on $[a, b]$, then g differentiable and

$$g'(x) = f(x) \quad \forall x \in (a, b).$$

Proof Let $x \in [a, b]$ and $h \in \mathbb{R}$ be such that $x + h \in [a, b]$. Then we have

$$g(x + h) - g(x) = \int_a^{x+h} f(t)dt - \int_a^x f(t)dt = \int_x^{x+h} f(t)dt.$$

Thus, since $|f|$ is integrable (Remark 3.1.7), we have

$$|g(x + h) - g(x)| \le \int_x^{x+h} |f(t)|dt \le M|h|,$$

where $M > 0$ is such that $|f(x)| \le M$ for all $x \in [a, b]$. Recall that such M exists, as Riemann integrability is defined only for bounded functions. Thus,

$$|g(x + h) - g(x)| \to 0 \quad \text{as} \quad h \to 0,$$

showing that g is continuous at x.

Next assume that f is continuous on $[a, b]$. Then, by mean value theorem, there exists ξ_h between x and $x + h$ such that

$$\frac{1}{h} \int_x^{x+h} f(t)dt = f(\xi_h).$$

Since f is continuous at x, we have $f(\xi_h) \to f(x)$ as $h \to 0$. Hence

$$\lim_{h \to 0} \frac{g(x+h) - g(x)}{h} = \lim_{h \to 0} f(\xi_h) = f(x).$$

Thus, $g'(x)$ exists and $g'(x) = f(x)$.   ∎

In view of Theorem 3.3.2,

> Every continuous function has an antiderivative.

Proof (**An alternative proof for Theorem** 3.3.1 **when** f **continuous**)

Let φ be the indefinite integral of f, i.e., $\varphi(x) = \int_a^x f(t)dt$, $x \in [a, b]$. Then we have $g'(x) = f(x) = \varphi'(x)$ for all $x \in [a, b]$, i.e., $g'(x) - \varphi'(x) = 0$ for all $x \in [a, b]$. Hence, $g - \varphi$ is a constant function. Hence, in view of the Theorem 3.3.2, we have

$$g(b) - g(a) = \varphi(b) - \varphi(a) = \int_a^b f(t)dt.$$

This completes the proof.   ∎

Theorem 3.3.2 leads to the following existence theorem in differential equation:

Theorem 3.3.3 *Given a continuous function f on $[a, b]$, there exists a function g which is continuous on $[a, b]$ and differentiable on (a, b) such that*

$$\frac{dg}{dx} = f(x) \quad \forall x \in (a, b) \quad and \quad g(a) = 0.$$

3.4   Some Consequences

In this section we derive some results as consequences of mean value theorems and fundamental theorems.

Theorem 3.4.1   (Integral as Riemann sum) *Suppose f is a continuous function on $[a, b]$. Then for every partition P of $[a, b]$, there exists a set T of tags on P such that*

$$S(P, f, T) = \int_a^b f(x)\, dx.$$

Proof Let $P = \{x_i : i = 0, 1, \ldots, k\}$ be a partition of $[a, b]$. Since f is continuous, by the mean value theorem (Theorem 3.2.1), there exists $\xi_i \in [x_{i-1}, x_i]$ such that

$$\int_{x_{i-1}}^{x_i} f(x)\, dx = f(\xi_i)(x_i - x_{i-1}), \quad i = 1, \ldots, k.$$

Hence, taking $T = \{\xi_i : i = 1, \ldots, k\}$,

$$S(P, f, T) = \sum_{i=1}^{k} f(\xi_i)(x_i - x_{i-1}) = \sum_{i=1}^{k} \int_{x_{i-1}}^{x_i} f(x)\, dx = \int_a^b f(x)\, dx.$$

This completes the proof.   ■

Next theorem is a consequence of Theorem 3.3.2 and the formula for the derivative of composition of two differentiable functions.

Theorem 3.4.2 *Let* $f : [a, b] \to \mathbb{R}$ *be a continuous function. Let*

$$F(x) := \int_a^{\varphi(x)} f(t)dt, \quad x \in [\alpha, \beta],$$

where $\varphi : [\alpha, \beta] \to \mathbb{R}$ *is a continuous function which is differentiable on* (α, β) *such that* $\varphi(x) \in (a, b)$ *for all* $x \in (\alpha, \beta)$. *Then*

$$F'(x) = f(\varphi(x))\varphi'(x) \quad \text{for all} \ \ x \in (\alpha, \beta).$$

Proof Note that,
$$F(x) = g(\varphi(x)) \quad \text{for} \ \ x \in [\alpha, \beta],$$

where

$$g(y) = \int_a^y f(t)dt, \quad y \in [a, b].$$

Since f is continuous on $[a, b]$, by Theorem 3.3.2,

$$g'(y) = f(y) \quad \text{for all} \ \ y \in (a, b).$$

Thus,
$$F'(x) = g'(\varphi(x))\varphi'(x) = f(\varphi(x))\varphi'(x) \quad \text{for all} \ \ x \in (\alpha, \beta).$$

This complete the proof.   ■

Example 3.4.1 Let us find an expression for the function

$$\frac{d}{dx} \int_0^{3x-2} \sqrt{1+t^2}\, dt.$$

Let $\varphi(x) := 3x - 2$ for $x \in [\alpha, \beta]$ and let $[a, b]$ be such that $3x - 2 \in (a, b)$ for all $x \in (\alpha, \beta)$. Then, by Theorem 3.4.2, we have

$$\frac{d}{dx} \int_0^{3x-2} \sqrt{1+t^2}\, dt = [\sqrt{1 + (3x-2)^2}]3 = 3\sqrt{9x^2 - 12x + 5}. \qquad \Diamond$$

We have seen in Theorem 3.1.9 that the product of two integrable functions is integrable. However, we did not give any formula for the computation of the integral of the product. Now, under some additional condition on one of the functions, we give a formula which would facilitate the computation of the integral of the product.

Theorem 3.4.3 (Product formula) *Suppose f and g are continuous functions defined on $[a, b]$. If f is differentiable in (a, b) and its derivative f' has integrable extension to $[a, b]$, then*

$$\int_a^b f(x)g(x)\, dx = [f(x)G(x)]_a^b - \int_a^b f'(x)G(x)\, dx,$$

where G is an antiderivative of g on (a, b).

Proof Since g is continuous on $[a, b]$, it has an antiderivative, say G, i.e., G is differentiable and $G' = g$ on (a, b). Then we have

$$(fG)' = fg + f'G \quad \text{on} \quad (a, b).$$

Hence, using fundamental theorem (Theorem 3.3.1),

$$[f(x)G(x)]_a^b = \int_a^b [f(x)G(x)]'\, dx$$

$$= \int_a^b f(x)g(x)\, dx + \int_a^b f'(x)G(x)\, dx.$$

Thus, $\displaystyle\int_a^b f(x)g(x)\,dx = [f(x)G(x)]_a^b - \int_a^b f'(x)G(x)\,dx.$ ∎

Another form of the product formula is the following.

Theorem 3.4.4 (Product formula) *Suppose f and g are continuous functions on $[a, b]$, which are differentiable in (a, b). Also, assume that both f' and g' have Riemann integrable extensions to $[a, b]$. Then*

$$\int_a^b f(x)g'(x)\,dx = [f(x)g(x)]_a^b - \int_a^b f'(x)g(x)\,dx.$$

Proof Proof follows as in the proof of Theorem 3.4.4, by replacing g there by g'. ∎

If we use new variables u and v which relate to x by

$$u = f(x), \quad v = g(x), \quad a \le x \le b,$$

then the product formula in Theorem 3.4.4 can be written, in short, as

$$\int_a^b u\,dv = \Big[uv\Big]_a^b - \int_a^b v\,du.$$

Thus, under appropriate conditions, denoting the antiderivative of g by G, we have the following formulae:

$$\int_a^b f(x)g(x)\,dx = [f(x)G(x)]_a^b - \int_a^b f'(x)G(x)\,dx$$

$$\int_a^b f(x)g'(x)\,dx = [f(x)g(x)]_a^b - \int_a^b f'(x)g(x)\,dx$$

$$\int_a^b u\,dv = [uv]_a^b - \int_a^b v\,du.$$

Example 3.4.2 Let us find the value of the integral $\int_0^{\pi/2} \sin^n x\,dx$ for $n \in \mathbb{N}$. Clearly,

$$\int_0^{\pi/2} \sin x\,dx = [-\cos x]_0^{\pi/2} = -[0 - 1] = 1,$$

and for $n = 2$,

$$\int_0^{\pi/2} \sin^2 x \, dx = \frac{1}{2} \int_0^{\pi/2} (1 - \cos 2x) \, dx = \frac{1}{2} \int_0^{\pi/2} dx = \frac{\pi}{4}.$$

For $n \geq 3$, we write $\int_0^{\pi/2} \sin^n x \, dx = \int_0^{\pi/2} \sin^{n-1} x \sin x \, dx$, so that by product formula, we obtain

$$\int_0^{\pi/2} \sin^{n-1} x \sin x \, dx = [\sin^{n-1} x (-\cos x)]_0^{\pi/2}$$

$$- \int_0^{\pi/2} (n-1) \sin^{n-2} x \cos x (-\cos x) \, dx$$

$$= \int_0^{\pi/2} (n-1) \sin^{n-2} x \cos^2 x \, dx$$

$$= \int_0^{\pi/2} (n-1) \sin^{n-2} x (1 - \sin^2 x) \, dx.$$

Hence, if we denote $\alpha_n := \int_0^{\pi/2} \sin^n x \, dx$ for $n \in \mathbb{N}$, then we obtain

$$\alpha_n = (n-1)\alpha_{n-2} - (n-1)\alpha_n,$$

i.e.,

$$\alpha_n = \frac{n-1}{n} \alpha_{n-2}.$$

In particular,

$$\alpha_3 = \frac{2}{3} \alpha_1 = \frac{2}{3}, \quad \alpha_4 = \frac{3}{4} \alpha_2 = \frac{3}{4} \frac{\pi}{4}.$$

More generally, for $k \in \mathbb{N}$,

$$\alpha_{2k} = \frac{\pi}{2} \cdot \frac{1}{2} \cdot \frac{3}{4} \cdot \cdots \cdot \frac{2k-1}{2k} = \frac{\pi (2k)!}{2^{2k+1} (k!)^2},$$

$$\alpha_{2k+1} = \frac{2}{3} \frac{4}{5} \cdots \frac{2k}{2k+1} = \frac{2^{2k} (k!)^2}{(2k+1)!} = \frac{\pi}{2(2k+1)\alpha_{2k}}.$$ ◊

Example 3.4.3 Let us evaluate the integral $\int_a^b e^x \sin x \, dx$. By product formula,

$$\int_a^b e^x \sin x \, dx = [e^x(-\cos x)]_a^b + \int_a^b e^x \cos x \, dx,$$

$$\int_a^b e^x \cos x \, dx = [e^x \sin x]_a^b - \int_a^b e^x \sin x \, dx.$$

Hence,

$$\int_a^b e^x \sin x \, dx = \frac{1}{2}\left([e^x(-\cos x)]_a^b + [e^x \sin x]_a^b\right)$$

$$= \frac{1}{2}[e^x(\sin x - \cos x)]_a^b.$$

In particular,

$$\int_0^{\pi/2} e^x \sin x \, dx = \frac{1}{2}[e^x(\sin x - \cos x)]_0^{\pi/2} = \frac{1}{2}[e^{\pi/2} + 1]. \qquad \Diamond$$

Theorem 3.4.5 (Change of variable formula) *Let $f : [a, b] \to \mathbb{R}$ be a continuous function. Let $\psi : [\alpha, \beta] \to \mathbb{R}$ be a continuous function which is differentiable on (α, β) such that $\psi(x) \in [a, b]$ for every $x \in [\alpha, \beta]$ with $\psi(\alpha) = a$ and $\psi(\beta) = b$. Then,*

$$\int_a^b f(x) \, dx = \int_\alpha^\beta f(\psi(t))\psi'(t) dt.$$

Proof Let F be an antiderivative of f, i.e., $F'(x) = f(x)$. Then taking $G(t) = F(\psi(t))$ for $t \in [\alpha, \beta]$, we have

$$G'(t) = F'(\psi(t))\psi'(t) = f(\psi(t))\psi'(t), \quad t \in [\alpha, \beta].$$

Hence, by fundamental theorem,

$$\int_\alpha^\beta f(\psi(t))\psi'(t) dt = \int_\alpha^\beta G'(t) dt = G(\beta) - G(\alpha) = F(\psi(\beta)) - F(\psi(\alpha)).$$

Hence,

$$\int_{\alpha}^{\beta} f(\psi(t))\psi'(t)dt = F(b) - F(a) = \int_{a}^{b} f(x)\,dx.$$

This completes the proof.   ∎

The formula in Theorem 3.4.5 can be written as:

$$\int_{\alpha}^{\beta} f(\psi(t))\psi'(t)dt = \int_{\psi(\alpha)}^{\psi(\beta)} f(x)\,dx.$$

The formula in Theorem 3.4.5 can be used to find the integral of certain complicated functions by expressing it in the form

$$\int_{\alpha}^{\beta} f(\psi(t))\psi'(t)dt$$

by using appropriate functions f and ψ.

Example 3.4.4 Let us find $\int_{0}^{\pi/2} \sin x \cos^2 dx$. Writing

$$\sin x \cos^2 = f(\psi(x))\psi'(x),$$

with $\psi(x) = -\cos x$ and $f(y) = y^2$, Theorem 3.4.5 gives

$$\int_{0}^{\pi/2} \sin x \cos^2 x dx = \int_{0}^{\pi/2} f(\psi(x))\psi'(x)dx = \int_{\psi(0)}^{\psi(\pi/2)} f(y)dy.$$

But,

$$\int_{\psi(0)}^{\psi(\pi/2)} f(y)dy = \int_{\psi(0)}^{\psi(\pi/2)} y^2 dy = \left[\frac{y^3}{3}\right]_{-1}^{0} = \frac{1}{3}.$$

Thus, $\int_{0}^{\pi/2} \sin x \cos^2 dx = 1/3$.   ◊

The method of expressing the integral $\int_{\alpha}^{\beta} f(\psi(t))\psi'(t)dt$ as $\int_{a}^{b} f(x)dx$ can be described as follows:

1. Consider the *change of variable* $x = \psi(t)$ so that $\dfrac{dx}{dt} = \psi'(t)$.
2. Write the above relation *formally* as $dx = \psi'(t)dt$.

3. Replace the expression integral $\int_\alpha^\beta f(\psi(t))\psi'(t)dt$ by $\int_{\psi(\alpha)}^{\psi(\beta)} f(x)dx$.

In view of the above remarks, let us look at another example.

Example 3.4.5 We would like to find the value of $\int_{-1}^1 \frac{dx}{(1+x^2)^2}$. Write $x = \tan t$ so that $dx = \sec^2 t\, dt$ and

$$\frac{dx}{(1+x^2)^2} = \frac{\sec^2 t\, dt}{(1+\tan^2 t)^2} = \frac{\sec^2 t\, dt}{(\sec^2 t)^2} = \cos^2 t\, dt = \frac{1}{2}(\cos 2t + 1)dt.$$

Since $\tan(\pi/4) = 1$ and $\tan(-\pi/4) = -1$, we obtain

$$\int_{-1}^1 \frac{dx}{(1+x^2)^2} = \int_{-\pi/4}^{\pi/4} \frac{\sec^2 t\, dt}{(1+\tan^2 t)^2}.$$

But,

$$\int_{-\pi/4}^{\pi/4} \frac{\sec^2 t\, dt}{(1+\tan^2 t)^2} = \int_{-\pi/4}^{\pi/4} \cos^2 t\, dt = \int_{-\pi/4}^{\pi/4} \frac{1}{2}(\cos 2t + 1)dt.$$

Thus,

$$\int_{-1}^1 \frac{dx}{(1+x^2)^2} = \frac{1}{2}\left[\frac{\sin 2t}{2} + t\right]_{-\pi/4}^{\pi/4} = \frac{1}{2} + \frac{\pi}{2}. \qquad \diamond$$

Theorem 3.4.6 (Taylor's formula - Cauchy integral form) *Let I be an open interval, $x_0 \in I$ and f has $n + 1$ continuous derivatives in I. Then for every $x \in I$,*

$$f(x) = \sum_{k=0}^n \frac{f^{(k)}(x_0)}{k!}(x - x_0)^k + \frac{1}{n!}\int_{x_0}^x f^{(n+1)}(t)(x - t)^n dt.$$

Proof Let $x \in I$ and $x \neq x_0$. By Fundamental theorem-I (Theorem 3.3.1), we have

$$f(x) = f(x_0) + \int_{x_0}^x f'(t)dt.$$

Hence, by integration by parts, we have

$$\int_{x_0}^{x} f'(t)dt = -\int_{x_0}^{x} f'(t)\frac{d}{dt}(x-t)dt$$

$$= \left[-f'(t)(x-t)\right]_{x_0}^{x} + \int_{x_0}^{x} f''(t)(x-t)dt$$

$$= f'(x_0)(x-x_0) + \int_{x_0}^{x} f''(t)(x-t)dt.$$

Thus, theorem holds for $n = 1$. Now, let $m < n$ and assume that theorem holds for $n = m - 1$, i.e.,

$$f(x) = \sum_{k=0}^{m-1} \frac{f^{(k)}(x_0)}{k!}(x-x_0)^k + \frac{1}{(m-1)!}\int_{x_0}^{x} f^{(m)}(t)(x-t)^{m-1}dt.$$

Then we have

$$\int_{x_0}^{x} f^{(m)}(t)(x-t)^{m-1}dt = -\frac{1}{m}\int_{x_0}^{x} f^{(m)}(t)\frac{d}{dt}(x-t)^m dt$$

$$= \frac{1}{m}f^{(m)}(x_0)(x-x_0)^m$$

$$+\frac{1}{m}\int_{x_0}^{x} f^{(m+1)}(t)(x-t)^m dt.$$

Thus, the theorem holds for $n = m$ as well.   ∎

Exercise 3.4.1 Derive the Taylor's formula in Theorem 2.3.23 from Theorem 3.4.6 by applying the generalized mean value theorem (Theorem 3.2.2).   ◁

3.5 Some Applications

In this section we shall make use of the properties of integrals to derive formulae for the area of certain regions in the plane, arc length of curves, volume of certain type of solid domains in the *three-dimensional space*, area of certain surfaces in the space, and also to obtain formulae for certain physical quantities such as centre of gravity of a material line, centre of gravity of certain planar region, moment of inertia of a material line and moment of inertia of certain planar region.

First let us define the notion of a *curve* in the plane.

Definition 3.5.1 By a **curve** in the plane we mean a function

$$t \mapsto \gamma(t) := (\varphi(t), \psi(t))$$

defined on a closed and bounded interval $[a, b]$ taking values in $\mathbb{R}^2$, where $\varphi : [a, b] \to \mathbb{R}$ and $\psi : [a, b] \to \mathbb{R}$ are continuous functions.   ◊

Given a curve $t \mapsto \gamma(t) := (\varphi(t), \psi(t))$ defined on $[a, b]$, we may also say that the set

$$C := \{\gamma(t) : a \leq t \leq b\}$$

is a curve, and it is given by the *parametric form*

$$x = \varphi(t), \quad y = \psi(t) \quad \text{for } t \in [a, b].$$

Note that, as t varies from a to b, the point $\gamma(t)$ moves from $\gamma(a)$ to $\gamma(b)$ along C. The point $\gamma(a)$ is called the *initial point* of C and $\gamma(b)$ is called the *terminal point* of C.

We may observe that, if $f : [a, b] \to \mathbb{R}$ is a continuous function, then it defines a curve, namely, $t \mapsto (t, f(t))$. We say that this curve is given by the equation

$$y = f(x), \quad x \in [a, b].$$

3.5.1 Computing Area Under the Graph of a Function

Using Cartesian coordinates

Motivated from the geometric interpretation of the Riemann integral of a positive function, we define area of certain regions in the plane.

Definition 3.5.2 Suppose a curve is given by an equation

$$y = f(x), \quad a \leq x \leq b,$$

where $f : [a, b] \to \mathbb{R}$ is a continuous function with $f(x) \geq 0$ for all $x \in [a, b]$. Then the **area of the region** bounded by the graph of f, namely, $\{(x, f(x)) : a \leq x \leq b\}$, the x-axis, and the ordinates at $x = a$ and $x = b$ is defined as the integral of f over $[a, b]$, that is,

$$\int_a^b f(x) \, dx,$$

which is also written as $\int_a^b y \, dx$.   ◊

Sometimes, the computation of the integral becomes easier if we can express the curve in parametric form. For example, suppose the curve is given in parametric form as

$$x = \varphi(t), \quad y = \psi(t), \quad \alpha \le t \le \beta,$$

such that $a = \varphi(\alpha)$, $b = \psi(\beta)$ and φ is differentiable on $[\alpha, \beta]$ with φ' continuous on $[\alpha, \beta]$. Then the area under the curve takes the form

$$\int_{\alpha}^{\beta} \psi(t)\varphi'(t)dt.$$

If f takes both positive and negative values, but changes sign only at a finite number of points, then the area bounded by the curve, the x-axis, and the ordinates at $x = a$ and $x = b$, is given by

$$\int_{a}^{b} |f(x)|\, dx.$$

Generalizing Definition 3.5.2, we have the following definition.

Definition 3.5.3 Suppose $f : [a, b] \to \mathbb{R}$ and $g : [a, b] \to \mathbb{R}$ are continuous functions such that $f(x) \le g(x)$ for all $x \in [a, b]$. Then the **area of the region** bounded by the graphs of f and g, and the ordinates at $x = a$ and $x = b$ is given by

$$A := \int_{a}^{b} [g(x) - f(x)]\, dx. \qquad\qquad \Diamond$$

A slight variant of the Definition 3.5.2 is the following.

Definition 3.5.4 Suppose f and g are continuous (real valued) functions defined on an interval I such that $f(x) \le g(x)$ for all $x \in I$. Suppose that the curves

$$y = f(x), \quad y = g(x) \quad \text{for} \quad x \in I \qquad\qquad (*)$$

intersect only at two points, say $f(a) = g(a)$ and $f(b) = g(b)$ with $a < b$ for some $a, b \in I$. Then the **area of the region** bounded by the curves in $(*)$ is defined by the integral

$$\int_{a}^{b} [g(x) - f(x)]\, dx. \qquad\qquad \Diamond$$

Example 3.5.1 We find the area of the region bounded by the curves defined by

$$y = \sqrt{x}, \quad y = x^2, \quad x \geq 0 :$$

Note that the points of intersection of the curves are at $x = 0$ and $x = 1$. Also, $\sqrt{x} \geq x^2$ for $0 \leq x \leq 1$. Hence, the required area is

$$\int_0^1 \left(\sqrt{x} - x^2 \right) \, dx = \left[\frac{x^{3/2}}{3/2} - \frac{x^3}{3} \right]_0^1 = \frac{1}{3}. \qquad \diamond$$

Example 3.5.2 Let us find the area of the region bounded by the straight line $y = x$ and the parabola $y = ax^2$, $a > 0$. Note that the required region is in the first quadrant of the plane, and the limits of integration are obtained by finding the intersection of the curves $y = x$ and $y = ax^2$, that is by solving $x = ax^2$. Thus, $x = 0$ and $x = 1/a$ are the limits of integration. Thus, the required area is given by

$$\int_0^{1/a} [x - ax^2] dx = \left[\frac{x^2}{2} - a\frac{x^3}{3} \right]_0^{1/a} = \frac{1}{6a^2}. \qquad \diamond$$

Example 3.5.3 We find the area of the region bounded by the *ellipse*

$$\frac{x^2}{a^2} + \frac{y^2}{b^2} = 1.$$

Since the region is symmetric with the coordinate axes, the required area is 4 times the area under the curve in the first quadrant. We shall compute this area in two different ways:

(i) The equation of the curve in the first quadrant is given by

$$y = b\sqrt{1 - \frac{x^2}{a^2}}, \quad 0 \leq x \leq a.$$

Hence, the required area is

$$A := 4 \int_0^a y \, dx = \frac{4b}{a} \int_0^a \sqrt{a^2 - x^2} \, dx.$$

By the change of variable $x = a \sin t$, we obtain

$$\frac{4b}{a} \int_0^{\pi/2} a^2 \cos^2 t \, dt = 4ab \int_0^{\pi/2} \frac{1 + \cos 2t}{2} dt = \pi ab.$$

(ii) Note that the ellipse can be represented in parametric form as

$$x(t) = a \cos t, \quad y(t) = b \sin t, \quad 0 \le t \le 2\pi.$$

Thus, using this parametrization, the required area is given by

$$4 \int_0^{\pi/2} y(t) x'(t) dt = 4 \int_0^{\pi/2} (b \sin t)(-a \sin t) \, dt$$

$$= 2ab \int_0^{\pi/2} (1 - \cos 2t) \, dt$$

$$= \pi ab.$$

Thus, the area of the region is πab. ◊

Example 3.5.4 We find the area bounded by one arch of the *cycloid*

$$x = a(t - \sin t), \quad y = a(1 - \cos t).$$

One arch of the cycloid is obtained by varying t over the interval $[0, 2\pi]$. Thus, the required area is

$$\int_0^{2\pi a} y \, dx = \int_0^{2\pi} y(t) x'(t) \, dt = \int_0^{2\pi} a^2 (1 - \cos t)^2 \, dt = 3\pi a^2.$$ ◊

Using polar coordinates

We may recall that a point in the plane $\mathbb{R}^2$, represented in Cartesian coordinates as (x, y) can also be represented in polar coordinates as (ρ, θ) such that

$$x = \rho \cos \theta, \quad y = \rho \sin \theta \quad \text{for} \quad \rho \ge 0, \quad 0 \le \theta < 2\pi.$$

Suppose a curve is given in polar coordinates as

$$\rho = \varphi(\theta), \quad \alpha \le \theta \le \beta,$$

where $\varphi : [\alpha, \beta] \to \mathbb{R}$ is a continuous function. We would like to find an expression for the area of the region bounded by the graph of φ and the rays $\theta = \alpha$ and $\theta = \beta$. For this, first consider a partition of $[\alpha, \beta]$, say $P : \alpha = \theta_0 < \theta_1 < \cdots < \theta_k = \beta$ and a set $\{\xi_i\}$ of tags on P. Then the area of the region bounded by the graph of φ and the rays $\theta = \theta_{i-1}$ and $\theta = \theta_i$ would be close to the quantity

$$\frac{1}{2}[\varphi(\xi_i)(\theta_i - \theta_{i-1})]\varphi(\xi_i).$$

Thus, the required area must be close to the

$$\sum_{i=1}^{k} \frac{1}{2}[\varphi(\xi_i)]^2(\theta_i - \theta_{i-1}).$$

Note that

$$\lim_{\mu(P)\to 0} \sum_{i=1}^{k} \frac{1}{2}[\varphi(\xi_i)]^2(\theta_i - \theta_{i-1}) = \frac{1}{2}\int_{\alpha}^{\beta} \rho^2 \, d\theta.$$

In view of this we have the following definition.

Definition 3.5.5 The **area of the region** bounded by the graph of φ and the rays $\theta = \alpha$ and $\theta = \beta$ is defined as

$$\frac{1}{2}\int_{\alpha}^{\beta} \rho^2 \, d\theta. \qquad \Diamond$$

As in the case of Cartesian coordinates, the above definition can be used to find the area of the certain regions in the plane using polar coordinates. Let us consider a few examples.

Example 3.5.5 We find the area bounded by a circle of radius a. Without loss of generality, we may assume that the centre of the circle is the origin. Then, the circle can be represented in polar coordinates as

$$\rho = a, \quad 0 \le \theta \le 2\pi.$$

Hence the required area is

$$A := \frac{1}{2}\int_{0}^{2\pi} \rho^2 \, d\theta = \pi a^2. \qquad \Diamond$$

Example 3.5.6 We find the area bounded by the *lemniscate*

$$\rho = a\sqrt{\cos 2\theta}.$$

Note that the lemniscate has two identical loops, and the curve traces one complete loop when θ varies from $-\pi/4$ to $\pi/4$. Thus, the required area is

$$2\left[\frac{1}{2}\int_{-\pi/4}^{\pi/4}\rho^2\,d\theta\right] = a^2\int_{-\pi/4}^{\pi/4}\cos 2\theta\,d\theta = a^2. \qquad \Diamond$$

3.5.2 Computing the Arc Length

Suppose a curve $C : \gamma(t) := (x(t), y(t))$, $a \le t \le b$, is given. In order to compute the length of C, we first consider a polygonal approximation of the curve. What we mean by that is the following:

Corresponding to a partition $P : a = t_0 < t_1 < \cdots < t_k = b$ of $[a, b]$, consider the length of the polygonal line obtained by joining the points $\gamma(t_0), \gamma(t_1), \ldots, \gamma(t_k)$, i.e., the quantity

$$\ell_P(C) := \sum_{i=1}^{k}\sqrt{(x(t_i) - x(t_{i-1}))^2 + (y(t_i) - y(t_{i-1}))^2}.$$

Using the above quantity, we define the length of C as follows:

Definition 3.5.6 The **length of the curve** C is defined by

$$\ell(C) := \sup_P \ell_P(C),$$

where the supremum is taken over all partitions P of $[a, b]$. $\qquad \Diamond$

In order to compute the quantity $\ell(C)$, we assume that the curve $C : \gamma(t) := (x(t), y(t))$, $t \in [a, b]$, is *smooth* in the sense that the functions $t \mapsto x(t)$ and $t \mapsto y(t)$ are differentiable in the open interval (a, b) and their derivatives $t \mapsto x'(t)$ and $t \mapsto y'(t)$ are continuous. Then, for a given a partition $P : a = t_0 < t_1 < \cdots < t_k = b$ of $[a, b]$, by mean value theorem, there exist $\xi_i, \eta_i \in (t_{i-1}, t_i)$ such that

$$x(t_i) - x(t_{i-1}) = x'(\xi_i)(t_i - t_{i-1}),$$

$$y(t_i) - y(t_{i-1}) = y'(\eta_i)(t_i - t_{i-1})$$

for $i = 1, 2, \ldots, k$. Thus, the quantity $\ell_P(C)$ takes the form

$$\ell_P(C) = \sum_{i=1}^{k} \sqrt{x'(\xi_i)^2 + y'(\eta_i)^2}(t_i - t_{i-1}).$$

By our assumptions on the functions $x(t)$ and $y(t)$, the function

$$f(t) := \sqrt{x'(t)^2 + y'(t)^2}, \quad a \le t \le b,$$

is integrable over $[a, b]$. Hence, if we take a sequence (P_n) of partitions on $[a, b]$, say

$$P_n : a = t_0^{(n)} < t_1^{(n)} < \cdots < t_{k_n}^{(n)} = b$$

such that $\mu(P_n) \to 0$ as $n \to \infty$, then

$$S(P_n, f, T_n) := \sum_{i=1}^{k_n} \sqrt{x'(c_i^{(n)})^2 + y'(c_i^{(n)})^2}(t_i^{(n)} - t_{i-1}^{(n)})$$

$$\to \int_a^b \sqrt{x'(t)^2 + y'(t)^2}\, dt$$

as $n \to \infty$, where $T_n := \{c_i^{(n)}\}_{i=1}^{k_n}$ is any set of tags on P_n. Corresponding to the partition P_n, we have

$$\ell_{P_n}(C) = \sum_{i=1}^{k_n} \sqrt{x'(\xi_i^{(n)})^2 + y'(\eta_i^{(n)})^2}(t_i^{(n)} - t_{i-1}^{(n)})$$

for some $\xi_i^{(n)}, \eta_i^{(n)} \in (t_{i-1}^{(n)}, t_i^{(n)})$ for $i = 1, \ldots, k_n$. Therefore, by the continuity of the functions $x'(t)$ and $y'(t)$, we obtain (Exercise)

$$|S(P_n, f, T_n) - \ell_{P_n}(C)| \to 0 \quad \text{as} \quad n \to \infty.$$

Hence, we can conclude that

$$\ell(C) = \sup_P \ell_P(C) = \lim_{n \to \infty} \ell_{P_n}(C) = \lim_{n \to \infty} S(P_n, f, T_n).$$

Thus,

$$\ell(C) = \int_a^b \sqrt{\left(\frac{dx}{dt}\right)^2 + \left(\frac{dy}{dt}\right)^2} \, dt.$$

Suppose $s(\tau)$ is the length of the arc which is part of the curve C from the point $\gamma(a)$ to the point $\gamma(\tau)$. Then from the above formula, we have

$$s(\tau) = \int_a^\tau \sqrt{x'(t)^2 + y'(t)^2} \, dt, \quad a < \tau \le b.$$

Example 3.5.7 Let us compute the length of the arc of the circle

$$x = r \cos\theta, \quad y = r \sin\theta, \quad 0 \le \theta \le 2\pi$$

when θ varies from $\theta = \alpha$ to $\theta = \beta$. Using the formula given above, the required length is

$$\ell(C) = \int_\alpha^\beta \sqrt{x'(\theta)^2 + y'(\theta)^2} \, d\theta$$

$$= \int_\alpha^\beta \sqrt{r^2 \sin^2\theta + r^2 \cos^2\theta} \, d\theta$$

$$= r(\beta - \alpha).$$

In particular, the circumference of the circle is $r(2\pi - 0) = 2\pi r$. ◊

Example 3.5.8 Let us find the length of the ellipse

$$x = a \cos\theta, \quad y = b \sin\theta, \quad 0 \le \theta \le 2\pi,$$

where $a > b$. The required length is

$$\ell(C) := 4 \int_0^{\pi/2} \sqrt{\left(\frac{dx}{d\theta}\right)^2 + \left(\frac{dy}{d\theta}\right)^2} \, d\theta$$

$$= 4 \int_0^{\pi/2} \sqrt{a^2 \sin^2 \theta + b^2 \cos^2 \theta} \, d\theta$$

$$= 4 \int_0^{\pi/2} \sqrt{a^2(1 - \cos^2 \theta) + b^2 \cos^2 \theta} \, d\theta$$

$$= 4 \int_0^{\pi/2} \sqrt{a^2 - (a^2 - b^2) \cos^2 \theta} \, d\theta.$$

Thus,

$$\ell(C) = 4a \int_0^{\pi/2} \sqrt{1 - \beta^2 \cos^2 \theta} \, d\theta, \quad \beta := \frac{\sqrt{a^2 - b^2}}{a}.$$

The above integral cannot be computed using standard methods, unless $\beta = 0$, i.e., $b = a$ in which case the ellipse is the circle. But, the integral can be approximately computed numerically.   ◊

Example 3.5.9 We find the length of the *astroid*:

$$x = a \cos^3 t, \quad y = a \sin^3 t.$$

We observe that the astroid consists of four loops, each of length

$$\int_0^{\pi/2} \sqrt{x'(t)^2 + y'(t)^2} \, dt = \int_0^{\pi/2} \sqrt{9a^2 \cos^4 t \sin^2 t + 9a^2 \sin^4 t \cos^2 t} \, dt$$

$$= 3a \int_0^{\pi/2} \sqrt{\cos^2 t \sin^2 t} \, dt$$

$$= \frac{3a}{2}.$$

Hence, the length of the astroid is $6a$.   ◊

Using Cartesian coordinates

If the curve C is given by an equation

$$y = f(x), \quad a \le x \le b,$$

where f is a continuous function on $[a, b]$, then we may write

$$C : \gamma(t) := (t, f(t)), \quad a \le t \le b.$$

In this case, the length of the curve C is given by

$$\ell(C) = \int_a^b \sqrt{[1 + f'(t)^2]} dt,$$

i.e.,

$$\ell(C) = \int_a^b \sqrt{1 + \left(\frac{dy}{dx}\right)^2} dx.$$

Example 3.5.10 Let us find the circumference of a circle of radius r using the above formula in Cartesian coordinates: Without loss of generality assume that the centre of the circle is the origin, i.e., the circle is given by $x^2 + y^2 = r^2$. The required length is

$$\ell(C) := 4 \int_0^r \sqrt{1 + \left(\frac{dy}{dx}\right)^2} dx, \quad y = \sqrt{r^2 - x^2}.$$

Thus,

$$\ell(C) := 4r \int_0^r \frac{dx}{\sqrt{r^2 - x^2}} = 2\pi r. \qquad \Diamond$$

Remark 3.5.1 Curves in parametric form can be assumed to be piecewise smooth, i.e., having unique tangents except possibly at a finite number of points. Note that if a curve is given in parametric form as

$$x = x(t), \quad y = y(t), \quad \alpha \le t \le \beta,$$

then it has unique tangent at $(x(t_0), y(t_0))$, if $x'(t_0)$, $y'(t_0)$ exist and $|x'(t_0)|^2 + |y'(t_0)|^2 \ne 0$. $\qquad \Diamond$

Using polar coordinates

Suppose a curve is given in polar coordinates as

$$\rho = \varphi(\theta), \quad \alpha \le \theta \le \beta,$$

where $\varphi : [\alpha, \beta] \to \mathbb{R}$ is a continuous function. Since

$$x = \rho \cos \theta, \quad y = \rho \sin \theta, \quad \alpha \le \theta \le \beta,$$

we have

$$\ell(C) = \int_\alpha^\beta \sqrt{\left(\frac{dx}{d\theta}\right)^2 + \left(\frac{dy}{d\theta}\right)^2} \, d\theta.$$

Note that

$$\frac{dx}{d\theta} = \rho' \cos \theta + \rho(-\sin \theta), \quad \frac{dy}{d\theta} = \rho' \sin \theta + \rho \cos \theta.$$

Hence, the **length of the curve** C is given by

$$\ell(C) = \int_\alpha^\beta \sqrt{\rho^2 + \left(\frac{d\rho}{d\theta}\right)^2} \, d\theta.$$

Example 3.5.11 We find the length of the cardioid $\rho = a(1 + \cos \theta)$. The required length $\ell(C)$ is given by

$$\ell(C) = \int_0^{2\pi} \sqrt{\rho^2 + \rho'^2} \, d\theta.$$

Since

$$\rho^2 = a^2(1 + \cos \theta)^2, \quad \rho'^2 = a^2 \sin^2 \theta,$$

we have

$$\ell(C) = \sqrt{2}a \int_0^{2\pi} \sqrt{1 + \cos \theta} \, d\theta = 2a \int_0^{2\pi} \left| \cos \frac{\theta}{2} \right| \, d\theta = 8a. \qquad \diamond$$

3.5.3 Computing Volume of a Solid

Suppose that a three-dimensional object, a solid, lies between two parallel planes $x = a$ and $x = b$ with $a < b$. Let $\alpha(x)$ be the area of the cross section of the solid at the point x, with cross section being parallel to the yz-plane. We assume that the function $\alpha(x)$, $x \in [a, b]$, is continuous.

Consider a partition $P : a = x_0 < x_1 < \ldots < x_k = b$ of the interval $[a, b]$. Let $\xi_i \in [x_{i-1}, x_i]$ for $i = 1, \ldots, k$. Then the quantity $\sum_{i=1}^{k} \alpha(\xi_i)(x_i - x_{i-1})$ can be thought of as the volume of a slice of the solid with width Δx_i. Note that

$$\lim_{\mu(P) \to 0} \sum_{i=1}^{k} \alpha(\xi_i)(x_i - x_{i-1}) = \int_{a}^{b} \alpha(x) \, dx.$$

In view of this relation we have the following definition.

Definition 3.5.7 The **volume of the solid** with cross-sectional area $\alpha(x)$ at the point x with x varies over the interval $[a, b]$ is defined by the integral

$$V := \int_{a}^{b} \alpha(x) \, dx. \qquad\qquad \Diamond$$

Example 3.5.12 Let us compute the volume of the solid enclosed by the ellipsoid

$$\frac{x^2}{a^2} + \frac{y^2}{b^2} + \frac{z^2}{c^2} = 1.$$

For a fixed $x \in [-a, a]$, the boundary of the cross section at x is given by the equation

$$\frac{y^2}{b^2} + \frac{z^2}{c^2} = 1 - \frac{x^2}{a^2}.$$

This equation can be written as

$$\frac{y^2}{\phi(x)^2} + \frac{z^2}{\psi(x)^2} = 1,$$

where

$$\phi(x) = b\sqrt{1 - \frac{x^2}{a^2}}, \quad \psi(x) = c\sqrt{1 - \frac{x^2}{a^2}},$$

so that it represents an ellipse. Hence, the cross-sectional area $\alpha(x)$ at x is given by

$$\alpha(x) = \pi \phi(x) \psi(x) = \pi bc\left(1 - \frac{x^2}{a^2}\right).$$

Therefore, the required volume V is given by

$$V = \int_{-a}^{a} \alpha(x)\, dx = \pi bc \int_{-a}^{a} \left(1 - \frac{x^2}{a^2}\right) dx = \frac{4}{3}\pi abc.$$

In particular, the volume of the solid bounded by the sphere of radius a is $\frac{4}{3}\pi a^3$. ◊

3.5.4 Computing the Volume of Solid of Revolution

Suppose a solid is obtained by revolving a curve

$$y = f(x), \quad x \in [a, b]$$

with x-axis as the axis of revolution. We would like to find the volume of the solid. In this case the area of the cross section at x is given by

$$\alpha(x) = \pi y^2 = \pi[f(x)]^2, \quad a \le x \le b.$$

Thus, in view of Definition 3.5.7, we have the following definition.

Definition 3.5.8 The **volume of the solid of revolution** obtained by revolving the curve $y = f(x)$, $a \le x \le b$ about the x-axis is given by

$$V := \pi \int_{a}^{b} [f(x)]^2\, dx.$$

◊

Analogously, we have the following definition.

Definition 3.5.9 The **volume of the solid of revolution** obtained by revolving the curve $x = g(y)$, $c \le y \le d$, about the y-axis is given by

$$V := \pi \int_{c}^{d} [g(y)]^2\, dy.$$

◊

What about volume of the solid of revolution of a curve with a specific straight line, say L, as the axis of revolution?

This can also be found, provided the curve is such that, the perpendicular at each point on the line intersects the curve at most at one point.

Let $\varphi(x, y)$ be the perpendicular distance from a point (x, y) on the curve C to the line L. Then, at the point (x, y), the cross-sectional area of the solid of revolution of C about L is given by

$$\pi [\varphi(x, y)]^2.$$

Now, suppose that the curve C and the line L are given in parametric form as

$$C : x = x(t), \quad y = y(t)$$
$$L : x = \ell_1(t), \quad y = \ell_2(t)$$

for $t \in [a, b]$. Note that $\ell_1(t)$ and $\ell_2(t)$ are linear in t, and an elementary segment on L is of the form

$$\sqrt{\ell_1'(t)^2 + \ell_2'(t)^2} \, \Delta t.$$

Thus, we have the following definition.

Definition 3.5.10 If the curve C and the line L have parametric representations

$$C : x = x(t), \quad y = y(t),$$
$$L : x = \ell_1(t), \quad y = \ell_2(t)$$

for $t \in [a, b]$. Then the **volume of the solid of revolution** of C about the line L is given by the integral

$$V := \pi \int_a^b [\varphi(x(t), y(t))]^2 \sqrt{\ell_1'(t)^2 + \ell_2'(t)^2} \, dt,$$

where $\varphi(\alpha, \beta)$ is the perpendicular distance from the point (α, β) on the curve C to the line L.   ◇

From the above formula in the above definition, we can deduce the expressions in Definition 3.5.8 and Definition 3.5.9 as special cases:

(i) Suppose C is given by the equation $y = f(x)$, $a \le x \le b$, and L is the x-axis. Then we may take

$$C : x = t, \quad y = f(t) \quad \text{and} \quad L : x = t, \quad y = 0$$

for $t \in [a, b]$. Then

$$\varphi(x(t), y(t)) = \varphi(t, f(t)) = |f(t)|, \quad \ell_1'(t)^2 + \ell_2'(t)^2 = 1$$

so that

$$V := \pi \int_a^b [\varphi(x(t), y(t))]^2 \sqrt{\ell_1'(t)^2 + \ell_2'(t)^2} dt = \pi \int_a^b [f(t)]^2 dt.$$

(ii) Suppose C is given by the equation $x = g(y)$, $c \leq y \leq d$, and L is the y-axis. Then we may take

$$C : x = g(t), \quad y = t \quad \text{and} \quad L : x = 0, \quad y = t$$

for $t \in [c, d]$. Then

$$\varphi(x(t), y(t)) = \varphi(g(t), t) = |g(t)|, \quad \ell_1'(t)^2 + \ell_2'(t)^2 = 1$$

so that

$$V := \pi \int_a^b [\varphi(x(t), y(t))]^2 \sqrt{\ell_1'(t)^2 + \ell_2'(t)^2} dt = \pi \int_c^d [g(t)]^2 dt.$$

Example 3.5.13 Let us compute the volume of the solid of revolution of the curve $y = x^2$ about x-axis for $-a \leq x \leq a$. The required volume is

$$V := \pi \int_{-a}^a y^2 \, dx = \pi \int_{-a}^a x^4 \, dx = \frac{2}{5} a^5. \qquad \diamondsuit$$

Example 3.5.14 We compute the volume of the solid of revolution of the *catenary*

$$y = \frac{a}{2} \left(e^{x/a} + e^{-x/a} \right)$$

about x-axis for $0 \leq x \leq b$. Hence,

$$\int_0^b y^2 dx = \int_0^b \frac{a^2}{4} \left(e^{x/a} + e^{-x/a} \right)^2 \, dx$$

$$= \frac{a^2}{4} \int_0^b \left(e^{2x/a} + e^{-2x/a} + 2 \right)$$

$$= \frac{a^3}{8} \left(e^{2b/a} - e^{-2b/a} \right) + \frac{\pi a^2 b}{2}.$$

The required volume is

$$V = \frac{\pi a^3}{8} \left(e^{2b/a} - e^{-2b/a}\right) + \frac{\pi a^2 b}{2}. \qquad \diamond$$

Example 3.5.15 Let us compute the volume of the solid of revolution of line segment $C : 0 \leq x \leq 1$ about the line $L : y = x$. Consider the parametrization of C and L as

$$C : x(t) = t, \ y(t) = 0; \quad L : \ell_1(t) = \frac{t}{2}, \ \ell_2(t) = \frac{t}{2}, \quad 0 \leq t \leq 1.$$

The perpendicular distance $\varphi(x, y)$ from the point (x, y) on the curve to the line L is given by

$$\varphi(x, y) = \frac{x}{\sqrt{2}}.$$

Thus,

$$\pi \int_0^1 [\varphi(x(t), y(t))]^2 \sqrt{\ell_1'(t)^2 + \ell_2'(t)^2} dt = \pi \int_0^1 \frac{t^2}{2} \sqrt{\frac{1}{4} + \frac{1}{4}} \, dt = \frac{\pi}{6\sqrt{2}}.$$

Thus, the volume is $\pi/6\sqrt{2}$. $\qquad \diamond$

3.5.5 Computing the Area of Surface of Revolution

Suppose a solid is obtained by revolving a curve $C : y = f(x)$, $a \leq x \leq b$, with x-axis as axis of revolution. We would like to find the area of the surface of the solid. We assume that $f(x) \geq 0$ for every $x \in [a, b]$.

Let us consider a partition $P : a = x_0 < x_1 < \ldots < x_k = b$ of the interval $[a, b]$. For each $i = 1, \ldots, k$, let $\xi_i \in [x_{i-1}, x_i]$ and let Δs_i be the length of the portion of the curve C when x varies over $[x_{i-1}, x_i]$. Then the quantity

$$2\pi f(\xi_i) \Delta s_i$$

would be *approximately* equal to the lateral surface area of the cylindrical piece with base as the cross section at ξ_i and height Δs_i. Hence the area of surface of revolution of the curve C is *approximately* equal to

$$\sum_{i=1}^k 2\pi f(\xi_i) \Delta s_i.$$

Since $\Delta s_i \approx \sqrt{1 + [f'(\xi_i)]^2}\,\Delta x_i$, where $\Delta x_i := x_i - x_{i-1}$ for $i = 1, \ldots, n$, we have

$$\sum_{i=1}^{k} 2\pi f(\xi_i)\Delta s_i \approx \sum_{i=1}^{k} 2\pi f(\xi_i)\sqrt{1 + [f'(\xi_i)]^2}\,\Delta x_i.$$

Note that

$$\lim_{\mu(P)\to 0} \sum_{i=1}^{k} 2\pi f(\xi_i)\sqrt{1 + [f'(\xi_i)]^2}\,\Delta x_i = 2\pi \int_a^b y\sqrt{1 + \left(\frac{dy}{dx}\right)^2}\,dx.$$

Here, y is the perpendicular distance of the point (x, y) on the curve from the x-axis, and the expression

$$\sqrt{1 + \left(\frac{dy}{dx}\right)^2}\,dx$$

corresponds to the elementary arc length of the curve.

In view of the above observation, we have the following definition.

Definition 3.5.11 The area of the **surface of revolution** S of the curve

$$C : y = f(x), \quad a \le x \le b$$

with x-axis as the axis of revolution is defined as

$$S := 2\pi \int_a^b y\sqrt{1 + \left(\frac{dy}{dx}\right)^2}\,dx. \qquad \Diamond$$

Suppose we are interested in finding the area of the surface of revolution of a curve C with a specific straight line L as the axis of revolution. Let

$$C : \gamma(t) := (x(t), y(t)), \quad a \le t \le b,$$

be the parametric representation of C. In this case, the expression corresponding to the elementary arc length of the curve is given by

$$\sqrt{x'(t)^2 + y'(t)^2}\,dt.$$

Let $p(t)$ be the perpendicular distance from a point $(x(t), y(t))$ on the curve C to the line L. Then, we obtain the required area as

$$S = 2\pi \int_a^b p(t)\sqrt{x'(t)^2 + y'(t)^2}\,dt.$$

The following are the two special cases:

(i) Suppose C is given by the equation $y = f(x)$, $a \leq x \leq b$, and L is the x-axis. Then,

$$C : \gamma(x) := (x, f(x)), \quad a \leq x \leq b,$$

is a parametrization of C and $f(x)$ is the perpendicular distance from the point $(x, f(x))$ on the curve C to the x-axis. Thus, we obtain

$$S = 2\pi \int_a^b f(x)\sqrt{1 + f'(x)^2}\,dx.$$

Note that this is the same as in Definition 3.5.11, as it should be.

(ii) Suppose C is given by the equation $x = g(y)$, $c \leq y \leq d$, and L is the y-axis. In this case, we can consider the parametrization of the curve as

$$C : \gamma(y) := (g(y), y), \quad c \leq y \leq d.$$

Then the perpendicular distance from the point $(g(y), y)$ on the curve C to the y-axis is $g(y)$, so that

$$S = 2\pi \int_c^d g(y)\sqrt{1 + g'(y)^2}\,dy.$$

Example 3.5.16 We find the area of the surface of revolution of the parabola

$$y^2 = 2px, \quad 0 \leq x \leq a \quad \text{for} \quad p > 0$$

with x-axis as the axis of revolution. The required area S is given by the formula

$$S = 2\pi \int_0^a y\sqrt{1 + \left(\frac{dy}{dx}\right)^2}\,dx,$$

where $y = \sqrt{2px}$. Thus,

$$S = 2\pi \int_0^a \sqrt{2px} \sqrt{1 + \frac{p}{2x}}\, dx$$

$$= 2\pi \sqrt{p} \int_0^a \sqrt{p + 2x}\, dx$$

$$= 2\pi \sqrt{p} \frac{2}{3} \left[(2x + p)^{3/2} \frac{1}{2} \right]_0^a$$

$$= \frac{2\pi \sqrt{p}}{3} \left[(2a + p)^{3/2} - p^{3/2} \right]. \qquad \Diamond$$

3.5.6 Centre of Gravity

In this subsection, we define the centre of gravity of a *material line* and a *material planar region* enclosed by certain curves.

First let us consider material particles on the plane at points

$$(x_1, y_1),\ (x_2, y_2),\ \ldots,\ (x_n, y_n)$$

with masses $m_1, m_2, \ldots m_n$, respectively. Then the **centre of gravity** of the system of these particles is at the point (x_0, y_0), where

$$x_0 := \frac{\sum_{i=1}^n x_i m_i}{\sum_{i=1}^n m_i}, \qquad y_0 := \frac{\sum_{i=1}^n y_i m_i}{\sum_{i=1}^n m_i}.$$

Centre of gravity of a material line in the plane

We would like to find a formula for the centre of gravity of a *material line L* in the plane. If $M(X, r)$ is the mass of an arc of the line containing the point X with length r, then the *point density* of this line at the point X is defined by

$$\gamma(X) := \lim_{r \to 0} \frac{M(X, r)}{r}.$$

Now, suppose L is given by the equation

$$y = f(x), \quad x \in [a, b].$$

In order to find the centre of gravity of L, we first consider a partition $P : a = x_0 < x_1 < \ldots < x_k$ of $[a, b]$, and points

$$X_i := (\xi_i, f(\xi_i)) \quad \text{for} \quad i = 1, \ldots, n,$$

on L, where $\xi_i \in [x_{i-1}, x_i]$, $i = 1, \ldots, n$. Then the centre of gravity of the system of material points $X_1, \ldots, X_n$ is the point $(x_0(P), y_0(P))$, where

$$x_0(P) = \frac{\sum_{i=1}^n \xi_i m_i}{\sum_{i=1}^n m_i}, \quad y_0(P) := \frac{\sum_{i=1}^n f(\xi_i) m_i}{\sum_{i=1}^n m_i}, \tag{$*$}$$

where, m_i is the mass of the material point X_i, that is, $m_i = \gamma_i \Delta s_i$ with γ_i as the point density at the point X_i and Δs_i is the length of the arcs joining (x_{i-1}, y_{i-1}) to (x_i, y_i). Here $y_i = f(x_i)$ for $i = 1, \ldots, n$. Note that $m_i = \gamma_i \Delta s_i$ is the approximate mass of the arc joining (x_{i-1}, y_{i-1}) to (x_i, y_i). We may consider the *centre of gravity* of L is at (x_0, y_0), where

$$x_0 := \lim_{\mu(P) \to 0} x_0(P), \quad y_0 := \lim_{\mu(P) \to 0} y_0(P).$$

Now, suppose that the function $x \mapsto \gamma(X) := \gamma(x, f(x))$ is continuous on $[a, b]$. Then, in view of the formula for $x_0(P)$ and $y_0(P)$ in $(*)$, we have

$$\lim_{\mu(P) \to 0} x_0(P) = \lim_{\mu(P) \to 0} \frac{\sum_{i=1}^n \xi_i m_i}{\sum_{i=1}^n m_i} = \frac{\int_a^b x \gamma(x, y) \sqrt{1 + \left(\frac{dy}{dx}\right)^2} \, dx}{\int_a^b \gamma(x, y) \sqrt{1 + \left(\frac{dy}{dx}\right)^2} \, dx},$$

$$\lim_{\mu(P) \to 0} y_0(P) = \lim_{\mu(P) \to 0} \frac{\sum_{i=1}^n f(\xi_i) m_i}{\sum_{i=1}^n m_i} = \frac{\int_a^b y \gamma(x, y) \sqrt{1 + \left(\frac{dy}{dx}\right)^2} \, dx}{\int_a^b \gamma(x, y) \sqrt{1 + \left(\frac{dy}{dx}\right)^2} \, dx},$$

Definition 3.5.12 Suppose that the material line L is given by the equation $y = f(x)$, $x \in [a, b]$, and for each point $X := (x, f(x))$ on L, $\gamma(x, f(x))$ is the point density such that the function $x \mapsto \gamma(X) := \gamma(x, f(x))$ is continuous on $[a, b]$. Then the **mass** of L is defined by

$$M := \int_a^b \gamma(x, y) \sqrt{1 + \left(\frac{dy}{dx}\right)^2} \, dx$$

and the **centre of gravity** of the line L is defined to be at the point (x_0, y_0), where

$$x_0 = \frac{1}{M} \int_a^b x \gamma(x, y) \sqrt{1 + \left(\frac{dy}{dx}\right)^2} \, dx,$$

$$x_0 = \frac{1}{M} \int_a^b y \gamma(x, y) \sqrt{1 + \left(\frac{dy}{dx}\right)^2} \, dx. \qquad \Diamond$$

Example 3.5.17 We find the centre of gravity of the semi-circular arc $x^2 + y^2 = a^2$, $y \geq 0$, assuming that the density of the material is constant. In this case,

$$y = f(x) := \sqrt{a^2 - x^2},$$

so that it follows that

$$\sqrt{1 + \left(\frac{dy}{dx}\right)^2} = \frac{a}{\sqrt{a^2 - x^2}}.$$

Since $\gamma(x, y)$ is constant,

$$x_C = 0 \qquad y_C = \frac{\int_{-a}^a y \sqrt{1 + \left(\frac{dy}{dx}\right)^2} \, dx}{\int_{-a}^a \sqrt{1 + \left(\frac{dy}{dx}\right)^2} \, dx} = \frac{2a}{\pi}. \qquad \Diamond$$

Centre of gravity of a material planar region

Next we consider the centre of gravity of a material planar region Ω bounded by two curves

$$y = f(x), \quad y = g(x) \quad \text{with} \quad f(x) \leq g(x) \quad a \leq x \leq b.$$

Suppose that the density of the material at the point X is $\gamma(X)$. This density is defined as follows: Suppose $M(X, r)$ is the mass of the circular region in Ω with centre at X and radius $r > 0$, and $\alpha(X, r)$ is the area of the same circular region. Then the *density* of the material at the point x is defined by

$$\gamma(X) := \lim_{r \to 0} \frac{M(X, r)}{\alpha(X, r)}.$$

Now, in order to find the *centre of gravity* of Ω, first consider a partition P : $a = x_0 < x_1 < \cdots < x_n = b$ of of the interval $[a, b]$. Let ξ_i be the mid-point of the interval $[x_{i-1}, x_i]$, that is, $\xi_i = \frac{x_{i-1}+x_i}{2}$ for $i = 1, \ldots, n$. Consider the rectangular boxes

$$R_i : \quad x_{i-1} \le x \le x_i, \quad f(\xi_i) \le y \le g(\xi_i)$$

for $i = 1, \ldots, n$. Note that, for each $i \in \{1, \ldots, n\}$,

$$X_i = \left(\xi_i, \frac{f(\xi_i) + g(\xi_i)}{2}\right),$$

is the mid-point of the rectangular box R_i. Of course, if $f(\xi_i) = g(\xi_i)$ for some $i \in \{1, \ldots, n\}$, then the corresponding R_i is a line segment. Assuming that the mass of the rectangular box R_i is concentrated at its mid-point, we can assume that this mass is approximately equal to

$$m_i = \gamma(X_i)[g(\xi_i) - f(\xi_i)]\Delta x_i,$$

where $\gamma(X_i)$ is the density of the material at X_i and $\Delta x_i = x_i - x_{i-1}$ for $i = 1, \ldots, n$. Now, the centre of gravity of the system of material points at $X_1, X_2, \ldots, X_n$ is

$$x_{0,P} := \frac{\sum_{i=1}^n \xi_i m_i}{\sum_{i=1}^n m_i}, \quad y_{0,P} := \frac{\sum_{i=1}^n \frac{f(\xi_i)+g(\xi_i)}{2} m_i}{\sum_{i=1}^n m_i}.$$

Then, the centre of gravity of Ω may be defined as

$$x_0 = \lim_{\mu(P)\to 0} x_{0,P}, \quad y_0 = \lim_{\mu(P)\to 0} y_{0,P}.$$

Denoting

$$\gamma(X_i) = \gamma\left(\xi_i, \frac{f(\xi_i) + g(\xi_i)}{2}\right) \text{ as } \tilde{\gamma}(\xi_i),$$

we have

$$x_{0,P} = \frac{\sum_{i=1}^n \xi_i \tilde{\gamma}(\xi_i)[g(\xi_i) - f(\xi_i)]\Delta x_i}{\sum_{i=1}^n \tilde{\gamma}(\xi_i)[g(\xi_i) - f(\xi_i)]\Delta x_i},$$

$$y_{0,P} = \frac{1}{2} \frac{\sum_{i=1}^n \tilde{\gamma}(\xi_i)[f(\xi_i) + g(\xi_i)][g(\xi_i) - f(\xi_i)]\Delta x_i}{\sum_{i=1}^n \tilde{\gamma}(\xi_i)[g(\xi_i) - f(\xi_i)]\Delta x_i}$$

$$= \frac{1}{2} \frac{\sum_{i=1}^n \tilde{\gamma}(\xi_i)[(g(\xi_i))^2 - (f(\xi_i))^2]\Delta x_i}{\sum_{i=1}^n \tilde{\gamma}(\xi_i)[g(\xi_i) - f(\xi_i)]\Delta x_i}.$$

Thus,

$$x_0 = \lim_{\mu(P) \to 0} x_{0,P} = \frac{\int_a^b x \tilde{\gamma}(x)[g(x) - f(x)] \, dx}{\int_a^b \tilde{\gamma}(x)[g(x) - f(x)] \, dx},$$

$$y_0 = \lim_{\mu(P) \to 0} y_{0,P} = \frac{1}{2} \frac{\int_a^b \tilde{\gamma}(x)[(g(x))^2 - (f(x)^2] \, dx}{\int_a^b \tilde{\gamma}(x)[g(x) - f(x)] \, dx}.$$

$$x_0 = \frac{\int_a^b x \tilde{\gamma}(x)[g(x) - f(x)] \, dx}{\int_a^b \tilde{\gamma}(x)[g(x) - f(x)] \, dx},$$

$$y_0 = \frac{1}{2} \frac{\int_a^b \tilde{\gamma}(x)[(g(x))^2 - (f(x))^2] \, dx}{\int_a^b \tilde{\gamma}(x)[g(x) - f(x)] \, dx}.$$

Example 3.5.18 We find the coordinates of the centre of gravity of the planar region bounded by the parabola $y^2 = a x$ cut off by the straight line $x = a$, assuming that the density to constant.

In this case

$$f(x) = -\sqrt{a x}, \quad g(x) = \sqrt{a x}, \quad 0 \le x \le a.$$

Hence the coordinates of the centre of gravity are

$$x_0 = \frac{\int_0^b x[g(x) - f(x)] \, dx}{\int_0^b [g(x) - f(x)] \, dx} = \frac{2 \int_0^b x \sqrt{a x} \, dx}{\int_0^b 2\sqrt{a x} \, dx} = \frac{3}{5} a.$$

$$y_0 = \frac{1}{2} \frac{\int_0^b [f(x) + g(x)][g(x) - f(x)] \, dx}{\int_0^b [g(x) - f(x)] \, dx} = 0. \qquad \diamond$$

Moment of Inertia

Suppose there are n material points in the plane. Let their masses be $m_1, m_2, \ldots m_n$ respectively. Suppose that these points are at distances $d_1, \ldots, d_n$ from a fixed point O. Then the **moment of inertia (M.I)** of the system of these points with respect to the point O is defined by the quantity:

$$I_O := \sum_{i=1}^{n} d_i^2 m_i.$$

If O is the origin, and $(x_1, y_1), (x_2, y_2), \ldots, (x_n, y_n)$ are the points, then

$$I_O := \sum_{i=1}^{n} (x_i^2 + y_i^2) m_i.$$

M.I. of a material line in the plane

Suppose a curve L is given by the equation $y = f(x)$, $a \le x \le b$. We assume that this curve is a *material line*. Suppose the density of the material at the point $X = (x, y)$ is $\gamma(X)$.

Now, in order to find the moment of inertia of L, we first consider a partition $P : a = x_0 < x_1 < \ldots < x_k = b$, and take points $\xi_i = [x_{i-1}, x_i], i = 1, \ldots, n$. Then we consider the moment of inertia of the system of material points at $(\xi_1, \eta_i), i = 1, \ldots, n$. Here, $\eta_i = f(\xi_i), i = 1, \ldots, n$.

$$I_{O,P} := \sum_{i=1}^{n} (\xi_i^2 + \eta_i^2) m_i.$$

Thus,

$$I_{O,P} := \sum_{i=1}^{n} (\xi_i^2 + \eta_i^2) \gamma_i \Delta s_i.$$

Here, Δs_i is the length of the arcs joining (x_{i-1}, y_{i-1}) to (x_i, y_i), and γ_i is the density at the point (ξ_i, η_i). Note that $\gamma_i \Delta s_i$ is the approximate mass of the arc joining $(x_{i-1}, f(x_{i-1}))$ to $(x_i, f(x_i))$. Now, assuming that the functions $f(x)$ and $\tilde{\gamma}(x) := \gamma(x, f(x))$ are continuous on $[a, b]$, the moment of inertia of L with respect to O is

$$I_O = \lim_{\mu(P) \to 0} I_{O,P} = \lim_{\mu(P) \to 0} \sum_{i=1}^{n} (\xi_i^2 + \eta_i^2) \gamma_i \Delta s_i.$$

Thus,

$$I_O = \int_a^b (x^2 + y^2) \gamma(x, y) \sqrt{1 + \left(\frac{dy}{dx}\right)^2} \, dx.$$

M.I. of a circular arc with respect to the centre

Suppose the given curve is a circular arc: $\rho = a, \alpha \le \theta \le \beta$. Following the arguments in the above paragraph, we compute the moment of inertia using polar coordinates:

The moment of inertia, in this, case is given by

$$I_O := \lim_{\mu(P) \to 0} \sum_{i=1}^{n} d_i^2 m_i,$$

where $d_i = a$, $m_i = \gamma_i a \Delta \theta_i$, for $i = 1, \ldots, n$, with $\gamma(\theta)$ being the point density. Hence,

$$I_O = \lim_{\mu(P) \to 0} \sum_{i=1}^{n} a^2 \gamma_i [a \Delta \theta_i].$$

Thus,

$$I_O = a^3 \int_{\alpha}^{\beta} \gamma(\theta) d\theta.$$

If $\gamma(\theta) = \gamma$, a constant, then

$$I_O = a^3 \int_{\alpha}^{\beta} \gamma(\theta) d\theta = (\beta - \alpha) \gamma a^3.$$

In particular, M.I of the circle $\rho = a$, $0 \leq \theta \leq 2\pi$, is

$$I_O = 2\pi \gamma a^3.$$

M.I. of inertia of a material sector in the plane

The region is $R : 0 \leq \rho \leq a$, $\alpha \leq \theta \leq \beta$ with constant density γ. To find the M.I. of R, we partition it by rays and circular arcs:

$$P : \alpha = \theta_0 < \theta_1 < \theta_2 < \ldots < \theta_n = \beta,$$
$$Q : 0 = \rho_0 < \rho_1 < \rho_2 < \ldots < \rho_m = a.$$

Consider the elementary region obtained by the above partition:

$$R_{ij} : \rho_{j-1} \leq \rho \leq \rho_j, \quad \theta_{i-1} \leq \theta \leq \theta_i.$$

Assume that the mass of this region R_{ij} is concentrated at the point $(\hat{\rho}_j, \hat{\theta}_i)$, where $\hat{\rho}_j \in [\rho_{j-1}, \rho_j]$, $\hat{\theta}_i \in [\theta_{i-1}, \theta_i]$. Then the M.I. of the material point at $(\hat{\rho}_j, \hat{\theta}_i)$ is $m_{ij}d_{ij}^2$ where m_{ij} is the mass of the region R_{ij} which is approximately equal to $[\hat{\rho}_j \Delta\theta_i \Delta\rho_j]\gamma$, and $d_{ij} = \hat{\rho}_j$. Thus the M.I. of the subsector $\theta_{i-1} \le \theta \le \theta_i$ is defined by

$$\lim_{\mu(Q)\to 0} \sum_{j=1}^{n} m_{ij}d_{ij}^2 = \lim_{\mu(Q)\to 0} \sum_{j=1}^{n} [\hat{\rho}_j \Delta\theta_i \Delta\rho_j]\gamma \hat{\rho}_j^2$$

$$= \lim_{\mu(Q)\to 0} \sum_{j=1}^{n} \left(\gamma \hat{\rho}_j^3 \Delta\rho_j\right) \Delta\theta_i$$

$$= \gamma \left(\int_0^a \rho^3 \, d\rho\right) \Delta\theta_i$$

$$= \frac{\gamma a^4}{4} \Delta\theta_i.$$

From this, it follows that, the moment of inertia of the sector $\alpha \le \theta \le \beta$ is given by

$$\text{M.I} = \frac{(\beta - \alpha)\gamma a^4}{4}.$$

In particular, moment of inertia of a circular disc is

$$\frac{\pi \gamma a^4}{2} = \frac{Ma^2}{2},$$

where $M = \pi a^2 \gamma$ is the mass of the disc.

Exercise 3.5.1 If M is the mass of a right circular homogeneous cylinder with base radius a, then show that its moment of inertia is $\frac{Ma^2}{2}$. ◁

3.6 Appendix

Theorem 3.6.1 *A bounded function* $f : [a, b] \to \mathbb{R}$ *is Riemann integrable if and only if there exists* γ *such that for every* $\varepsilon > 0$, *there exists a partition* P *of* $[a, b]$ *satisfying*

$$|S(P, f, T) - \gamma| < \varepsilon$$

for every tag T *on* P, *and in that case* $\gamma = \int_a^b f(x)dx$.

Proof Let $\varepsilon > 0$ be given. Suppose f is Riemann integrable. Then, by Theorem 3.1.3, there exists a partition P of $[a, b]$ such that $U(P, f) - L(P, f) < \varepsilon$. Since $L(P, f) \leq S(P, f, T) \leq U(P, f)$ for every tag T on P, we obtain

$$|S(P, f, T) - \gamma| < \varepsilon \quad \text{for every tag } T \text{ on } P.$$

Conversely, suppose there exists γ such that for every $\varepsilon > 0$, there exists a partition P of $[a, b]$ satisfying $|S(P, f, T) - \gamma| < \varepsilon$ for every tag T on P. So, assume that such a $\gamma \in \mathbb{R}$ exists. For a given $\varepsilon > 0$, let P be a partition of $[a, b]$ such that

$$\gamma - \varepsilon < S(P, f, T) < \gamma + \varepsilon \tag{1}$$

holds. Let

$$T_n' = \{t_{i,n}' : i = 1, \ldots, k\}, \qquad T_n'' = \{t_{i,n}'' : i = 1, \ldots, k\}$$

be tags on $P : a = x_0 < x_1 < \ldots < x_k = b$ such that

$$f(t_{i,n}') \to m_i, \qquad f(t_{i,n}'') \to M_i \quad \text{as} \quad n \to \infty.$$

Then

$$S(P, f, T_n') \to L(P, f), \qquad S(P, f, T_n'') \to U(P, f) \quad \text{as} \quad n \to \infty. \tag{2}$$

Also, from (1), we have

$$\gamma - \varepsilon < S(P, f, T_n') < \gamma + \varepsilon, \qquad \gamma - \varepsilon < S(P, f, T_n'') < \gamma + \varepsilon \quad \forall n \in \mathbb{N}.$$

Now, taking limit and using (2), we have

$$\gamma - \varepsilon < L(P, f) < \gamma + \varepsilon, \qquad \gamma - \varepsilon < U(P, f) < \gamma + \varepsilon.$$

Therefore, $U(P, f) - L(P, f) < \varepsilon$. By Theorem 3.1.3, f is integrable. ∎

As a consequence of the above theorem we have the following.

Corollary 3.6.2 *Suppose $f : [a, b] \to \mathbb{R}$ is a bounded function. If (P_n) is a sequence of partitions on $[a, b]$ such that $\{S(P_n, f, T_n)\}$ converges for every tag T_n on P_n for each $n \in \mathbb{N}$, then f is Riemann integrable, and*

$$\int_a^b f(x)dx = \lim_{n \to \infty} S(P_n, f, T_n).$$

Knowing a sequence (P_n) of partitions, how to assert the convergence of $\{S(P_n, f, T_n)\}$ for every tag T_n of P_n for each $n \in \mathbb{N}$?

In this regard, analogous to Theorem 3.1.16, we have the following result (See [6]) for its proof).

Theorem 3.6.3 *Suppose* $f : [a, b] \to \mathbb{R}$ *is a Riemann integrable (bounded) function. If* (P_n) *is a sequence of partitions on* $[a, b]$ *such that* $\mu(P_n) \to 0$ *as* $n \to \infty$, *then*

$$S(P_n, f, T_n) \to \int_a^b f(x)dx \quad as \quad n \to \infty$$

for every tag T_n *on* P_n *for each* $n \in \mathbb{N}$.

Now, we specify a large class of functions of practical importance which are Riemann integrable.

Theorem 3.6.4 *Suppose* $f : [a, b] \to \mathbb{R}$ *is bounded and piecewise continuous, i.e., there are at most a finite number of points in* $[a, b]$ *at which* f *is discontinuous. Then* f *is integrable.*

Proof Let $\varepsilon > 0$ be given. We have to show that there exists a partition P such that $U(P, f) - L(P, f) < \varepsilon$.

Suppose that $c \in (a, b)$ such that f is continuous on $[a, c)$ and $(c, b]$. Let $\delta > 0$ be such that $c + \delta < b$ and $c - \delta > a$. Let f_1, f_2, f_3 be restrictions of f to the intervals $[a, c - \delta]$, $[c + \delta, b]$ and $[c - \delta, c + \delta]$, respectively. Since f is continuous on $[a, c - \delta]$ and $[c + \delta, b]$, f is integrable on these intervals, so that there exist partitions P_1 on $[a, c - \delta]$ and P_2 on $[c + \delta, b]$ such that

$$U(P_1, f_1) - L(P_1, f_1) < \frac{\varepsilon}{3}, \quad U(P_2, f_2) - L(P_2, f_2) < \frac{\varepsilon}{3}.$$

Now, for any partition P_3 on $[c - \delta, c + \delta]$, we have

$$2m\delta \le L(P_3, f_3) \le U(P_3, f_3) \le 2M\delta,$$

where $M = \sup\{f(x) : x \in [a, b]\}$ and $m = \inf\{f(x) : x \in [a, b]\}$. So, if $\delta > 0$ is so small that $2(M - m)\delta < \varepsilon/3$, we have

$$U(P_3, f_3) - L(P_3, f_3) < \frac{\varepsilon}{3}.$$

Now, for the partition $P = P_1 \cup P_2 \cup P_3$ on $[a, b]$, we have

$$U(P, f) = U(P_1, f_1) + U(P_2, f_2) + U(P_3, f_3),$$

$$L(P, f) = L(P_1, f_1) + L(P_2, f_2) + L(P_3, f_3).$$

Hence,

$$U(P, f) - L(P, f) < \varepsilon.$$

Thus, we have shown that for every $\varepsilon > 0$, there exists a partition P such that $U(P, f) - L(P, f) < \varepsilon$, so that f is integrable.

The case of more than one (but finite number of) points of discontinuity, including discontinuity at the endpoints, can be handled analogously. ∎

3.7 Additional Exercises

1. Let f be a bounded function on $[a, b]$. Prove that, if there is a partition P of $[a, b]$ such that $L(P, f) = U(P, f)$, then f is a constant function.
2. If f and g are bounded functions on $[a, b]$ such that $f \leq g$, then prove that $\int_a^b f(x)dx \leq \int_a^b g(x)dx$.
3. Let $f : [a, b] \to \mathbb{R}$ be a bounded function. If P and Q are partitions of $[a, b]$ such that Q is a refinement of P, then show that

$$L(P, f) \leq L(Q, f) \leq U(Q, f) \leq U(P, f)$$

   for all $n \in \mathbb{N}$.
4. For a bounded function $f : I \to \mathbb{R}$, let $m_f := \inf_{x \in I} f(x)$ and $M_f := \sup_{x \in I} f(x)$. Prove that, if f and g are bounded functions on I, then

$$m_f + m_g \leq m_{f+g} \leq M_{f+g} \leq M_f + M_g.$$

5. Let $f : [a, b] \to \mathbb{R}$ be a bounded function. Show that

$$L(P, f) + L(P, g) \leq L(P, f + g) \leq U(P, f + g) \leq U(P, f) + U(P, g)$$

   for any partition P of $[a, b]$.
6. Suppose $f : [a, b] \to \mathbb{R}$ is Riemann integrable. Justify the following statements:

   (a) For every $\varepsilon > 0$, there exists a partition P of $[a, b]$ such that

$$0 \leq \int_a^b f(x)dx - L(P, f) < \varepsilon.$$

   (b) For every $\varepsilon > 0$, there exists a partition P of $[a, b]$ such that

$$0 \leq U(P, f) - \int_a^b f(x)dx < \varepsilon.$$

(c) For every $\varepsilon > 0$, there exists a partition P of $[a, b]$ such that

$$0 \le \left| S(P, f, T) - \int_a^b f(x)dx \right| < \varepsilon$$

for every tag T on P.

7. If f is either monotonically increasing or monotonically decreasing function on $[a, b]$, then prove that f is integrable. [Hint: Use Corollary 3.1.4.]
8. Prove that every piecewise constant function on $[a, b]$ is integrable.
9. Given a bounded function f and a partition P on $[a, b]$, find piecewise constant functions g and h such that

$$L(P, f) = \int_a^b g(x)dx, \quad U(P, f) = \int_a^b h(x)dx.$$

10. Prof Theorem 3.1.7 and Theorem 3.1.8 using Corollary 3.1.4.
11. Prove that every bounded piecewise continuous function on $[a, b]$ is integrable.
12. Let f be a bounded function on $[a, b]$ such that $|f|$ is integrable. Is it necessary that f is integrable?
13. Let f be a bounded function on $[a, b]$. Prove the following:

    (a) If f is continuous on (a, b), then f is integrable on $[a, b]$.
    (b) If g is a bounded function on (a, b) such that $g(x) = f(x)$ for all $x \in [a, b]$, then g is integrable on $[a, b]$ and $\int_a^b g(x)dx = \int_a^b f(x)dx$.
    (c) If h is a bounded function on $[a, b]$ which is continuous on $[a, c)$ and $(c, b]$ for some $c \in (a, b)$, then h is integrable on $[a, b]$ and $\int_a^b h(x)dx = \int_a^b f(x)dx$.

14. Let f be integrable on $[a, b]$ and $g(x) = \int_a^x f(t)dt$. Show the following:

    (a) If $f(x) \ge 0$ for all $x \in [a, b]$, then g is monotonically increasing.
    (b) If $f(x) \le 0$ for all $x \in [a, b]$, then g is monotonically decreasing.

15. Suppose f is integrable on $[a, b]$. Prove the following.

    (a) $\lim_{t \to b} \int_a^t f(x)dx = \int_a^b f(x)dx$.
    (b) If (a_n) and (b_n) are in $[a, b]$ which converge to a and b, respectively, then $\lim_{n \to \infty} \int_{a_n}^{b_n} f(x)dx = \int_a^b f(x)dx$.

16. Using Newton-Leibnitz formula, evaluate the following integrals.

$$(a) \int_0^{1/\sqrt{2}} \frac{dx}{\sqrt{1 - x^2}}, \quad (b) \int_0^1 \sqrt{3x + 1}dx,$$

$$(c) \int_1^2 (x^3 + x + 1)dx, \quad (d) \int_0^{2\pi/\omega} \cos^2(\omega t)dt, \quad \omega > 0.$$

17. Verify the conclusions in Theorem 3.3.2 for the following functions.

$$(a)\, f(x) = |x|, \quad x \in [-1, 1], \, (b)\, f(x) = \begin{cases} -1, & -1 \le x \le 0, \\ 1, & 0 < x \le 1 \end{cases}$$

$$(c)\, f(x) = \begin{cases} x, & 0 \le x \le 1, \\ 1, & 1 < x \le 2 \end{cases}, \quad (d)\, f(x) = \begin{cases} x, & 0 \le x \le 1, \\ x - 1, & 1 < x \le 2 \end{cases}$$

18. Antiderivative of a Riemann integrable function is Riemann integrable, and difference of any two antiderivatives of a given function is a constant -- Why?

19. Find the following limits by interpreting the sequence as a Riemann sum and then applying Newton-Leibnitz formula.

(a) $\displaystyle\lim_{n \to \infty} \frac{1}{n} [\sin(\pi/n) + \sin(2\pi/n) + \cdots + \sin(n\pi/n)].$

(b) $\displaystyle\lim_{n \to \infty} \left[\frac{1}{n} + \frac{1}{n + 1} + \cdots + \frac{1}{3n} \right].$

(c) $\displaystyle\lim_{n \to \infty} \left[\frac{1^2}{n^3} + \frac{2^2}{n^3} + \cdots + \frac{(n - 1)^2}{n^3} \right].$

(d) $\displaystyle\lim_{n \to \infty} \left[\frac{1}{\sqrt{n^2}} + \frac{1}{\sqrt{n^2 + n}} + \cdots + \frac{1}{\sqrt{n^2 + (n - 1)n}} \right].$

[*Hint:* Theorem 3.1.16 together with Theorem 3.3.1]

20. Let $g : [0, 1] \to \mathbb{R}$ be defined by

$$g(x) = \begin{cases} \sin(1/x), & 0 < x \le 1, \\ 0, & x = 0. \end{cases}$$

Show that g is differentiable at every point in $(0, 1)$, but g' does not have a Riemann integrable extension to $[0, 1]$.

21. Let $f : [0, 1] \to \mathbb{R}$ be defined by $f(x) = \begin{cases} x^2 \sin(1/x), & x \ne 0, \\ 0, & x = 0. \end{cases}$ Show that f is differentiable in $(0, 1)$ and f' has Riemann integrable extension to $[0, 1]$. Find $\int_a^b f'(x)dx$.

22. If f is continuous and bounded on an open interval (a, b), then it has Riemann integrable extension to $[a, b]$. Why?

23. Suppose f is defined on (a, b). If f_1 and f_2 are Riemann integrable extensions of a function f to $[a, b]$, then $\int_a^b f_1(x)dx = \int_a^b f_2(x)dx$. Why?

24. Justify the statement: Given a continuous function f on $[a, b]$ and $c \in \mathbb{R}$, there exists a function g which is continuous on $[a, b]$ and differentiable on (a, b) such that $g' = f$ on (a, b) and $g(a) = c$.

25. State the required properties of the function $f : [a, b] \to \mathbb{R}$ such that $\frac{d}{dx}(\int_a^x f(t)\,dt) = \int_a^x \frac{d}{dt} f(t)\,dt$.

26. Find explicit expressions for the following.

$$(a)\ \frac{d}{dx} \int_1^{e^x} \ln t \, dt \qquad (b)\ \frac{d}{dx} \int_0^{\sin x} \frac{dt}{\sqrt{1-t^2}}$$

$$(a)\ \frac{d}{dx} \int_1^{x^2} \frac{1}{1+t^3} \, dt \quad (b)\ \frac{d}{dx} \int_{x^2}^{x} \sqrt{\sqrt{1+t^2}}$$

27. Evaluate the following integral by using appropriate change of variable, and properties of integrals.

$$(a)\ \int_0^1 \sqrt{1-x^2}\,dx \quad (b)\ \int_0^1 xe^{x^2}\,dx \quad (c)\ \int_0^1 x^3 e^{x^2}\,dx$$

$$(d)\ \int_1^4 \frac{x^2}{\sqrt{1+x^3}}\,dx \quad (e)\ \int_1^4 \frac{\sin\sqrt{x}}{\sqrt{x}}\,dx \quad (f)\ \int_0^1 x\sqrt{1+x^2}\,dx$$

28. Prove the following:

$$(a)\ \int_0^1 x^m (1-x)^n\,dx = \int_0^1 x^n (1-x)^m\,dx \text{ for any } m, n \in \mathbb{N}.$$

$$(b)\ \int_a^b f(x)\,dx = \int_a^b f(a+b-x)\,dx.$$

29. Suppose f is continuous on $[a, b]$ and φ and ψ are differentiable on an open interval J such that $\varphi(x), \psi(x) \in (a, b)$ for all $x \in J$. Prove that the function g, defined by $g(x) = \int_{\psi(x)}^{\varphi(x)} f(t)\,dt$, $x \in J$, is differentiable and $g'(x) = f(\varphi(x))\varphi'(x) - f(\psi(x))\psi'(x)$ for all $x \in J$.

30. Let $f(x) = \int_1^x (1/t)\,dt$ for $x > 0$. Justify the following statements:

(a) $f(x) < 0$ for $0 < x < 1$ and $f(x) > 0$ for $x > 1$,
(b) f is continuous and strictly increasing on $(0, \infty)$,
(c) f is differentiable and $f'(x) = 1/x$ for $x > 0$.

31. Let $f(x) = \int_0^x \frac{dt}{\sqrt{1-t^2}}$ for $-1 < x < 1$. Show that

(a) f is continuous and strictly increasing on $(-1, 1)$,

(b) f is differentiable and $f'(x) = 1/\sqrt{1 - x^2}$ on $(-1, 1)$,

(c) $f(\sin x) = x$ for all $x \in (-1, 1)$.

32. Show that, if f is twice differentiable and f' and f'' are continuous in an open interval containing $[a,]$, then

$$\int_a^b f'(x)dx + \int_a^b xf''(x)dx = [bf'(b) - af'(a)].$$

33. Drive Theorem 3.4.2 using Theorem 3.4.5. [*Hint:* Observe that $\int_a^{\varphi(x)} f(t)dt = \int_a^{\varphi(\alpha)} f(t)dt + \int_{\varphi(\alpha)}^{\varphi(x)} f(t)dt$.]

34. Suppose f and ψ are as in Theorem 3.4.5. Assume that $\psi'(x) \neq 0$ for all $x \in (\alpha, \beta)$. If φ is the inverse of ψ defined on the range of ψ and if φ' exists, then show that

$$\int_\alpha^\beta f(\psi(t))dt = \int_{\psi(\alpha)}^{\psi(\beta)} f(x)\varphi'(x)dx.$$

[*Hint:* Write $f(x) = (f \circ \psi)(\varphi(x))$ and use Theorem 3.4.5.]

35. Use Problem 34 to evaluate the following integrals.

$$(a) \int_1^4 \frac{\sqrt{t}}{1 + \sqrt{t}} dt \quad (b) \int_1^3 \frac{dt}{t\sqrt{1 + t}}$$

Geometric and mechanical applications

36. Find the area of the portion of the circle $x^2 + y^2 = 1$ which lies inside the parabola $y^2 = 1 - x$.

[*Hint:* Area enclosed by the circle in the second and third quadrant and the area enclosed by the parabola in the first and fourth quadrant. The required area is $\frac{\pi}{2} + 2\int_0^1 \sqrt{1 - x} \, dx$. Ans: $\frac{\pi}{2} + \frac{4}{3}$.]

37. Find the area common to the cardioid $\rho = a(1 + \cos\theta)$ and the circle $\rho = \frac{3a}{2}$.

[*Hint:* The points of intersections of the given curves are given by $1 + \cos\theta = \frac{3}{2}$, i.e., for $\theta = \pm\frac{\pi}{3}$. Hence the required area is $2\left[\frac{1}{2}\int_0^{\pi/3}(\frac{3a}{2})^2 \, d\theta + \frac{1}{2}\int_{\pi/3}^{\pi} a^2(1 + \cos\theta)^2 \, d\theta\right]$. Ans: $\frac{7}{4}\pi - \frac{9\sqrt{3}}{8}$.]

38. For $a, b > 0$, find the area included between the parabolas given by $y^2 = 4a(x + a)$ and $y^2 = 4b(b - x)$.

[*Hint*: Points of intersection of the curves is given by $a(x + a) = b(b - x)$, i.e., $x = \frac{b^2 - a^2}{a+b} = b - a$; $y = 2\sqrt{ab}$. The required area is $2 \times \left[\int_{-a}^{b-a} \sqrt{4a(x + a)} + \int_{b-a}^{b} \sqrt{4b(b - x)}\, dx \right]$. *Ans*: $\frac{8}{3}\sqrt{ab}(a + b)$.]

39. Find the area of the loop of the curve $r^2 \cos\theta = a^2 \sin 3\theta$
    [*Hint*: $r = 0$ for $\theta = 0$ and $\theta = \pi/3$, and r is maximum for $\theta = \pi/6$. The area is $\int_0^{\pi/3} \frac{r^2}{2}\, d\theta$.]

40. Find the area of the region bounded by the curves $x - y^3 = 0$ and $x - y = 0$.
    [*Hint*: Points of intersections of the curves are at $x = 0, 1, -1$. The area is $2\int_0^1 (x^{1/3} - x)dx$. *Ans*: 1/2]

41. Find the area of the region that lies inside the circle $r = a\cos\theta$ and outside the cardioid $r = a(1 - \cos\theta)$.
    [*Hint*: Note that the circle is the one with centre at $(0, a/2)$ and radius $a/2$. The curves intersect at $\theta = \pm\pi/3$. The required area is $\int_{-\pi/3}^{\pi/3} (r_1^2 - r_2^2)d\theta$, where $r_1 = a\cos\theta$, $r_2 = a(1 - \cos\theta)$. *Ans*: $\frac{a^3}{3}(3\sqrt{3} - \pi)$]

42. Find the area of the loop of the curve $x = a(1 - t^2)$, $y = at(1 - t^2)$ for $-1 \le t \le 1$.
    [*Hint*: $y = 0$ for $t \in \{-1, 0, 1\}$, and y negative for $-1 \le t \le 0$ and positive for $0 \le t \le 1$. Also, $y^2 = x^2(a - x)/a$ so that the curve is symmetric w.r.t. the x-axis. Area is $2\int_0^a y dx = 2\int_1^0 y(t)x'(t)dt$. *Ans*: $8a^2/15$]

43. Find the length of an arch of the cycloid $x = a(t - \sin t)$, $y = a(1 - \cos t)$.
    [*Hint*: The curve cuts the x-axis at $x = a$ and $x = 2\pi a$ for $t = 0$ and $t = 2\pi$, respectively. $\ell(C) = \int_0^{2\pi} \sqrt{[x'(t)]^2 + [y'(t)]^2}dt$. *Ans*: 8a.]

44. For $a > 0$, find the length of the loop of the curve $3a y^2 = x(x - a)^2$.
    [*Hint*: The curve cuts the x-axis at $x = a$, and the curve is symmetric w.r.t. the x-axis. Thus the required area is $2\int_0^a \sqrt{1 + (\frac{dy}{dx})^2}\, dx$. Since $6ayy' = (x - a)(3x - a)$, $1 + y'^2 = \frac{(3x+a)^2}{12ax}$. *Ans*: $\frac{4a}{\sqrt{3}}$.]

45. Find the length of the curve $r = \frac{2}{1+\cos\theta}$, $0 \le \theta \le \pi/2$.
    [*Hint*: $\ell(C) = 2\int_0^{\pi/4} \sec^3\theta d\theta$. *Ans*: $\sqrt{2} + \ln(\sqrt{2} + 1)$.]

46. Find the volume of the solid obtained by revolving the curve $y = 4\sin 2x$, $0 \le x \le \pi/2$, about y-axis.
    [*Hint*: Writing $y = 4\sin 2x$ for $0 \le x \le \pi/4$ and $y = 4\sin 2u$ for $\pi/4 \le u \le \pi/2$, $V = \int_0^4 (u^2 - x^2)dy = \pi \int_{\pi/4}^{\pi/2} u^2(8\cos 2u)du - \pi \int_0^{\pi/4} x^2(8\cos 2x)dx$. Also, note that the curve is symmetric w.r.t. the line $x = \pi/4$. Hence, the volume is $\pi \int_0^{\pi/4} [(\frac{\pi}{4} - x)^2 - x^2]dy$. *Ans*: $2\pi^2$.]

47. Find the area of the surface obtained by revolving a loop of the curve $9ax^2 = y(3a - y)^2$ about y-axis.
    [*Hint*: $x = 0$ iff $y = 0$ or $y = 3a$. $A = 2\pi \int_0^{3a} x\sqrt{1 + (\frac{dx}{dy})^2}\, dx$. *Ans*: $3\pi a^2$.]

48. Find the area of the surface obtained by revolving about x-axis, an arc of the catenary $y = c\cosh(x/c)$ between $x = -a$ and $x = a$ for $a > 0$.
    [*Hint*: The area is $2\pi \int_{-a}^a y\sqrt{1 + y'^2}\, dx = 2\pi c \int_{-a}^a \cosh^2 \frac{x}{c}\, dx$. *Ans*: $\pi c[2a + c\sinh\frac{2a}{c}]$.]

49. The lemniscate $\rho^2 = a^2 \cos 2\theta$ revolves about the line $\theta = \frac{\pi}{4}$. Find the area of the surface of the solid generated.
    [*Hint*: The required surface area is $4\pi \int_{-\pi/4}^{\pi/4} h\sqrt{\rho^2 + \rho'^2}\, d\theta$, where $h := \rho \sin \left(\frac{\pi}{4} - \theta\right)$, $\rho = a\sqrt{\cos 3\theta}$ so that $\rho^2 + \rho'^2 = \frac{a^2}{\cos 2\theta}$. *Ans*: $4\pi a^2$.]

50. Find the volume of revolution of the cardioid $\rho = a(1 + \cos \theta)$ about the initial line. [*Ans*: $\frac{8}{3}$.]

Chapter 4
Improper Integrals

In Chap. 3, we defined the concept of definite integral, more precisely, the Riemann integral, only for those functions which are bounded and defined on closed and bounded intervals. In this chapter, we extend the notion of the integral in a natural way when some of the above requirements are not satisfied. The resulting integrals are known as improper integrals.

4.1 Definitions

4.1.1 Integrals over Unbounded Intervals

First we consider integrals of functions defined over intervals of the forms $[a, \infty)$, $(-\infty, b]$ and $(-\infty, \infty)$. Recall that the Riemann integral was defined for bounded functions defined on intervals of the form $[a, b]$ where $a, b \in \mathbb{R}$ with $a < b$.

Definition 4.1.1 (Integral over $[a, \infty)$) Suppose f is a real valued function defined on $[a, \infty)$ for some $a \in \mathbb{R}$ and f is integrable on $[a, t]$ for all $t > a$. If

$$\lim_{t \to \infty} \int_a^t f(x)\, dx$$

exists as a real number, then we say that the **improper integral** of f over $[a, \infty)$, denoted by $\int_a^\infty f(x) dx$, exists and it is defined by

M. T. Nair, *Calculus of One Variable*,
https://doi.org/10.1007/978-3-030-88637-0_4

$$\int\limits_{a}^{\infty} f(x)dx = \lim_{t\to\infty} \int\limits_{a}^{t} f(x)\, dx.$$ $\Diamond$

Definition 4.1.2 (Integral over $(-\infty, b]$) Suppose f is a real valued function defined on $(-\infty, b]$ for some $b \in \mathbb{R}$ and f is integrable on $[t, b]$ for all $t < b$, . If

$$\lim_{t\to-\infty} \int\limits_{t}^{b} f(x)\, dx$$

exists as a real number, then we say that the **improper integral** of f over $(-\infty, b]$, denoted by $\int_{-\infty}^{b} f(x)\, dx$, exists and it is defined by

$$\int\limits_{-\infty}^{b} f(x)\, dx = \lim_{t\to-\infty} \int\limits_{t}^{b} f(x)\, dx.$$ $\Diamond$

Suppose $f : \mathbb{R} \to \mathbb{R}$ is such that the improper integrals $\int_{-\infty}^{c} f(x)\, dx$ and $\int_{c}^{\infty} f(x)\, dx$ in the sense of Definitions 4.1.1 and 4.1.2, respectively, exist for some $c \in \mathbb{R}$. Then, it can be shown that (Exercise), for any $d \in \mathbb{R}$, the improper integrals $\int_{-\infty}^{d} f(x)\, dx$ and $\int_{d}^{\infty} f(x)\, dx$ exist in the sense of Definitions 4.1.1 and 4.1.2, respectively, and

$$\int\limits_{-\infty}^{c} f(x)\, dx + \int\limits_{c}^{\infty} f(x)\, dx = \int\limits_{-\infty}^{d} f(x)\, dx + \int\limits_{d}^{\infty} f(x)\, dx.$$

Thus, the sum $\int_{-\infty}^{c} f(x)\, dx + \int_{c}^{\infty} f(x)\, dx$ is independent of the choice of $c \in \mathbb{R}$. In view of this, we have the following definition (Figs. 4.1 and 4.2).

Definition 4.1.3 Suppose $f : \mathbb{R} \to \mathbb{R}$ is such that the improper integrals $\int_{-\infty}^{c} f(x)\, dx$ and $\int_{c}^{\infty} f(x)\, dx$ exist in the sense of Definitions 4.1.1 and 4.1.2, respectively, for some $c \in \mathbb{R}$. Then we say that the **improper integral** of f over $(-\infty, \infty)$, denoted by $\int_{-\infty}^{\infty} f(x)\, dx$, exists and it is defined by

$$\int\limits_{-\infty}^{\infty} f(x)\, dx = \int\limits_{-\infty}^{c} f(x)\, dx + \int\limits_{c}^{\infty} f(x)\, dx.$$ $\Diamond$

Fig. 4.1 Improper integral
over $[a, \infty)$

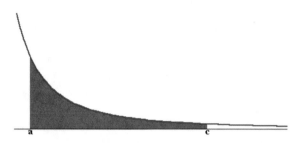

Fig. 4.2 Improper integral
over $(-\infty, a]$

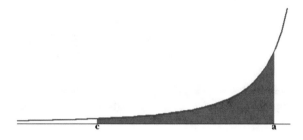

Remark 4.1.1 We may observe that the existence of

$$\lim_{t \to \infty} \int_{-t}^{t} f(x)\,dx$$

does not, in general, imply the existence of $\int_{-\infty}^{\infty} f(x)\,dx$. To see this, consider the
function $f : \mathbb{R} \to \mathbb{R}$ defined by

$$f(x) = x, \quad x \in \mathbb{R}.$$

Then we have $\int_{-t}^{t} f(x)\,dx = 0$ for every $t \in \mathbb{R}$, but the integrals $\int_{-\infty}^{c} f(x)\,dx$ and
$\int_{c}^{\infty} f(x)\,dx$ do not exist for any $c \in \mathbb{R}$.   ◊

$$\lim_{t \to \infty} \int_{-t}^{t} f(x)\,dx \text{ exists} \quad \not\Rightarrow \quad \int_{-\infty}^{\infty} f(x)\,dx$$

Remark 4.1.2 If $\lim\limits_{t \to \infty} \int_{-t}^{t} f(x)\,dx$ exists, then its value is called the *Cauchy principal
value* of the integral of f. This particular integral will be useful while studying
integrals of complex valued functions of a complex variable.   ◊

Next we consider integrals of functions defined over infinite integrals of the form
(a, ∞) and $(-\infty, b)$.

Definition 4.1.4 Suppose f is a real valued function defined on (a, ∞) and the improper integral $\int_t^\infty f(x)dx$ exists in the sense of Definition 4.1.1 for all $t > a$. If

$$\lim_{t \to a^+} \int_t^\infty f(x)\, dx$$

exists, then we say that the **improper integral** of f over (a, ∞), denoted by $\int_a^\infty f(x)\, dx$, exists, and it is defined by

$$\int_a^\infty f(x)\, dx = \lim_{t \to a^+} \int_t^\infty f(x)\, dx. \qquad \qquad \Diamond$$

Definition 4.1.5 Suppose f is a real valued function defined on $(-\infty, b)$ and the improper integral $\int_{-\infty}^t f(x)dx$ exists in the sense of Definition 4.1.2 for all $t < b$. If

$$\lim_{t \to b^-} \int_{-\infty}^t f(x)\, dx$$

exists, then we say that the **improper integral** of f over $(-\infty, b)$, denoted by $\int_{-\infty}^b f(x)\, dx$, exists, and it is defined by

$$\int_{-\infty}^b f(x)\, dx = \lim_{t \to b^-} \int_{-\infty}^t f(x)\, dx. \qquad \qquad \Diamond$$

Remark 4.1.3 In the case of (a, ∞), the function may not be defined at the point a or may be unbounded on $(a, a + \delta)$ for some $\delta > 0$ so that we cannot talk about the Riemann integral over $[a, a + \delta]$ for $\delta > 0$. Analogous remark holds for functions defined on $(-\infty, b)$. $\qquad \qquad \Diamond$

4.1.2 Improper Integrals over Bounded Intervals

Now, we consider improper integral of a function f which is defined on a bounded interval of the form either $(a, b]$ or $[a, b)$ and use them to define for the case of $[a, b] \setminus J_0$, where J_0 is a finite subset of $[a, b]$ (Figs. 4.3 and 4.4).

Fig. 4.3 Improper integral over $(a, b]$

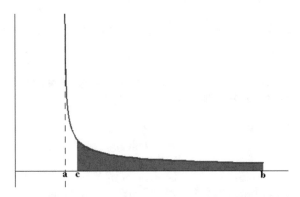

Fig. 4.4 Improper integral over $[a, b)$

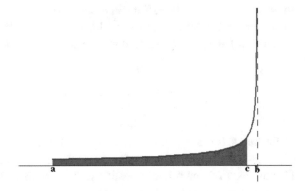

Definition 4.1.6 Suppose $f : (a, b] \to \mathbb{R}$ is such that it is integrable on $[t, b]$ for every $t \in (a, b)$. If

$$\lim_{t \to a^+} \int_t^b f(x) \, dx$$

exists, then we say that the **improper integral of f over** $(a, b]$, denoted by $\int_a^b f(x) \, dx$, exists, and it is defined by

$$\int_a^b f(x) \, dx = \lim_{t \to a^+} \int_t^b f(x) \, dx. \qquad \diamondsuit$$

Definition 4.1.7 Suppose $f : [a, b) \to \mathbb{R}$ is such that it is integrable on $[a, t]$ for every $t \in (a, b)$. If

$$\lim_{t \to b^-} \int_a^t f(x)\, dx$$

exists, then we say that the **improper integral of** f **over** $[a, b)$, denoted by $\int_a^b f(x)\, dx$, exists and it is defined by

$$\int_a^b f(x)\, dx = \lim_{t \to b^-} \int_a^t f(x)\, dx. \qquad \qquad \Diamond$$

Definition 4.1.8 Suppose f is a real valued function defined on the set $[a, c) \cup (c, b]$ for some c with $a < c < b$. If the improper integrals $\int_a^c f(x)\, dx$ and $\int_c^b f(x)\, dx$ exist in the sense of Definitions 4.1.7 and 4.1.6, respectively, then we say that the **improper integral of** f **over** $[a, b]$, denoted by $\int_a^b f(x)\, dx$, exists and it is defined by

$$\int_a^b f(x)\, dx = \int_a^c f(x)\, dx + \int_c^b f(x)\, dx. \qquad \qquad \Diamond$$

Definition 4.1.9 Suppose f is a real valued function defined on (a, b). If for some $c \in (a, b)$, the improper integrals $\int_a^c f(x)\, dx$ and $\int_c^b f(x)\, dx$ exist in the sense of Definitions 4.1.6 and 4.1.7, respectively, then we say that the **improper integral of** f **over** (a, b), denoted by $\int_a^b f(x)\, dx$, exists and it is defined by

$$\int_a^b f(x)\, dx = \int_a^c f(x)\, dx + \int_c^b f(x)\, dx. \qquad \qquad \Diamond$$

More generally we have the following definition.

Definition 4.1.10 Suppose f is defined on $[a, b] \setminus J_0$, where J_0 is a finite subset of $[a, b]$, say $J_0 = \{x_1, \ldots x_k\}$. Suppose f is integrable over every closed and bounded interval contained in $[a, b] \setminus J_0$. Assume further that the improper integrals

$$\int_{x_{i-1}}^{x_i} f(x)dx, \quad i = 1, \ldots, k+1,$$

exist, where $x_0 = a$ and $x_{k+1} = b$. Then we say that the **improper integral of f over** $[a, b]$, denoted by $\int_a^b f(x)\, dx$, exists and it is defined by

$$\int_a^b f(x)\, dx = \sum_{i=1}^{k+1} \int_{x_{i-1}}^{x_i} f(x)\, dx. \qquad \Diamond$$

Terminology: If an improper integral exists, then we also say that the *improper integral converges*; otherwise we say that the *improper integral diverges*.

Remark 4.1.4 In the case of improper integrals over a non-closed bounded interval of the form $(a, b]$, the function may not be defined at the point a or may be unbounded on $(a, a + \delta)$ for some $\delta > 0$. If the function is not defined at a, then we can assign some value to it at a, say $f(a) = \alpha$ for some $\alpha \in \mathbb{R}$. Now, if the redefined function is not bounded in $[a, b]$, then we cannot talk about the Riemann integral over $[a, b]$. Analogous statement holds for the case of improper integrals over $[a, b)$ or (a, b) or $[a, b] \setminus J_0$, where J_0 is a finite set. $\qquad \Diamond$

Remark 4.1.5 Suppose f is defined on a set of the form $[a, b] \setminus J_0$, where J_0 is a finite set. In case f is bounded on $[a, b] \setminus J_0$ and can be extended to $[a, b]$ such that the extended function $\tilde{f}$ is integrable over $[a, b]$, then it can be shown that (Exercise) the improper integral $\int_a^b f(x)\, dx$ exists and

$$\int_a^b f(x)\, dx = \int_a^b \tilde{f}(x)\, dx,$$

where the integral on the right-hand side is the Riemann integral of $\tilde{f}$. $\qquad \Diamond$

Let us illustrate the above remark by an example.

Example 4.1.1 Let $f(x) = \frac{\sin x}{x}$, $0 < x \le 1$. We know that $\tilde{f}$ defined by

$$\tilde{f}(x) = \begin{cases} f(x), & x \ne 0, \\ 1, & x = 0 \end{cases}.$$

is a continuous extension of f to $[0, 1]$, and hence, $\int_0^1 \tilde{f}(x) dx$ exists. Now, the question is whether,

$$\lim_{t \to 0+} \int_t^1 \frac{\sin x}{x}\, dx = \int_0^1 \tilde{f}(x) dx.$$

Since $\tilde{f}(x) = \frac{\sin x}{x}$ for $t \le x \le 1$ for any $t \in (0, 1)$, we have

$$\int_0^1 \tilde{f}(x)dx - \int_t^1 \frac{\sin x}{x}dx = \int_0^t \tilde{f}(x)dx.$$

But, since $|\tilde{f}(t)| \le 1$ for all $t \in [0, 1]$,

$$\left| \int_0^t \tilde{f}(x)dx \right| \le t \to 0 \quad \text{as} \quad t \to 0.$$

Thus, we have shown that

$$\lim_{t \to 0+} \int_t^1 \frac{\sin x}{x}dx = \int_0^1 \tilde{f}(x)dx.$$

Thus, the improper integral $\int_0^1 \frac{\sin x}{x}dx$ exists and

$$\int_0^1 \frac{\sin x}{x}dx = \int_t^1 \tilde{f}(x)dx. \qquad \qquad \Diamond$$

4.1.3   Typical Examples

Example 4.1.2   Consider the improper integral $\int_1^\infty \frac{1}{x} dx$. Note that

$$\int_1^t \frac{1}{x} dx = [\ln x]_1^t = \ln t \to \infty \quad \text{as} \quad t \to \infty.$$

Hence, $\int_1^\infty \frac{1}{x} dx$ diverges. $\qquad \qquad \Diamond$

Example 4.1.3   Consider the improper integral $\int_1^\infty \frac{1}{x^2} dx$. Note that

$$\int_1^t \frac{1}{x^2} dx = \left[-\frac{1}{x} \right]_1^t = 1 - \frac{1}{t} \to 1 \quad \text{as} \quad t \to \infty.$$

Hence, $\int_1^\infty \frac{1}{x^2}\,dx$ converges and

$$\int_1^\infty \frac{1}{x^2}\,dx = 1. \qquad \diamond$$

Example 4.1.4 For $p \neq 1$, consider the improper integral $\int_1^\infty \frac{1}{x^p}\,dx$. In this case, we have

$$\int_1^t \frac{1}{x^p}\,dx = \left[\frac{x^{-p+1}}{-p+1}\right]_1^t = \frac{t^{-p+1} - 1}{-p+1}.$$

Note that,

$$p > 1 \quad \Rightarrow \quad \frac{t^{-p+1} - 1}{-p+1} \to \frac{1}{p-1} \quad \text{as} \quad t \to \infty,$$

and

$$p < 1 \quad \Rightarrow \quad \frac{t^{-p+1} - 1}{-p+1} \to \infty \quad \text{as} \quad t \to \infty,$$

The above observations combined with Example 4.1.2 show that

$$\int_1^\infty \frac{1}{x^p}\,dx \quad \begin{cases} \text{converges for } p > 1, \\ \text{diverges for } p \leq 1, \end{cases}$$

and

$$\int_1^\infty \frac{1}{x^p}\,dx = \frac{1}{p-1} \quad \text{for} \quad p > 1. \qquad \diamond$$

$$\int_1^\infty \frac{1}{x^p}\,dx \text{ exsits iff } p > 1 \text{ and in that case } \int_1^\infty \frac{1}{x^p}\,dx = \frac{1}{p-1}$$

Example 4.1.5 Note that

$$\int_0^t e^{-x}\,dx = \left[-e^{-x}\right]_0^t = 1 - e^{-t} \to 1 \quad \text{as} \quad t \to \infty.$$

Hence, $\int_0^\infty e^{-x}dx$ converges and

$$\int\limits_0^\infty e^{-x}dx = 1. \qquad \Diamond$$

Example 4.1.6 The integral $\int_0^\infty \dfrac{1}{1+x^2}dx$ converges: Note that for $t > 0$,

$$\int\limits_0^t \frac{1}{1+x^2}dx = \left[\tan^{-1}(x)\right]_0^t = \tan^{-1}(t) - \tan^{-1}(0) = \tan^{-1}(t).$$

Since $\tan^{-1}(t) \to \pi/2$ as $t \to \infty$ we obtain

$$\int\limits_1^\infty \frac{1}{1+x^2}dx = \lim_{t\to\infty} \int\limits_1^t \frac{1}{1+x^2}dx = \frac{\pi}{2}. \qquad \Diamond$$

Example 4.1.7 The integral $\int_{-\infty}^\infty \frac{1}{1+x^2}dx$ converges: Note that for $t > 0$.

$$\int\limits_{-t}^0 \frac{1}{1+x^2}dx = \int\limits_0^t \frac{1}{1+x^2}dx.$$

Hence, as in last example,

$$\lim_{t\to\infty} \int\limits_{-t}^0 \frac{1}{1+x^2}dx = \lim_{t\to\infty} \int\limits_0^t \frac{1}{1+x^2}dx = \frac{\pi}{2}.$$

Thus, both $\int_{-\infty}^0 \frac{1}{1+x^2}dx$ and $\int_0^\infty \frac{1}{1+x^2}dx$ exist, and each of them is equal to $\pi/2$. Hence,

$$\int\limits_{-\infty}^\infty \frac{1}{1+x^2}dx = \int\limits_{-\infty}^0 \frac{1}{1+x^2}dx + \int\limits_0^\infty \frac{1}{1+x^2}dx = \pi. \qquad \Diamond$$

Example 4.1.8 **(i)** The integral $\int_0^1 \frac{1}{x} \, dx$ is improper as the function $f(x) = 1/x$, $0 < x \le 1$ is unbounded. We note, for $0 < t < 1$,

$$\int_t^1 \frac{1}{x} \, dx = [\log x]_t^1 = \log 1 - \log t$$

$$= -\log t = \log\left(\frac{1}{t}\right).$$

Thus $\lim_{t \to 0^+} \int_t^1 \frac{1}{x} dx = \infty$, so that $\int_0^1 \frac{1}{x} \, dx$ diverges.

Since $\int_a^b \frac{1}{x} \, dx$ exists as a Riemann integral for any $a > 0$, $b > a$, we can conclude that $\int_0^b \frac{1}{x} \, dx$ diverges for every $b > 0$.

(ii) Clearly, for $p \le 0$, the integral $\int_0^1 \frac{1}{x^p} \, dx$ exists as a Riemann integral. Next, consider the case $p > 0$. In this case, $\int_0^1 \frac{1}{x^p} \, dx$ is an improper integral, as the function $f(x) = 1/x^p$, $0 < x \le 1$ is unbounded. We observe that, for $0 < t < 1$,

$$\int_t^1 \frac{1}{x^p} \, dx = \left[\frac{x^{-p+1}}{-p+1}\right]_t^1 = \frac{1 - t^{-p+1}}{-p+1}.$$

Note that,

$$p > 1 \quad \Rightarrow \quad \frac{1 - t^{-p+1}}{-p+1} = \frac{t^{-p+1} - 1}{p-1} \to \infty \quad \text{as} \quad t \to 0,$$

and

$$p < 1 \quad \Rightarrow \quad \frac{1 - t^{-p+1}}{-p+1} = \frac{1 - t^{1-p}}{1-p} \to \frac{1}{1-p} \quad \text{as} \quad t \to 0.$$

These observations combined with (i) shows that

$$\int_0^1 \frac{1}{x^p} \, dx \quad \begin{cases} \text{converges for } p < 1, \\ \text{diverges for } p \ge 1, \end{cases}$$

and

$$\int_0^1 \frac{1}{x^p} \, dx = \frac{1}{1-p} \quad \text{for} \quad p < 1.$$

Since $\int_a^b \frac{1}{x^p} \, dx$ exists as a Riemann integral for any $a > 0$, $b > a$ and $p \neq 0$, we can conclude that, for every $b > 0$, $\int_0^b \frac{1}{x^p} \, dx$ converges for $p < 1$ and diverges for $p > 1$. ◊

$$\int_0^1 \frac{1}{x^p} \, dx \text{ exists iff } p < 1 \text{ and in that case } \int_0^1 \frac{1}{x^p} \, dx = \frac{1}{1-p}$$

Example 4.1.9 Let $a < b$ and $p > 0$. We shall discuss the convergence of the improper integrals

$$\int_a^b \frac{dx}{(b-x)^p} \quad \text{and} \quad \int_a^b \frac{dx}{(x-a)^p}.$$

(i) We observe that

$$\lim_{t \to b-} \int_a^t \frac{dx}{(b-x)^p} \text{ exists} \iff \lim_{t \to b-} \int_{b-t}^{b-a} \frac{ds}{s^p} \text{ exists}$$

$$\iff \lim_{\varepsilon \to 0} \int_\varepsilon^{b-a} \frac{ds}{s^p} \text{ exists}$$

We know that $\lim_{\varepsilon \to 0} \int_\varepsilon^{b-a} \frac{ds}{s^p}$ does not exist for $p = 1$. Thus, $\int_a^b \frac{dx}{(b-x)^p}$ diverges if $p = 1$. Next, let $p \neq 1$ and $0 < \varepsilon < b - a$. Then, we have

$$\int_\varepsilon^{b-a} \frac{ds}{s^p} = \left[\frac{s^{1-p}}{1-p} \right]_\varepsilon^{b-a} = \frac{(b-a)^{1-p}}{1-p} - \frac{\varepsilon^{1-p}}{1-p}.$$

Since $\lim \varepsilon \to 0 \, \varepsilon^{1-p}/(1-p)$ exists as a real number if and only if $p < 1$,

$$\lim_{t \to b-} \int_a^t \frac{dx}{(b-x)^p} \text{ exists} \iff p < 1,$$

and in that case

$$\int_a^b \frac{dx}{(b-x)^p} = \frac{(b-a)^{1-p}}{1-p}.$$

(ii) Modifying the arguments above, it can also be seen that, for $p > 0$, the improper integral $\int_a^b \frac{dx}{(x-a)^p}$ converges if and only if $p < 1$. ◊

Remark 4.1.6 For the computation of improper integrals, whenever they exist, we essentially used the fundamental theorem of integration over intervals over which the integral exist as Riemann integrals and then took limits. This motivates us in considering the following procedure:

Suppose f is defined on an open interval (a, b), where a can be $-\infty$ and b can be ∞. Suppose f is integrable on every closed and bounded interval contained in (a, b) and there exists a function g which is differentiable on (a, b) and

$$g'(x) = f(x) \quad \text{for all} \quad x \in (a, b).$$

Assume further that

$$\beta := \lim_{x \to b-} g(x) \quad \text{and} \quad \alpha := \lim_{x \to a+} g(x) \quad \text{exist.}$$

Then, it can be shown that the improper integral $\int_a^b f(x)dx$ is given by

$$\int_a^b f(x)dx = \beta - \alpha.$$

For example, to compute

$$\int_a^b \frac{dx}{(b-x)^p} \quad \text{for} \quad p < 1,$$

first note that

$$\int \frac{dx}{(b-x)^p} = g(x) := -\frac{(b-x)^{-p+1}}{-p+1}.$$

Since $g(x) \to \beta := 0$ as $x \to b^{-1}$ and $g(x) \to \alpha := g(a)$ as $x \to a^+$, we obtain

$$\int_a^b \frac{dx}{(b-x)^p} = \beta - \alpha = \frac{(b-a)^{1-p}}{1-p} \quad \text{for} \quad p < 1. \qquad \Diamond$$

Exercise 4.1.1 Suppose $f : [a, \infty) \to \mathbb{R}$ be such that it is integrable on $[a, b]$ for every $b > a$. Show that $\int_a^\infty f9x)dx$ converges if and only if for $\varepsilon > 0$, there exists $\beta_0 > a$ such that $|\int_\alpha^\beta f(x)\,dx| < \varepsilon$ for all $\alpha, \beta \geq \beta_0$. ◁

4.2 Tests for Integrability

4.2.1 Integrability by Comparison

We state a result which will be useful in asserting the existence of certain improper integrals by comparing it with certain other improper integrals.

Notation 4.2.1 Suppose J is an interval of either finite or infinite length. Suppose f is defined on J, except possibly at a finite number of points in J. We denote the improper integral of f over J by

$$\int_J f(x)\, dx. \qquad \Diamond$$

For example, if $J = [a, b]$, then f may not be defined at a or at b or at some point $c \in (a, b)$, and then the corresponding improper integrals, by definition, are

$$\lim_{t \to a} \int_t^b f(x)\, dx, \quad \lim_{t \to b} \int_a^t f(x)\, dx,$$

$$\lim_{t \to c^-} \int_a^t f(x)\, dx + \lim_{t \to c^+} \int_t^b f(x)\, dx,$$

respectively.

We omit the proof of the following two theorems as they can be proved using the corresponding results for definite integrals and then taking limits.

Theorem 4.2.1 (Comparison test) *Let J be an interval, f and g be defined on $J \setminus J_0$, where J_0 is a finite set. If $0 \le f(x) \le g(x)$ for all $x \in J \setminus J_0$ and if $\int_J g(x)\, dx$ exists, then $\int_J f(x)\, dx$ exists and*

$$\int_J f(x) dx \le \int_J g(x) dx.$$

Theorem 4.2.2 *Let J be an interval and f be defined on $J \setminus J_0$, where J_0 is a finite set. If $\int_J |f(x)|\, dx$ exists, then $\int_J f(x)\, dx$ exists. and*

$$\left| \int_J f(x)\, dx \right| \le \int_J |f(x)|\, dx.$$

Definition 4.2.1 Let J be an interval and f be defined on $J \setminus J_0$, where J_0 is a finite set. Then f is said to be **absolutely integrable** over J, if $\int_J |f(x)|\, dx$ is convergent. $\quad \Diamond$

Example 4.2.1 We have already seen that the improper integral $\int_0^\infty \frac{1}{1+x^2} dx$ converges. This can also be seen as follows:

We know that the Riemann integral $\int_0^1 \frac{1}{1+x^2} dx$ exists. Also, since

$$\frac{1}{1+x^2} \le \frac{1}{x^2} \quad \text{for all} \quad x \ge 1,$$

and since $\int_1^\infty \frac{1}{x^2} dx$ converges, by Theorem 4.2.1, $\int_1^\infty \frac{1}{1+x^2} dx$ also converges. Similarly, we can assert the convergence of the improper integrals $\int_{-\infty}^0 \frac{1}{1+x^2} dx$ and $\int_{-\infty}^\infty \frac{1}{1+x^2} dx$. ◇

Example 4.2.2 Let us check the convergence of

$$\int_1^\infty \frac{dx}{1+e^x}.$$

We have already seen that $\int_1^\infty \frac{dx}{x^2}$ converges and its value is 1. Now, for $x \ge 1$ we have

$$\frac{1}{1+e^x} \le \frac{2}{x^2}$$

so that by Theorem 4.2.1 $\int_1^\infty \frac{dx}{1+e^x}$ also converges. ◇

Example 4.2.3 Let $p > 1$. Since

$$\left| \frac{\sin x}{x^p} \right| \le \frac{1}{x^p}, \left| \frac{\cos x}{x^p} \right| \le \frac{1}{x^p}$$

for $x > 0$, it follows from Example 4.1.4 and Theorem 4.2.2 that the improper integrals

$$\int_1^\infty \frac{\sin x}{x^p} dx \quad \text{and} \quad \int_1^\infty \frac{\cos x}{x^p} dx$$

converge for all $p > 1$. ◇

In fact $\int_1^\infty \frac{\sin x}{x^p} dx$ and $\int_1^\infty \frac{\cos x}{x^p} dx$ converge for all $p > 0$ as we see in the next example.

Example 4.2.4 Let $p > 0$. Then for $t > 0$,

$$\int_1^t \frac{\sin x}{x^p} \, dx = \left[\frac{1}{x^p}(-\cos x) \right]_1^t - p \int_1^t \frac{1}{x^{p+1}} \cos x \, dx$$

$$= \left[\cos 1 - \frac{\cos t}{t^p} \right] - p \int_1^t \frac{\cos x}{x^{p+1}} \, dx.$$

By the result in Example 4.2.3, $\int_1^\infty \frac{\cos x}{x^{p+1}} \, dx$ converges for all $p > 0$. Also, $\frac{\cos t}{t^p} \to 0$ as $t \to \infty$. Hence,

$$\int_1^\infty \frac{\sin x}{x^p} \, dx \quad \text{converges for all} \quad p > 0.$$

Similarly, we see that

$$\int_1^\infty \frac{\cos x}{x^p} \, dx \quad \text{converges for all} \quad p > 0. \qquad \Diamond$$

Example 4.2.5 Clearly, for $p \le 0$, $\int_0^1 \frac{\sin x}{x^p} \, dx$ and $\int_0^1 \frac{\cos x}{x^p} \, dx$ exist as Riemann integrals. So, let $p > 0$. Then the relations

$$\left| \frac{\sin x}{x^p} \right| = \left| \frac{\sin x}{x} \right| \frac{1}{x^{p-1}} \le \frac{1}{x^{p-1}}, \quad \left| \frac{\cos x}{x^p} \right| \le \frac{1}{x^p}$$

and the results in Example 4.1.8 and Theorem 4.2.1 imply that

$$\int_0^1 \frac{\sin x}{x^p} \, dx \quad \text{converges for all} \quad p < 2$$

and

$$\int_0^1 \frac{\cos x}{x^p} \, dx \quad \text{converges for all} \quad p < 1. \qquad \Diamond$$

$$\int_1^\infty \frac{\sin x}{x^p}\, dx \text{ converges for all } p > 0$$

$$\int_1^\infty \frac{\cos x}{x^p}\, dx \text{ converges for all } p > 0$$

$$\int_0^\infty \frac{\sin x}{x^p}\, dx \text{ converges for } 0 < p < 2$$

$$\int_0^\infty \frac{\cos x}{x^p}\, dx \text{ converges for } 0 < p < 1$$

Example 4.2.6 Since $\frac{\sin x}{x}$ is decreasing on $(0, 1]$,

$$\frac{\sin x}{x^p} = \frac{\sin x}{x} \frac{1}{x^{p-1}} \ge \frac{\sin 1}{x^{p-1}} \quad \forall x \in (0, 1].$$

Also, since $\int_0^1 \frac{1}{x^{p-1}}\, dx$ diverges for $p - 1 \ge 1$, i.e., for $p \ge 2$, by comparison test (Theorem 4.2.1), we have

$$\int_0^1 \frac{\sin x}{x^p}\, dx \quad \text{diverges for all} \quad p \ge 2. \qquad \diamond$$

$$\int_0^1 \frac{\sin x}{x^p}\, dx \text{ converges if and only if } p < 2$$

4.2.2 Integral Test for Series of Numbers

Recall that, in Example 1.2.9, we have proved the convergence of the series $\sum_{n=1}^{\infty} \frac{1}{n^p}$ for $p > 1$ by using the fact that the sequence (a_n) with

$$a_n = \int_1^n \frac{dx}{x^p}, \quad n \in \mathbb{N},$$

converges. Now, we obtain a general procedure for showing the convergence or divergence of a series of positive terms by using the convergence or divergence of an appropriate improper integral. First we observe the following result (Exercise):

Suppose f is defined on $[a, \infty)$ with values in $[0, \infty)$. Then $\int_a^\infty f(x)dx$ converges if and only if $\lim_{n \to \infty} \int_a^n f(x)dx$ exists.

Theorem 4.2.3 (Integral test) *Let f be a continuous positive and decreasing function defined on $[1, \infty)$. Then*

$$\sum_{n=1}^{\infty} f(n) \ \ converges \ \ \Longleftrightarrow \ \ \int_1^{\infty} f(x)\, dx \ \ converges.$$

Proof We observe that, for each $k \in \mathbb{N}$,

$$k \le x \le k+1 \ \Rightarrow \ f(k+1) \le f(x) \le f(k)$$

$$\Rightarrow \ f(k+1) \le \int_k^{k+1} f(x)dx \le f(k).$$

Hence,

$$\sum_{k=1}^{n} f(k+1) \le \sum_{k=1}^{n} \int_k^{k+1} f(x)dx \le \sum_{k=1}^{n} f(k),$$

i.e.,

$$\sum_{k=1}^{n} f(k+1) \le \int_1^{n+1} f(x)dx \le \sum_{k=1}^{n} f(k),$$

Hence, by comparison test for sequences, and the statement preceding the theorem we obtain, the series $\sum_{n=1}^{\infty} f(n)$ converges if and only if the improper integral $\int_1^{\infty} f(x)\, dx$ converges. ∎

Remark 4.2.1 From the above theorem it follows that if f is a continuous positive and decreasing function defined on $[1, \infty)$ such that $\int_1^{\infty} f(x)\, dx$ converges, then $f(n) \to 0$ as $n \to \infty$. In fact, under the above assumption, it can also shown that

$$f(x) \to \infty \quad \text{as } x \to \infty.$$

Can we conclude the same if we drop the assumption that f is a decreasing? The answer is in the negative. To see this consider the continuous function $f : [1, \infty) \to \mathbb{R}$ defined by

$$f(x) = \begin{cases} 1 - n^2|x - 2n| & \text{if } |x - 2n| \le 1/n^2, \ n \in N, \\ 0 & \text{otherwise.} \end{cases}$$

We observe that $|x - 2n| \leq 1/n^2$ if and only if $2n - 1/n^2 \leq x \leq 2n + 1/n^2$ so that

$$\int_1^\infty f(x)\,dx = \sum_{n=1}^\infty \int_{2n-1/n^2}^{2n+1/n^2} f(x)\,dx = \sum_{n=1}^\infty \frac{1}{n^2}.$$

Thus, $\int_1^\infty f(x)\,dx$ converges, but $f(x) \not\to 0$ as $x \to \infty$, since $f(2n) = 1$ for all $n \in \mathbb{N}$.   ◇

Exercise 4.2.1 Let $f : [1, \infty) \to \mathbb{R}$ be defined by

$$f(x) = \begin{cases} n - n^4|x - 2n| & \text{if } |x - 2n| \leq 1/n^3, \ n \in N, \\ 0 & \text{otherwise.} \end{cases}$$

Show that f is a continuous function such that $\int_1^\infty f(x)\,dx$ converges and $f(x) \to \infty$ as $x \to \infty$.   ◁

4.2.3 Integrability Using Limits

Now we prove a result which facilitates the assertion of convergence and divergence of improper integrals.

Theorem 4.2.4 *Suppose*

1. $f(x) \geq 0$, $g(x) > 0$ *for all* $x \in [a, \infty)$,
2. $\int_a^b f(x)dx$ *and* $\int_a^b g(x)dx$ *exist for every* $b > a$,
3. $\frac{f(x)}{g(x)} \to \ell$ *as* $x \to \infty$ *for some* $\ell \geq 0$.

Then the following hold.

(i) *If* $\ell > 0$, *then* $\int_a^\infty f(x)dx$ *converges* $\Longleftrightarrow$ $\int_a^\infty g(x)dx$ *converges.*
(ii) *If* $\ell = 0$ *and* $\int_a^\infty g(x)dx$ *converges, then* $\int_a^\infty f(x)dx$ *converges.*

Further, if $\frac{f(x)}{g(x)} \to \infty$ *as* $x \to \infty$ *and* $\int_a^\infty f(x)dx$ *converges, then* $\int_a^\infty g(x)dx$ *converges.*

Proof Suppose $\dfrac{f(x)}{g(x)} \to \ell$ as $x \to \infty$ for some $\ell \geq 0$.
   (i) Suppose $\ell \neq 0$. Then $\ell > 0$, and for $\varepsilon > 0$ with $\ell - \varepsilon > 0$, there exists $x_0 \geq a$ such that

$$\ell - \varepsilon < \frac{f(x)}{g(x)} < \ell + \varepsilon \quad \forall x \geq x_0.$$

Hence

$$(\ell - \varepsilon)g(x) < f(x) < (\ell + \varepsilon)g(x) \quad \forall x \geq x_0.$$

Consequently, $\int_{x_0}^{\infty} f(x)dx$ converges iff $\int_{x_0}^{\infty} g(x)dx$ converges. Also, $\int_a^{x_0} f(x)dx$ and $\int_a^{x_0} g(x)dx$ exist, by assumption.

(ii) Suppose $\ell = 0$. Then for $\varepsilon > 0$, there exists $x_0 \geq a$ such that

$$\frac{f(x)}{g(x)} < \varepsilon \quad \forall x \geq x_0.$$

Thus, $f(x) < \varepsilon g(x)$ for all $x \geq x_0$. Hence, convergence of $\int_{x_0}^{\infty} g(x)dx$ implies the convergence of $\int_{x_0}^{\infty} f(x)dx$.

Next, suppose that $\dfrac{f(x)}{g(x)} \to \infty$ as $x \to \infty$. Then for any $M > 0$, there exists $x_0 \geq a$ such that

$$\frac{f(x)}{g(x)} \geq M \quad \forall x \geq x_0.$$

Hence

$$0 \leq g(x) \leq \frac{1}{M} f(x) \quad \forall x \geq x_0.$$

Consequently, if $\int_{x_0}^{\infty} f(x)dx$ converges, then $\int_{x_0}^{\infty} g(x)dx$ converges. As $\int_a^{x_0} f(x)dx$ and $\int_a^{x_0} g(x)dx$ exist, the proof is over. ∎

Example 4.2.7 The integral

$$\int\limits_0^{\infty} \frac{1+x}{1+x^3} dx \quad \text{converges:}$$

To see this, first note that

$$\frac{(1+x)(1+x^2)}{1+x^3} = \frac{1+x+x^2+x^3}{1+x^3} \to 1 \quad \text{as} \quad x \to \infty.$$

Hence, by Theorem 4.2.4,

$$\int\limits_0^{\infty} \frac{1+x}{1+x^3} dx \quad \text{converges} \iff \int\limits_0^{\infty} \frac{1}{1+x^2} dx \quad \text{converges.}$$

We have already seen that $\int_0^{\infty} \frac{1}{1+x^2} dx$ converges. Hence, $\int_0^{\infty} \frac{1+x}{1+x^3} dx$ converges. ◊

Exercise 4.2.2 Suppose f and g are non-negative continuous functions on J. Then, prove that

$$\int\limits_J f(x)dx \text{ exists} \iff \int\limits_J g(x)dx \text{ exists}$$

in the following cases:

1. $J = (a, b]$ and $\lim\limits_{x \to a} \frac{f(x)}{g(x)} = \ell$ and $\ell > 0$.
2. $J = [a, b)$ and $\lim\limits_{x \to b} \frac{f(x)}{g(x)} = \ell$ and $\ell > 0$.
3. $J = [a, \infty)$ and $\lim\limits_{x \to \infty} \frac{f(x)}{g(x)} = \ell$ and $\ell > 0$.
4. $J = (-\infty, b]$ and $\lim\limits_{x \to -\infty} \frac{f(x)}{g(x)} = \ell$ and $\ell > 0$.

In 1–4 above, prove that, if $\ell = 0$ and $\int_J g(x)dx$ exists, then $\int_J f(x)dx$ exists.   ◁

4.3   Gamma and Beta Functions

Gamma and beta functions are certain improper integrals which appear in many applications.

4.3.1   Gamma Function

We show that for $x > 0$, the improper integral

$$\Gamma(x) := \int_0^\infty t^{x-1} e^{-t} \, dt$$

converges. The function $\Gamma(x)$, $x > 0$, is called the **gamma function**.

Let $x > 0$. Note that

$$t^{x-1} e^{-t} \leq t^{x-1} \quad \forall t > 0$$

and by the result in Example 4.1.8(ii), the improper integral $\int_0^1 t^{x-1} \, dt$ converges. Hence, by comparison test (Theorem 4.2.1),

$$\int_0^1 t^{x-1} e^{-t} \, dt \quad \text{converges.}$$

Also, we observe that

$$\frac{t^{x-1} e^{-t}}{t^{-2}} \to 0 \quad \text{as} \quad t \to \infty,$$

and by the result in Example 4.1.3, the integral $\int_1^\infty t^{-2} dt$ converges. Hence, by Theorem 4.2.4, $\int_1^\infty t^{x-1} e^{-t} \, dt$ converges. Thus,

$$\Gamma(x) := \int_0^\infty t^{x-1} e^{-t}\, dt = \int_0^1 t^{x-1} e^{-t}\, dt + \int_1^\infty t^{x-1} e^{-t}\, dt$$

converges for every $x > 0$. Recall from Example 4.1.5 that $\int_0^\infty e^{-t} dt = 1$. Hence, we have

$$\Gamma(1) = 1. \tag{1}$$

Also, using integration by parts, we obtain

$$\Gamma(x+1) = x\Gamma(x) \quad \forall x > 1. \tag{2}$$

Indeed, for $x > 1$, we have

$$\int t^x e^{-t}\, dt = -t^x e^{-t} + \int x t^{x-1} e^{-t} dt.$$

Since $t^x e^{-t} \to 0$ as $t \to 0$ and also as $t \to \infty$, we obtain

$$\Gamma(x+1) = \int_0^\infty t^x e^{-t}\, dt = x \int_0^\infty t^{x-1} e^{-t} dt = x\Gamma(x).$$

Combining (1) and (2), we obtain

$$\Gamma(n+1) = n! \quad \forall n \in \mathbb{N}.$$

We may also observe that

$$\Gamma(1/2) = \int_0^\infty t^{-1/2} e^{-t}\, dt.$$

Using the change of variable, $y = t^{1/2}$, we have $dy = \frac{1}{2t^{1/2}} dt$, it can be seen that

$$\int_0^\infty t^{-1/2} e^{-t}\, dt = 2 \int_0^\infty e^{-y^2}\, dy.$$

It is known (can be proved using the method of calculus of two variables) that

$$\int_0^\infty e^{-x^2} dx = \frac{\sqrt{\pi}}{2}.$$

Hence, we have

$$\Gamma(1/2) = \sqrt{\pi}.$$

This, together with (2) gives

$$\Gamma\left(\frac{2n+1}{2}\right) = \frac{1 \times 3 \times \cdots \times (2n-1)}{2^n}\sqrt{\pi}.$$

4.3.2 Beta Function

We show that for $x > 0$, $y > 0$, the improper integral

$$\beta(x, y) := \int_0^1 t^{x-1}(1-t)^{y-1}\, dt$$

converges. The function $\beta(x, y)$ for $x > 0$, $y > 0$ is called the **beta function**.

Clearly, the above integral is *proper* for $x \geq 1$ and $y \geq 1$. In order to treat the remaining cases, we consider the integrals

$$\int_0^{1/2} t^{x-1}(1-t)^{y-1}\, dt, \quad \int_{1/2}^1 t^{x-1}(1-t)^{y-1}\, dt.$$

We note that if $0 < t \leq 1/2$, then $(1-t)^{y-1} \leq 2^{1-y}$ so that

$$t^{x-1}(1-t)^{y-1} \leq t^{x-1}2^{1-y} \leq 2t^{x-1} \quad \text{for} \ \ 0 < t < \frac{1}{2}.$$

Now, $\int_0^{1/2} t^{x-1}\, dt$ converges, by the result in Example 4.1.8(ii). Hence, by comparison test (Theorem 4.2.1), $\int_0^{1/2} t^{x-1}(1-t)^{y-1}\, dt$ converges.

To show the convergence of the second integral, consider the change of variable $s = 1 - t$. Then

$$\int_{1/2}^1 t^{x-1}(1-t)^{y-1}\, dt = \int_0^{1/2} s^{y-1}(1-s)^{x-1}\, ds.$$

We have shown that the integral $\int_0^{1/2} s^{y-1}(1-s)^{x-1}\, ds$ converges. Thus,

$$\beta(x, y) := \int_0^1 t^{x-1}(1-t)^{1-y}\, dt, x > 0,\ y > 0$$

converges for every $x > 0,\ y > 0$.

Beta function and gamma function are related by the relation,

$$\beta(x, y) = \frac{\Gamma(x)\Gamma(y)}{\Gamma(x+y)}$$

the proof of which is too much involved (see, e.g., [6]).

4.4 Additional Exercises

1. Justify the following statements:
   (a) If $\int_a^\infty f(x)\, dx$ exists, then $\int_c^\infty f(x)\, dx$ exists for any $c > a$.
   (b) If $\int_{-\infty}^b f(x)\, dx$ exists, then $\int_{-\infty}^c f(x)\, dx$ exists for any $c < b$.
   (c) If f is defined on $(-\infty, \infty)$ and integrable on every closed and bounded interval, and if $\int_{-\infty}^c f(x)\, dx$ and $\int_c^\infty f(x)\, dx$ exist for some $c \in \mathbb{R}$, then $\int_{-\infty}^d f(x)\, dx$ and $\int_d^\infty f(x)\, dx$ exist for any $d \in \mathbb{R}$.

2. Suppose $f : \mathbb{R} \to \mathbb{R}$ is an even function, that is, $f(-x) = f(x)$ for every $x \in \mathbb{R}$. Prove that $\int_{-\infty}^\infty f(x)dx$ exists if and only if $\int_0^\infty f(x)dx$ exists, and in that case

$$\int_{-\infty}^\infty f(x)dx = 2\int_0^\infty f(x)dx.$$

3. Suppose $f : \mathbb{R} \to \mathbb{R}$ is an odd function, that is, $f(-x) = -f(x)$ for every $x \in \mathbb{R}$. Prove that $\int_{-\infty}^0 f(x)dx$ exists if and only if $\int_0^\infty f(x)dx$ exists, and in that case

$$\int_{-\infty}^\infty f(x)dx = 0.$$

4. Justify the following statements:
   (a) Suppose f is defined on $[a, \infty)$ with values in $[0, \infty)$. Then $\int_a^\infty f(x)dx$ converges if and only if $\lim_{n\to\infty} \int_a^n f(x)dx$ exists.
   (b) Suppose f is defined on $(-\infty, b]$ with values in $[0, \infty)$. Then $\int_{-\infty}^b f(x)dx$ converges if and only if $\lim_{n\to\infty} \int_{-n}^b f(x)dx$ exists.

5. Justify the following statements:

   (a) Suppose f is defined on $[a, \infty)$. Then $\int_a^\infty f(x)dx$ converges if and only if for any sequence (b_n) in $\mathbb{R}$, $b_n \to \infty$ implies $\lim_{n \to \infty} \int_a^{b_n} f(x)dx$ exists.

   (b) Suppose f is defined in $(-\infty, b]$. Then $\int_{-\infty}^b f(x)dx$ converges if and only if for any sequence (a_n) in $\mathbb{R}$, $a_n \to -\infty$ implies $\lim_{n \to \infty} \int_{a_n}^b f(x)dx$ exists.

6. Suppose f is absolutely integrable over $(-\infty, \infty)$. Show that

$$\int_{-\infty}^{\infty} f(x)dx = \lim_{t \to \infty} \int_{-t}^{t} f(x)dx.$$

7. Suppose f is defined on $(a, b]$ or $[a, b)$ and it has an integrable extension $\tilde{f}$ to $[a, b]$. Prove that the improper integral $\int_a^b f(x)\,dx$ exists and $\int_a^b f(x)\,dx = \int_a^b \tilde{f}(x)\,dx$.

8. For $p < 1$, let $f_p(x) := \frac{\sin x}{x^p}$, $0 < x \le 1$. Show (without using the previous problem) that f_p has a continuous extension $\tilde{f}_p$ to $[0, 1]$ and the improper integral $\int_0^1 f_p(x)dx$ is equal to $\int_0^1 \tilde{f}_p(x)dx$.

9. Justify: $\int_1^2 \frac{x^2-1}{x-1}dx = \int_1^2 (x + 1)dx$.

10. Suppose $f \ge 0$ on $[a, b)$ and the integral $\int_a^t f(x)dx$ exists for every $t \in [a, b)$. If $\lim_{x \to b}(b - x)^\alpha f(x)$ converges for some $\alpha < 1$, then show that $\int_a^b f(x)dx$ also converges.

   [*Hint:* Observe that for any $\varepsilon > 0$, there exists $x_0 \in [a, b)$ such that the number $\beta := \lim_{x \to b}(b - x)^\alpha f(x)$ satisfies $0 \le f(x) \le \frac{\beta+\varepsilon}{(b-x)^\alpha}$ for all $x \in [x_0, b)$.]

11. For $a < b$ and $p < 1$, show that the improper integral $\int_a^b \frac{dx}{(x-a)^p}$ converges.

12. Show that the improper integral $\int_1^\infty \frac{\cos x}{x^p}\,dx$ converges for all $p > 0$.

13. Show that, for $p > 1$, $\int_0^1 \frac{dx}{x^{1/p}} = \frac{p}{p-1}$.

14. Show that, for $p > 1$, $\int_0^1 \frac{dx}{x^{1/p}} - \int_1^\infty \frac{dx}{x^{1/p}} = 1$.

15. Show that $\int_1^\infty \frac{dx}{x^p} \to 0$ as $p \to \infty$ and $\int_0^1 \frac{dx}{x^p} \to 1$ as $p \to 0$.

16. Show that $\lim_{t \to 1-} \int_0^1 \frac{dx}{x^p} = \infty$.

17. Does $\int_1^\infty \sin\left(\frac{1}{x^2}\right) dx$ converge? [*Hint:* Note that $\left|\sin\left(\frac{1}{x^2}\right)\right| \le \frac{1}{x^2}$.]

18. Does $\int_2^\infty \frac{\cos x}{x(\log x)^2}dx$ converge?

   [*Hint:* Observe $\left|\frac{\cos x}{x(\log x)^2}\right| \le \frac{1}{x(\log x)^2}$ and use the change of variable $t = \log x$.]

19. Does $\int_0^\infty \frac{\sin^2 x}{x^2}dx$ converge?

   [*Hint:* Observe $\frac{\sin^2 x}{x^2} \le \frac{1}{x^2}$ for $x \ge 1$ and $\frac{\sin^2 x}{x^2}$, $0 < x \le 1$ has a continuous extension on $[0, 1]$.]

20. Does $\int_0^1 \frac{\sin x}{x^2} dx$ converge?

   [*Hint*: Observe $\frac{\sin x}{x^2} = \left(\frac{\sin x}{x}\right) \frac{1}{x} \geq \frac{\sin 1}{1}$.]

21. Suppose $f \geq 0$ on $[a, b)$ and the integral $\int_a^t f(x)dx$ exists for every $t \in [a, b)$. If $\lim_{x \to b} (b-x)^\alpha f(x)$ converges for some $\alpha < 1$, then show that $\int_a^b f(x)dx$ also converges.

   [*Hint*: Observe that for any $\varepsilon > 0$, there exists $x_0 \in [a, b)$ such that the number $\beta := \lim_{x \to b} (b-x)^\alpha f(x)$ satisfies $0 \leq f(x) \leq \frac{\beta + \varepsilon}{(b-x)^\alpha}$ for all $x \in [x_0, b)$.]

22. Does $\displaystyle\int_{a_0}^\infty f(x)dx$ exists implies $\int_a^b f(x)dx \to 0$ as $a, b \to \infty$.

   [*Hint*: Observe: $\int_a^b f(x)dx = \int_{a_0}^b f(x)dx - \int_{a_0}^a f(x)dx$ for $a_0 > 0$ and take limits.]

23. Does $\int_0^\infty e^{-x^2} dx$ converge?

   [*Hint*: $e^{-x^2} \leq \frac{1}{x^2}$ for $1 \leq x \leq \infty$.]

24. Does $\int_2^\infty \frac{\sin(\log x)}{x} dx$ converge?

   [*Hint*: Change of variable $t = \log x$, and the fact that $\int_{\log 2}^\infty \sin t \, dt$ diverges.]

25. Does $\int_0^1 \ln x dx$ converge?

   [*Hint*: Change of variable $t = \log x$.]

Chapter 5
Sequence and Series of Functions

Suppose we have real valued functions $f_1, f_2, \ldots$ defined on an interval I. Then, for each $x \in I$, we have a sequence $(f_n(x))$ of real numbers. Suppose $(f_n(x))$ converges at each $x \in I$ to say, $f(x)$. Some of the natural questions that one would like to ask are the following:

(i) If each f_n is continuous, then is f continuous?

(ii) If each f_n is differentiable on $(a, b) \subseteq I$, then is f differentiable on (a, b), and if so, is it true that $f'(x) = \lim_{n \to \infty} f_n'(x)$ for each $x \in (a, b)$?

(iii) If each f_n is Riemann integrable on $[a, b] \subseteq I$, then is f Riemann integrable on $[a, b]$, and if so, is it true that $\int_a^b f(x)\,dx = \lim_{n \to \infty} \int_a^b f_n(x)\,dx$?

If $\sum_{n=1}^\infty f_n(x)$ converges for each $x \in I$ to $g(x) = \sum_{n=1}^\infty f_j(x)$ and if $g_n(x) = \sum_{j=1}^n f_j(x)$ for $n \in \mathbb{N}$ and for each $x \in I$, then questions (i), (ii), (iii) above can be asked about g_n and g in place of f_n and f, respectively.

The purpose of this chapter is to discuss the above questions and many other related issues. We shall also deal with the convergence of a special case of the series of functions, namely the power series.

5.1 Sequence of Functions

5.1.1 Pointwise Convergence and Uniform Convergence

Analogous to the concepts of sequence and series of numbers, we have the concepts of sequence and series of functions.

Definition 5.1.1 For each $n \in \mathbb{N}$, let f_n be a (real valued) function defined on an interval I. Then, we say that (f_n) is a **sequence of functions** defined on I. $\diamond$

© The Author(s), under exclusive license to Springer Nature Switzerland AG 2021
M. T. Nair, *Calculus of One Variable*,
https://doi.org/10.1007/978-3-030-88637-0_5

Definition 5.1.2 Let (f_n) be a sequence of functions defined on an interval I. We say that

(1) (f_n) **converges at a point** $x_0 \in I$, if the sequence $(f_n(x_0))$ of real numbers converges;
(2) (f_n) **converges pointwise** on I, if it converges at every $x \in I$.   ◊

Suppose (f_n) converges pointwise on I. Then, we can define a function $f : I \to \mathbb{R}$ by

$$f(x) = \lim_{n \to \infty} f_n(x), \quad x \in I.$$

Definition 5.1.3 Suppose (f_n) converges pointwise on I. Then, the function f defined by

$$f(x) = \lim_{n \to \infty} f_n(x), \quad x \in I,$$

is called the **pointwise limit** of (f_n), and we write this fact as

$$f_n \to f \quad \text{pointwise on} \quad I.$$ ◊

Exercise 5.1.1 Let (f_n) be a sequence of functions defined on an interval I such that $f_n \to f$ and $f_n \to g$ pointwise for some functions f and g on I. Then $f = g$.   ◁

We may observe that, for a given sequence (f_n) of functions defined on an interval I,

$f_n \to f$ pointwise on I if and only if for every $\varepsilon > 0$, and for each $x \in I$, there exists $N \in \mathbb{N}$
(in general, depends on both ε and x) such that $|f_n(x) - f(x)| < \varepsilon$ for all $n \geq N$.

For any given $\varepsilon > 0$, if we are able to find an $N \in \mathbb{N}$ which does not vary as x varies over I such that $|f_n(x) - f(x)| < \varepsilon$ for all $n \geq N$ and for all $x \in I$, then we say that (f_n) *converges uniformly* to f on I (Figs. 5.1 and 5.2).

More precisely, we have the following definition.

Definition 5.1.4 Let (f_n) be a sequence of functions defined on an interval I. We say that (f_n) **converges to** f **uniformly on** I if for every $\varepsilon > 0$ there exists $N \in \mathbb{N}$ such that

Fig. 5.1 Graph of f_n lies in a strip containing the graph of f

$f_n(x)$
$f(x)$

Fig. 5.2 $|f_n(x) - f(x)| < \varepsilon$
for all $x \in I$

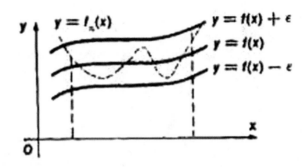

$$|f_n(x) - f(x)| < \varepsilon \qquad \forall n \geq N \quad \text{and} \quad \forall x \in I,$$

and in that case we write

$$f_n \to f \quad \text{uniformly on} \quad I. \qquad \qquad \Diamond$$

Remark 5.1.1 It is to be remarked that the Definitions 5.1.1–5.1.4 will also make sense if we take the domains of definition of the functions as any subset of $\mathbb{R}$ instead of an interval I. However, for the sake of simplicity of presentation and also for illustrating with examples, we have taken the domains of definition of the functions as intervals. $\qquad \Diamond$

Remark 5.1.2 We observe from the definition that if (f_n) converges uniformly to f on I, then (f_n) converges uniformly to f on every subinterval $I_0 \subseteq I$. $\qquad \Diamond$

Exercise 5.1.2 Verify the statement in the above remark. $\qquad \qquad \triangleleft$

It is obvious that uniform convergence implies pointwise convergence. However, the converse is not true. To see this, look at the following example.

Example 5.1.1 For each $n \in \mathbb{N}$, let

$$f_n(x) = x^n, \quad x \in (0, 1).$$

We have $f_n(x) \to 0$ for every $x \in (0, 1)$. Thus, (f_n) converges pointwise to the zero function on $(0, 1)$. However, this convergence is not uniform. To see this, let $\varepsilon > 0$ be given. If the convergence is uniform, then there exists $N \in \mathbb{N}$ such that $|x^n| < \varepsilon$ for all $x \in (0, 1)$ and for all $n \geq N$; in particular, $|x^N| < \varepsilon$ for all $x \in (0, 1)$, which is not possible if we had chosen $\varepsilon < 1$. This is illustrated in Fig. 1.1 below. $\qquad \Diamond$

In the next example we give two uniformly convergent sequence of functions (Fig. 5.3).

Fig. 5.3 $f_n(x) = x^n$,
$0 < x < 1$

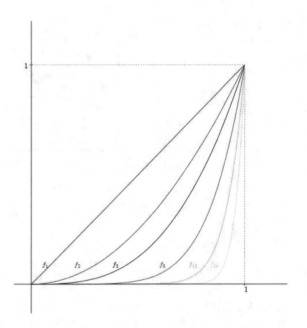

Example 5.1.2 (i) For each $n \in \mathbb{N}$, let

$$f_n(x) = x^n, \quad 0 \le x \le a$$

for some fixed $a \in (0, 1)$. We show that (f_n) converges to the zero function uniformly on $[0, a]$: Note that

$$|f_n(x)| = x^n \le a^n \quad \forall n \in \mathbb{N}, \ \forall x \in [0, a].$$

Let $\varepsilon > 0$ be given. Since $0 < a < 1$, there exists $N \in \mathbb{N}$ such that $0 \le a^n < \varepsilon$ for all $n \ge N$. Thus, we have proved that

$$|f_n(x)| < \varepsilon \quad \forall n \ge N, \ \forall x \in [0, a],$$

showing that (f_n) converges to the zero function uniformly.
  (ii) For each $n \in \mathbb{N}$, let

$$f_n(x) = \frac{\sin(nx)}{n}, \quad x \in \mathbb{R}.$$

Then we have

$$|f_n(x)| \le \frac{1}{n} \quad \forall n \in \mathbb{N}, \ \forall x \in \mathbb{R}.$$

Since $1/n \to 0$, it can be seen, as in the case of (i) above, that (f_n) converges to the zero function uniformly.   ◊

In the next theorem, we give a sufficient condition and also a necessary condition for uniform convergence of functions. Although, its proof is very easy, we supply the details also.

Theorem 5.1.1 *Let (f_n) be a sequence of functions defined on an interval I, and let f be also a function on I.*

(i) If there exists a sequence (α_n) of positive real numbers such that

$$|f_n(x) - f(x)| \leq \alpha_n \quad \forall n \in \mathbb{N}, \quad \forall x \in I,$$

with $\alpha_n \to 0$, then (f_n) converges uniformly to f.
(ii) If (f_n) converges uniformly to f on I, then for every sequence (x_n) in I, $|f_n(x_n) - f(x_n)| \to 0$.

Proof (i) Suppose there exists a sequence (α_n) of positive real numbers such that $\alpha_n \to 0$ and

$$|f_n(x) - f(x)| \leq \alpha_n \quad \forall n \in \mathbb{N}, \quad \forall x \in I.$$

Then for every $\varepsilon > 0$, there exists $N \in \mathbb{N}$ such that $\alpha_n < \varepsilon$ for all $n \geq N$. Hence,

$$|f_n(x) - f(x)| < \varepsilon \quad \forall n \geq N, \quad \forall x \in I.$$

(ii) Suppose (f_n) converges uniformly to f on I. Then for every $\varepsilon > 0$, there exists $N \in \mathbb{N}$ such that

$$|f_n(x) - f(x)| < \varepsilon \quad \forall n \geq N, \quad \forall x \in I.$$

In particular, for any sequence (x_n) in I, we have

$$|f_n(x_n) - f(x_n)| < \varepsilon \quad \forall n \geq N.$$

Thus, $|f_n(x_n) - f(x_n)| \to 0$.   ∎

In order to prove (f_n) does not converge uniformly to f on I, it is enough to find a sequence (x_n) in I such that $|f_n(x_n) - f(x_n)| \nrightarrow 0$

Let (f_n) be a sequence of functions defined on I, and let $f : I \to \mathbb{R}$. By Theorem 5.1.1 (i),

$$\sup_{x \in I} |f_n(x) - f(x)| \to 0 \quad \Rightarrow \quad f_n \to f \text{ uniformly} .$$

In fact, the converse is also true.

Theorem 5.1.2 *Let (f_n) be a sequence of functions defined on I, and let $f : I \to \mathbb{R}$. Then*

$$\sup_{x \in I} |f_n(x) - f(x)| \to 0 \iff f_n \to f \text{ uniformly .}$$

Proof We have already observed that if $\sup_{x \in I} |f_n(x) - f(x)| \to 0$, then $f_n \to f$ uniformly. Conversely, suppose $f_n \to f$ uniformly. Let $\varepsilon > 0$ be given. Then we know that there exists $N \in \mathbb{N}$ such that

$$|f_n(x) - f(x)| < \varepsilon \quad \forall n \geq N, \forall x \in I.$$

In particular,

$$\sup_{x \in I} |f_n(x) - f(x)| < \varepsilon \quad \forall n \geq N.$$

Hence, $\sup_{x \in I} |f_n(x) - f(x)| \to 0$. ∎

(f_n) converges uniformly to f $\iff$ $\sup_{x \in I} |f_n(x) - f(x)| \to 0$.

Let us use Theorem 5.1.1 to verify uniform convergence and non-uniform convergence of some sequences.

Example 5.1.3 For each $n \in \mathbb{N}$, let

$$f_n(x) = \frac{2nx}{1 + n^4 x^2}, \quad x \in [0, 1].$$

Using the relation $a^2 + b^2 \geq 2ab$ for any real numbers a and b, we have

$$0 \leq f_n(x) = \frac{2nx}{1 + n^4 x^2} \leq \frac{1}{n}.$$

Thus, by Theorem 5.1.1 (i), $f_n \to 0$ uniformly. ◇

Example 5.1.4 For each $n \in \mathbb{N}$, let

$$f_n(x) = \frac{1}{n^3} \log(1 + n^4 x^2), \quad x \in [0, 1].$$

Then we have

$$0 \leq f_n(x) \leq \frac{1}{n^3} \log(1 + n^4) = g(n) \quad \forall n \in \mathbb{N},$$

where $g(t) := \frac{1}{t^3} \log(1 + t^4)$ for $t > 0$. Note that, by L'Hospital's rule, we have

$$\lim_{t \to \infty} g(t) = \lim_{t \to \infty} \frac{4t^3}{3t^2(1 + t^4)} = 0.$$

In particular, $\lim_{n\to\infty} g(n) = 0$. Hence, by Theorem 5.1.1 (i), (f_n) converges uniformly to the zero function.   ◊

Example 5.1.5 For each $n \in \mathbb{N}$, let

$$f_n(x) = x^n, \quad x \in (0, 1).$$

Taking f as the zero function on $(0, 1)$, we have seen in Example 5.1.1 that $f_n \to f$ pointwise, but not uniformly. Now, the non-uniform convergence of (f_n) also follows by Theorem 5.1.1 (ii), by taking $x_n = n/(n+1)$ for $n \in \mathbb{N}$, since

$$|f_n(x_n) - f(x_n)| = f_n(x_n) = \left(\frac{n}{n+1}\right)^n \to \frac{1}{e}. \qquad ◊$$

Example 5.1.6 For each $n \in \mathbb{N}$, let

$$f_n(x) = \frac{nx}{1 + n^2 x^2}, \quad x \in [0, 1].$$

Note that $f_n(0) = 0$, and for $x \neq 0$, $f_n(x) \to 0$ as $n \to \infty$. Hence, (f_n) converges pointwise to the zero function. Since $f_n(1/n) = 1/2$ for all $n \in \mathbb{N}$, it follows from Theorem 5.1.1 (ii) that (f_n) does not converge uniformly on $[0, 1]$.   ◊

Example 5.1.7 For each $n \in \mathbb{N}$, let

$$f_n(x) = \tan^{-1}(nx), \quad x \in \mathbb{R}.$$

Note that $f_n(0) = 0$, and for $x \neq 0$, $f_n(x) \to f(x)$ as $n \to \infty$, where

$$f(x) = \begin{cases} \pi/2, & x > 0, \\ 0, & x = 0 \\ -\pi/2, & x < 0. \end{cases}$$

Hence, (f_n) converges pointwise to f. However, it does not converge uniformly to f on any interval containing 0. To see this, let I be an interval containing 0 and let $x_n = 1/n$ for $n \in \mathbb{N}$. Then we have

$$f_n(x_n) = \pi/4 \quad \text{and} \quad f(x_n) = \pi/2 \quad \forall n \in \mathbb{N}.$$

Hence, by Theorem 5.1.1 (ii), (f_n) does not converge uniformly.   ◊

Exercise 5.1.3 Show the non-uniform convergence of the sequence (f_n) in Example 5.1.7 by using $\varepsilon - N$ arguments.   ◁

Example 5.1.8 Let $\{r_1, r_2, \ldots\}$ be an enumeration of rational numbers in $[0, 1]$. For each $n \in \mathbb{N}$, let $f_n : [0, 1] \to \mathbb{R}$ be defined by

$$f_n(x) = \begin{cases} 0, & x \in \{r_1, \ldots, r_n\}, \\ 1, & x \notin \{r_1, \ldots, r_n\}. \end{cases}$$

We observe that for each $x \in [0, 1]$,

$$f_n(x) \to f(x) := \begin{cases} 0, & x \text{ rational}, \\ 1, & x \text{ irrational} \end{cases}$$

Thus, $f_n \to f$ pointwise. Now, note that $f_n(r_{n+1}) = 1$ and $f(r_{n+1}) = 0$ for all $n \in \mathbb{N}$ so that $|f_n(r_{n+1}) - f(r_{n+1})| \nrightarrow 0$. Hence, by Theorem 5.1.1 (ii), (f_n) does not converge to f uniformly.   ◊

Example 5.1.9 For each $n \in \mathbb{N}$, let

$$f_n(x) = \frac{1}{nx}, \quad x \in (0, 1).$$

Clearly, $f_n(x) \to 0$ for every $x \in (0, 1]$. Thus, taking $f = 0$, the zero function, $f_n \to f$ pointwise. But, (f_N) does not converge uniformly. This follows from Theorem 5.1.1 (ii), since $f_n(1/n) = 1$ for all $n \in \mathbb{N}$.   ◊

We observe that the sequence (f_n) of functions in Example 5.1.9 is such that f is bounded, whereas every f_n is unbounded. Such situation never arise for a uniformly convergent sequence (f_n) of functions, as Theorem 5.1.3 below shows. Before stating the theorem we introduce the following definition.

Definition 5.1.5 A sequence (f_n) of functions is said to be **eventually uniformly bounded** on I if there exists $M > 0$ and $N \in \mathbb{N}$ such that $|f_n(x)| \leq M$ for all $x \in I$ and for all $n \geq N$.   ◊

Theorem 5.1.3 *Suppose* $f_n \to f$ *uniformly on* I. *Then,* f *is bounded if and only if* (f_n) *is eventually uniformly bounded.*

Proof Let $\varepsilon > 0$ be given. Since $f_n \to f$ uniformly on I, there exists $N \in \mathbb{N}$ such that $|f_n(x) - f(x)| < \varepsilon$ for all $x \in I$ and $n \geq N$.

Now, suppose f is bounded on I. Then there exists $M > 0$ such that $|f(x)| \leq M$ for all $x \in I$. Hence,

$$|f_n(x)| \leq |f_n(x) - f(x)| + |f(x)| < \varepsilon + M \quad \forall x \in I, \ \forall n \geq N,$$

and thus, (f_n) is eventually uniformly bounded.

Conversely, suppose (f_n) is eventually uniformly bounded on I. Then there exists $M > 0$ and $N \in \mathbb{N}$ such that $|f_n(x)| \leq M$ for all $x \in I$ and for all $n \geq N$. Hence,

$$|f(x)| \le |f(x) - f_n(x)| + |f_n(x)| \le \varepsilon + M \quad \forall x \in I, \ \forall n \ge N,$$

and thus, f is bounded on I. ∎

Before closing this subsection, let us observe the following simple properties of the uniform convergence.

Theorem 5.1.4 *Let (f_n) and (g_n) be sequences of functions defined on I. If $f_n \to f$ and $g_n \to g$ uniformly for some functions f and g defined on I, then*

$$f_n + g_n \to f + g \quad uniformly.$$

Proof Note that for every $n \in \mathbb{N}$ and for all $x \in I$,

$$|(f_n(x) + g_n(x)) - (f(x) + g(x))| \le |f_n(x) - f(x)| + |g_n(x) - g(x)|. \quad (*)$$

Now, let $\varepsilon > 0$ be given. Since $f_n \to f$ and $g_n \to g$ uniformly, there exist $N_1, N_2 \in \mathbb{N}$ such that

$$|f_n(x) - f(x)| < \varepsilon \quad \forall x \in I, \ \forall n \ge N_1,$$

$$|g_n(x) - g(x)| < \varepsilon \quad \forall x \in I, \ \forall \ge N_2.$$

Hence, from $(*)$, we obtain

$$|(f_n(x) + g_n(x)) - (f(x) + g(x))| \le |f_n(x) - f(x)| + |g_n(x) - g(x)|$$
$$< 2\varepsilon$$

for all $n \ge N := \max\{N_1, N_2\}$ and for all $x \in I$. Thus, $f_n + g_n \to f + g$ uniformly. ∎

5.1.2 Uniform Convergence and Continuity

We have seen in Examples 5.1.5 and 5.1.7 that the limit function f is not continuous, although every f_n is continuous. The following theorem shows that such a situation will not arise if the convergence is uniform.

Theorem 5.1.5 *Suppose (f_n) is a sequence of continuous functions defined on an interval I such that it converges uniformly to a function f. Then f is continuous on I.*

Proof Suppose $x_0 \in I$. Then for any $x \in I$ and for any $n \in \mathbb{N}$,

$$|f(x) - f(x_0)| \le |f(x) - f_n(x)| + |f_n(x) - f_n(x_0)| + |f_n(x_0) - f(x_0)|. \quad (*)$$

Let $\varepsilon > 0$ be given. Since (f_n) converges to f uniformly, there exists $N \in \mathbb{N}$ such that

$$|f_n(x) - f(x)| < \varepsilon \quad \forall n \geq N, \ \forall x \in I.$$

Since f_N is continuous, there exists $\delta > 0$ such that

$$|f_N(x) - f_N(x_0)| < \varepsilon \quad \text{whenever} \quad |x - x_0| < \delta.$$

Hence from $(*)$, we have

$$|f(x) - f(x_0)| \leq |f(x) - f_N(x)| + |f_N(x) - f_N(x_0)| + |f_N(x_0) - f(x_0)|$$
$$< 3\varepsilon$$

whenever $|x - x_0| < \delta$. Thus, f is continuous at x_0. This is true for all $x_0 \in I$. Hence, f is a continuous function on I.   ∎

> If (f_n) is a sequence of continuous functions defined on an interval I which converges pointwise to a discontinuous function f on I, then (f_n) does not converge uniformly to f

5.1.3   Uniform Convergence and Integration

Suppose $f_n \to f$ pointwise on an interval I. If each f_n is Riemann integrable on $[a, b] \subseteq I$, then can we say that f is integrable? If f is integrable, do we have the convergence

$$\int_a^b f_n(x)\mathrm{d}x \to \int_a^b f(x)\mathrm{d}x?$$

The answer to the above question need not be in the affirmative as the following examples show.

Example 5.1.10 Let (f_n) and f be as in Example 5.1.8. That is, if $\{r_1, r_2, \ldots\}$ is an enumeration of rational numbers in $[0, 1]$, then for each $n \in \mathbb{N}$, $f_n : [0, 1] \to \mathbb{R}$ is defined by

$$f_n(x) = \begin{cases} 0, \ x \in \{r_1, \ldots, r_n\}, \\ 1, \ x \notin \{r_1, \ldots, r_n\}. \end{cases}$$

Then, for each $n \in \mathbb{N}$, f_n is continuous except at the points $r_1, \ldots, r_n$ so that f_n is integrable. Note that (f_n) converges pointwise to $f : [0, 1] \to \mathbb{R}$ defined by

$$f(x) := \begin{cases} 0, \ x \quad \text{rational}, \\ 1, \ x \quad \text{irrational}, \end{cases}$$

$\int_0^1 f_n(x)dx = 1$ for all $n \in \mathbb{N}$, but f is not integrable.   ◊

Example 5.1.11 For each $n \in \mathbb{N}$, let

$$f_n(x) = nx(1 - x^2)^n, \qquad 0 \le x \le 1.$$

Then we see that

$$\lim_{n \to \infty} f_n(x) = 0 \quad \forall x \in [0, 1].$$

Indeed, $f_n(0) = 0 = f_n(1)$ and for each $x \in (0, 1)$,

$$\frac{f_{n+1}(x)}{f_n(x)} = (1 - x^2)\left(\frac{n+1}{n}\right) \to (1 - x^2) \quad \text{as} \quad n \to \infty.$$

Since $(1 - x^2) < 1$ for $x \in (0, 1)$, we obtain $\lim_{n \to \infty} f_n(x) = 0$ for every $x \in [0, 1]$.
We note that,

$$\int_0^1 f_n(x)dx = \frac{n}{2n + 2} \to \frac{1}{2} \quad \text{as} \quad n \to \infty.$$

Thus, the limit of the integrals is not the integral of the limit.   ◊

We observe that in the above two examples, the convergence of (f_n) is not uniform:
(a) The non-uniform convergence of the sequence (f_n) in Example 5.1.10 has
been shown in Example 5.1.8
(b) Let (f_n) be as in Example 5.1.11. Note that,

$$f_n(1/n) = \left(1 - \frac{1}{n^2}\right)^n = \left(1 - \frac{1}{n}\right)^n \left(1 + \frac{1}{n}\right)^n.$$

We know that (see Example 2.3.23)

$$\lim_{n \to \infty} \left(1 - \frac{1}{n}\right)^n = \exp(-1), \quad \lim_{n \to \infty} \left(1 + \frac{1}{n}\right)^n = \exp(1).$$

Hence, we have $\lim_{n \to \infty} f_n(1/n) = 1$. In particular, with f as the zero function,
$|f_n(1/n) - f(1/n)| \to 1$ as $n \to \infty$. Hence, by Theorem 5.1.1 (ii), (f_n) does not converge to f uniformly.
Now, we show that the answer to the question raised in the beginning of this
subsection is in the affirmative if the convergence of (f_n) to f is uniform.

Theorem 5.1.6 *Let (f_n) be a sequence of integrable functions defined on an interval $[a, b]$. Suppose (f_n) converges uniformly on $[a, b]$ and f is its limit. Then f is
integrable, and if*

$$g_n(x) = \int_a^x f_n(t)dt, \quad g(x) = \int_a^x f(t)dt, \quad n \in \mathbb{N}$$

for $x \in [a, b]$ and $n \in \mathbb{N}$, then $g_n \to g$ uniformly on $[a, b]$. In particular,

$$\int_a^b f_n(x)dx \to \int_a^b f(x)dx.$$

Proof We know that $\alpha_n := \sup_{x \in [a,b]} |f_n(x) - f(x)| \to 0$ (Theorem 5.1.2). Since $f_n(x) - \alpha_n \leq f(x) \leq f_n(x) + \alpha_n$ for all $x \in [a, b]$ and for all $n \in \mathbb{N}$, by Theorem 3.1.14, f is integrable.

Next, let $\varepsilon > 0$ be given. Let $N \in \mathbb{N}$ be such that

$$|f_n(t) - f(t)| < \varepsilon \quad \forall n \geq N \quad \forall t \in [a, b].$$

Then we have

$$|g_n(x) - g(x)| \leq \int_a^x |f_n(t) - f(t)|dt \leq \varepsilon(b - a)$$

for all $n \geq N$ and for all $x \in [a, b]$. Hence, $g_n \to g$ uniformly. In particular, $g_n(b) \to g(b)$, that is,

$$\int_a^b f_n(x)dx \to \int_a^b f(x)dx.$$

Thus, the proof is complete.   ∎

The following example shows that the uniform convergence of (f_n) to f in the above theorem is not necessary for its conclusion.

Example 5.1.12 For $n \in \mathbb{N}$, let

$$f_n(x) = \frac{nx}{1 + nx}, \quad x \in [0, 1].$$

Note that $f_n(0) = 0$ for every $n \in \mathbb{N}$, and $f_n(x) \to 1$ for every $x \in (0, 1]$. Thus, (f_n) converges pointwise to the function f defined by

$$f(x) = \begin{cases} 0, & x = 0, \\ 1, & 0 < x \leq 1. \end{cases}$$

Since each f_n is continuous and f is not continuous on $[0, 1]$, the convergence of (f_n) to f is not uniform. This can also be seen by noting that

$$f(1/n) - f_n(1/n) = \frac{1}{2} \quad \forall n \in \mathbb{N}.$$

However (cf. Example 2.2.17 and Example 2.3.28),

$$\int_0^1 (f(x) - f_n(x))dx = \int_0^1 \frac{dx}{1 + nx} = \frac{1}{n} \log(1 + n) \to 0.$$

Thus, the conclusion in Theorem 5.1.6 holds, though the sequence (f_n) does not converge uniformly to f. ◇

5.1.4 Uniform Convergence and Differentiation

Suppose $f_n \to f$ pointwise on an open interval I and each f_n is differentiable on I. Then the following questions are natural:

1. Is f differentiable on I?
2. If f is differentiable on I, then do we have

$$\frac{d}{dx} f(x) = \lim_{n \to \infty} \frac{d}{dx} f_n(x)$$

for every $x \in I$?

The following examples show that the answers need not be in the affirmative even if the convergence of (f_n) to f is uniform.

Example 5.1.13 For each $n \in \mathbb{N}$, let

$$f_n(x) = \frac{\sin(nx)}{\sqrt{n}}, \quad x \in \mathbb{R}.$$

Note that

$$|f_n(x)| \le \frac{1}{\sqrt{n}}, \quad \forall n \in \mathbb{N},$$

and hence (f_n) converges to the zero function uniformly. Since

$$f_n'(x) = \sqrt{n} \cos(nx), \quad x \in \mathbb{R}, \ n \in \mathbb{N},$$

we have

$$f_n'(0) = \sqrt{n} \to \infty \quad \text{as} \ n \to \infty.$$

Therefore, (f_n') does not converge pointwise. ◇

Example 5.1.14 For each $n \in \mathbb{N}$, let

$$f_n(x) = \sqrt{\frac{1}{n^2} + x^2} \quad \text{for} \quad x \in \mathbb{R}.$$

Then we see that, $\lim_{n \to \infty} f_n(x) = |x|$ for all $x \in \mathbb{R}$. Note that the function $f(x) = |x|$, $x \in \mathbb{R}$, is not differentiable at 0. Thus, pointwise limit of a sequence of differentiable functions need not be even differentiable.

In this case, with $f(x) = |x|$ for $x \in \mathbb{R}$, the convergence (f_n) to f is uniform. Indeed, for every $x \in \mathbb{R}$,

$$\sqrt{\frac{1}{n^2} + x^2} - |x| = \frac{(\frac{1}{n^2} + x^2) - x^2}{\sqrt{\frac{1}{n^2} + x^2} + |x|} = \frac{1/n^2}{\sqrt{\frac{1}{n^2} + x^2} + |x|}$$

so that $|f_n(x) - f(x)| \leq 1/n$ for all $x \in \mathbb{R}$ and for all $n \in \mathbb{N}$. Hence, (f_n) converges to f uniformly.   ◊

Example 5.1.15 For $n \in \mathbb{N}$, let

$$f_n(x) = \frac{x}{1 + n^2 x^2}, \quad x \in (-1, 1).$$

Since

$$|f_n(x)| = \frac{|x|}{1 + n^2 x^2} \leq \frac{1}{2n} \quad \forall x \in (-1, 1), \ \forall n \in \mathbb{N},$$

(f_n) converges uniformly to the zero function $f \equiv 0$. Note that

$$f_n'(x) = \frac{(1 + n^2 x^2) - x(2n^2 x)}{(1 + n^2 x^2)^2} = \frac{1 - n^2 x^2}{(1 + n^2 x^2)^2}.$$

Note that $f_n'(0) = 1$, and for $x \neq 0$, $f_n'(x) \to 0$ as $n \to \infty$. Hence, the sequence (f_n') converges pointwise to the function g defined by

$$g(x) = \begin{cases} 1 \text{ if } x = 0, \\ 0 \text{ if } 0 < |x| < 1, \end{cases}$$

Thus, we have shown that $f_n \to f \equiv 0$ uniformly and $f_n' \to g$ pointwise, but, $f' \neq g$.   ◊

In Examples 5.1.13, 5.1.13 and 5.1.15 we have seen that (f_n) is a sequence of differentiable functions converging to a function f uniformly, but either

(a) f is not differentiable or
(b) f may be differentiable, but (f_n') does not converge pointwise, or

(c) f is differentiable and (f_n') converges pointwise to a function g, but f' is not equal to g.

In view of the above discussion, the following theorem is important.

Theorem 5.1.7 *Suppose* (f_n) *is a sequence of differentiable functions defined on an open interval I such that each f_n' is continuous, $(f_n(x_0))$ converges for some $x_0 \in I$, and (f_n') converges uniformly on I. Then (f_n) converges uniformly to a differentiable function f on I and*

$$\lim_{n \to \infty} f_n'(x) = f'(x) \quad \forall x \in I.$$

Proof Let $g(x) := \lim_{n \to \infty} f_n'(x)$ for $x \in I$, and $\alpha := \lim_{n \to \infty} f_n(x_0)$. Since the convergence of (f_n') to g is uniform, by Theorem 5.1.5, the function g is continuous. Now, let

$$\varphi_n(x) = \int_{x_0}^{x} f_n'(t)dt, \quad \varphi(x) = \int_{x_0}^{x} g(t)dt \quad \text{for} \ \ x \in I.$$

Then, using the arguments as in Theorem 5.1.6, (φ_n) converges to φ uniformly. Since g is continuous, by Theorem 3.3.2, φ is differentiable and $\varphi'(x) = g(x)$ for $x \in I$, and by Theorem 3.3.1,

$$\varphi_n(x) = \int_{x_0}^{x} f_n'(t)dt = f_n(x) - f_n(x_0).$$

Hence, we have

$$\lim_{n \to \infty} f_n(x) = \lim_{n \to \infty} [\varphi_n(x) + f_n(x_0)] = \varphi(x) + \alpha \quad \forall x \in I.$$

Let $f(x) = \varphi(x) + \alpha, x \in I$. Since $\varphi_n \to \varphi$ uniformly, it follows from the above that $f_n \to f$ uniformly, and f is differentiable. Note that

$$g(x) = \varphi'(x) = f'(x) \quad \forall x \in I.$$

Thus, (f_n') converges to f'. ∎

Let us illustrate the above theorem using the sequence of functions in Example 5.1.15, but defined on a different interval.

Example 5.1.16 For $n \in \mathbb{N}$, let

$$f_n(x) = \frac{x}{1 + n^2 x^2}, \quad x \in (1, 2).$$

As in Example 5.1.15, we see that

$$\text{(i)} \quad |f_n(x)| \le \frac{1}{2n} \quad \text{and} \quad \text{(ii)} \quad f_n'(x) = \frac{1 - n^2 x^2}{(1 + n^2 x^2)^2}$$

for all $x \in (1, 2)$ and for all $n \in \mathbb{N}$. The inequality (i) implies that $f_n \to 0$ uniformly on $(1, 2)$, and by (ii), we have

$$|f_n'(x)| \le \frac{1 + n^2 x^2}{(1 + n^2 x^2)^2} \le \frac{1}{1 + n^2 x^2} \le \frac{1}{1 + n^2}$$

for all $x \in (1, 2)$ and for all $n \in \mathbb{N}$. Hence, $f_n' \to 0$ uniformly on $(1, 2)$. Thus, in this case, we have we have $\lim_{n \to \infty} f_n'(x) = f'(x)$ for all $x \in (1, 2)$, as in the conclusion of Theorem 5.1.7.

We may observe that, in the above, in place of the interval $(1, 2)$, we could have taken any open interval (a, b) with $0 \notin [a, b]$. $\qquad \Diamond$

Remark 5.1.3 In Theorem 5.1.7, the requirement of uniform convergence of (f_n') is not a necessary condition. To see this, for each $n \in \mathbb{N}$, consider

$$f_n(x) = \frac{x^n}{n}, \quad x \in (0, 1).$$

Note that (f_n) converges uniformly to the zero function $f = 0$ and

$$f_n'(x) = x^{n-1} \quad \forall x \in (0, 1), \ \forall n \in \mathbb{N}.$$

In this case we have $\lim_{n \to \infty} f_n'(x) = f'(x)$ for every $x \in (0, 1)$. Note that the convergence of (f_n') is not uniform. $\qquad \Diamond$

5.2   Series of Functions

Series of functions is defined as in the case of series of numbers.

Definition 5.2.1 By a **series of functions** on an interval I, we mean an expression of the form

$$\sum_{n=1}^{\infty} f_n \quad \text{or} \quad \sum_{n=1}^{\infty} f_n(x),$$

where (f_n) is a sequence of functions defined on I. $\qquad \Diamond$

Definition 5.2.2 Given a series $\sum_{n=1}^{\infty} f_n$ of functions on an interval I, let

$$s_n(x) := \sum_{i=1}^{n} f_i(x), \quad x \in I.$$

Then s_n is called the nth **partial sum** of the series $\sum_{n=1}^{\infty} f_n$.   ◊

Definition 5.2.3 Let (f_n) be a sequence of functions defined on an interval I, and let s_n be the nth partial sum of the series $\sum_{n=1}^{\infty} f_n$. Then we say that the series

(1) $\sum_{n=1}^{\infty} f_n$ **converges at a point** $x_0 \in I$ if (s_n) converges at x_0,
(2) $\sum_{n=1}^{\infty} f_n$ **converges pointwise on** I if (s_n) converges pointwise on I,
(3) $\sum_{n=1}^{\infty} f_n$ **converges uniformly on** I if (s_n) converges uniformly on I,
(4) $\sum_{n=1}^{\infty} f_n$ **converges absolutely on** I if $\sum_{n=1}^{\infty} |f_n(x)|$ converges for every $x \in I$. ◊

Recall from Chap. 1 that if (a_n) is a sequence of real numbers, then convergence of $\sum_{n=1}^{\infty} |a_n|$ implies the convergence of $\sum_{n=1}^{\infty} a_n$. From this, it follows that, absolute convergence of a series of functions implies its pointwise convergence.

The proof of the following theorem is easy and left as an exercise.

Theorem 5.2.1 (Comparison test) *Suppose (f_n) and (g_n) are sequences of functions defined on an interval I such that*

$$|f_n(x)| \le |g_n(x)| \quad \forall x \in I, \quad \forall n \in \mathbb{N}.$$

If the series $\sum_{n=1}^{\infty} g_n(x)$ converges absolutely on I, then $\sum_{n=1}^{\infty} f_n(x)$ converges absolutely on I.

The first theorem below is a consequence of Theorems 5.1.5 and 5.1.6, and the second theorem follows from Theorem 5.1.7.

Theorem 5.2.2 *Suppose (f_n) is a sequence of continuous functions on I. If $\sum_{n=1}^{\infty} f_n(x)$ converges uniformly on I, say to $f(x)$, then f is continuous on I, and for $[a, b] \subseteq I$,*

$$\int_a^b f(x)dx = \sum_{n=1}^{\infty} \int_a^b f_n(x)dx.$$

Theorem 5.2.3 *Suppose (f_n) is a sequence of continuously differentiable functions on an open interval I. If $\sum_{n=1}^{\infty} f_n'(x)$ converges uniformly on I, and if $\sum_{n=1}^{\infty} f_n(x)$ converges at some point $x_0 \in I$, then $\sum_{n=1}^{\infty} f_n(x)$ converges uniformly to a differentiable function on I, and*

$$\frac{d}{dx}\left(\sum_{n=1}^{\infty} f_n(x)\right) = \sum_{n=1}^{\infty} f_n'(x).$$

Exercise 5.2.1 Show that, in Theorem 5.2.2, the continuity of the functions f_n can be replaced by Riemann integrability of f_n and f.   ◁

Definition 5.2.4 Given a series $\sum_{n=1}^{\infty} f_n$ of functions defined on an open interval I, if f_n are differentiable on I, then the series $\sum_{n=1}^{\infty} f_n'$ is called the **derived series** of $\sum_{n=1}^{\infty} f_n$.   ◊

5.2.1 Dominated Convergence

Now, we consider a useful sufficient condition to check uniform convergence. First a definition.

Definition 5.2.5 We say that $\sum_{n=1}^{\infty} f_n$ is a **dominated series** if there exists a sequence (α_n) of positive real numbers such that

$$|f_n(x)| \le \alpha_n \quad \forall x \in I, \quad \forall n \in \mathbb{N},$$

and the series $\sum_{n=1}^{\infty} \alpha_n$ converges.   ◊

By Theorem 5.2.1, every dominated series converges absolutely.

Theorem 5.2.4 (Weierstrass test) *Every dominated series converges uniformly and absolutely.*

Proof Let $\sum_{n=1}^{\infty} f_n$ be a dominated series defined on an interval I, and let (α_n) be a sequence of positive reals such that $|f_n(x)| \le \alpha_n$ for all $n \in \mathbb{N}$ and for all $x \in I$, $\sum_{n=1}^{\infty} \alpha_n$ converges.

Let $s_n(x) = \sum_{i=1}^{n} f_i(x)$ and $\sigma_n = \sum_{k=1}^{n} \alpha_k$ for $n \in \mathbb{N}$. Then for $n > m$,

$$|s_n(x) - s_m(x)| = \left| \sum_{i=m+1}^{n} f_i(x) \right| \le \sum_{i=m+1}^{n} |f_i(x)| \le \sum_{i=m+1}^{n} \alpha_i = \sigma_n - \sigma_m.$$

Since $\sum_{n=1}^{\infty} \alpha_n$ converges, the sequence (σ_n) is a Cauchy sequence. Now, let $\varepsilon > 0$ be given, and let $N \in \mathbb{N}$ be such that

$$|\sigma_n - \sigma_m| < \varepsilon \quad \forall n, m \ge N.$$

Hence, from the relation $|s_n(x) - s_m(x)| \le \sigma_n - \sigma_m$, we have

$$|s_n(x) - s_m(x)| < \varepsilon \quad \forall n, m \ge N, \forall x \in I.$$

This, in particular, implies that $(s_n(x))$ is a Cauchy sequence at each $x \in I$. Hence, $(s_n(x))$ converges for each $x \in I$. Let $f(x) = \lim_{n \to \infty} s_n(x)$, $x \in I$. Then, we have

$$|f(x) - s_m(x)| = \lim_{n \to \infty} |s_n(x) - s_m(x)| \le \varepsilon \quad \forall m \ge N, \forall x \in I.$$

Thus, the series $\sum_{n=1}^{\infty} f_n$ converges uniformly to f on I. By Theorem 5.2.1, it converges absolutely as well.   ∎

Remark 5.2.1 Theorem 5.2.4 is usually known as **Weierstrass M-test**, as in standard books on Calculus, the numbers α_n in Theorem 5.2.4 are usually denoted by M_n.   ◊

Example 5.2.1 The series $\sum_{n=1}^{\infty} \frac{\cos nx}{n^2}$ and $\sum_{n=1}^{\infty} \frac{\sin nx}{n^2}$ are dominated series, since

$$\left|\frac{\cos nx}{n^2}\right| \le \frac{1}{n^2}, \quad \left|\frac{\sin nx}{n^2}\right| \le \frac{1}{n^2} \quad \forall n \in \mathbb{N}$$

and $\sum_{n=1}^{\infty} \frac{1}{n^2}$ is convergent.   ◊

Example 5.2.2 The series $\sum_{n=0}^{\infty} x^n$ is a dominated series on $[-\rho, \rho]$ for $0 < \rho < 1$, since $|x^n| \le \rho^n$ for all $n \in \mathbb{N}$, and $\sum_{n=0}^{\infty} \rho^n$ is convergent.   ◊

Example 5.2.3 Consider the series $\sum_{n=1}^{\infty} \frac{x}{n(1+nx^2)}$ on $\mathbb{R}$. Note that

$$\frac{|x|}{n(1 + nx^2)} \le \frac{1}{n}\left(\frac{1}{2\sqrt{n}}\right),$$

and $\sum_{n=1}^{\infty} \frac{1}{n^{3/2}}$ converges. Thus, the given series is dominated series.   ◊

Example 5.2.4 Consider the series $\sum_{n=1}^{\infty} \frac{x}{1+n^2x^2}$ for $x \in [c, \infty)$, $c > 0$. Note that

$$\frac{x}{1 + n^2x^2} \le \frac{x}{n^2x^2} \le \frac{1}{n^2x} \le \frac{1}{n^2c}$$

and $\sum_{n=1}^{\infty} \frac{1}{n^2}$ converges. Thus, the given series is a dominated series on $[c, \infty)$ for any $c > 0$.   ◊

Example 5.2.5 The series $\sum_{n=1}^{\infty} (xe^{-x})^n$ is dominated on $[0, \infty)$: To see this, note that

$$\left(xe^{-x}\right)^n = \frac{x^n}{e^{nx}} \le \frac{x^n}{(nx)^n/n!} = \frac{n!}{n^n}$$

and the series $\sum_{n=1}^{\infty} \frac{n!}{n^n}$ converges (Fig. 5.4).   ◊

Example 5.2.6 The series $\sum_{n=1}^{\infty} x^{n-1}$ is not uniformly convergent on $(0, 1)$, in particular, not dominated on $(0, 1)$. This is seen as follows: Note that

$$s_n(x) := \sum_{k=1}^{n} x^{k-1} = \frac{1 - x^n}{1 - x} \to f(x) := \frac{1}{1 - x} \quad \text{as} \quad n \to \infty.$$

Also, we observe that

$$f(x) - s_n(x) = \frac{x^n}{1 - x}.$$

Taking $x_n := n/(n+1)$, we have $x_n^n \to 1/e$ and $1 - x_n \to 0$ so that

$$f(x_n) - s_n(x_n) = \frac{x_n^n}{1 - x_n} \to\to \infty.$$

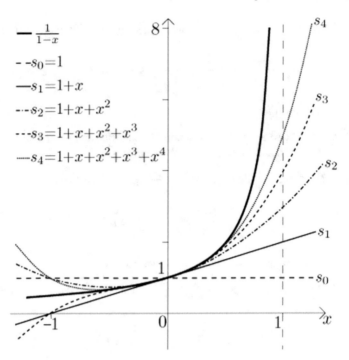

Fig. 5.4 $s_n(x) = \sum_{k=1}^{n} x^{k-1}, \quad 0 < x < 1$

Hence, the convergence of the series is not uniform on $(0, 1)$. However, we have seen that the series is dominated on $(0, a]$ for $0 < a < 1$.   ◇

Example 5.2.7 The series $\sum_{n=1}^{\infty} (1 - x)x^{n-1}$ is not uniformly convergent on $[0, 1]$; in particular, not dominated on $[0, 1]$. This is seen as follows: Note that

$$s_n(x) := \sum_{k=1}^{n} (1 - x)x^{k-1} = \begin{cases} 1 - x^n & \text{if } x \neq 1 \\ 0 & \text{if } x = 1. \end{cases}$$

In particular,
$$s_n(x) = 1 - x^n \quad \forall x \in [0, 1), \quad \forall n \in \mathbb{N}.$$

By Example 5.1.5, we know that $(s_n(x))$ converges to $f(x) \equiv 1$ pointwise, but not uniformly on $[0, 1)$.   ◇

Remark 5.2.2 Note that if a series $\sum_{n=1}^{\infty} f_n$ converges uniformly to a function f on an interval I, then we must have

$$\beta_n := \sup_{x \in I} |s_n(x) - f(x)| \to 0 \quad \text{as} \quad n \to \infty.$$

Here, s_n is the nth partial sum of the series. Conversely, if $\beta_n \to 0$, then the series is uniformly convergent. Thus, if $\sum_{n=1}^{\infty} f_n$ converges to a function f on I, and if $\sup_{x \in I} |s_n(x) - f(x)| \not\to 0$ as $n \to \infty$, then we can infer that the convergence is not uniform.

As an illustration, consider the Example 5.2.7. In this case, we have

$$|s_n(x) - f(x)| = \begin{cases} x^n & \text{if } x \neq 1 \\ 0 & \text{if } x = 1, \end{cases}$$

where $f(x) = \begin{cases} 1 & \text{if } x \neq 1 \\ 0 & \text{if } x = 1. \end{cases}$ Hence, $\sup_{|x| \leq 1} |s_n(x) - f(x)| = 1$. Moreover, the limit function f is not continuous. Hence, the non-uniform convergence also follows from Theorem 5.2.2. ◊

Exercise 5.2.2 Suppose (f_n) is a sequence of functions on I such that $a_n : = \sup_{x \in I} |f_n(x)| < \infty$ for all $n \in \mathbb{N}$. Show that $\sum_{n=1}^{\infty} f_n$ is a dominated series if and only if $\sum_{n=1}^{\infty} a_n$ converges. ◁

Next example shows that in Theorem 5.2.3, the condition that the *derived series* converges uniformly is not a necessary condition for the conclusion.

Example 5.2.8 Consider the series $\sum_{n=0}^{\infty} x^n$. We know that it converges to $1/(1-x)$ for $|x| < 1$. It can be seen that the derived series $\sum_{n=1}^{\infty} nx^{n-1}$ converges uniformly for $|x| \leq \rho$ for any $0 < \rho < 1$. This follows since $\sum_{n=1}^{\infty} n\rho^{n-1}$ converges. Hence, by Theorem 5.2.3,

$$\frac{1}{(1-x)^2} = \frac{d}{dx} \frac{1}{1-x} = \sum_{n=1}^{\infty} nx^{n-1} \quad \text{for } |x| \leq \rho.$$

The above relation is true for x in any *open* interval $I \subseteq (-1, 1)$ because we can choose ρ sufficiently close to 1 such that $I \subseteq [-\rho, \rho]$. Hence, we have

$$\frac{1}{(1-x)^2} = \sum_{n=1}^{\infty} nx^{n-1} \quad \text{for } |x| < 1.$$

Recall (cf. Example 5.2.6) that the given series is not uniformly convergent. Also, using Theorem 5.1.7, it follows (How?) that the derived series $\sum_{n=1}^{\infty} nx^{n-1}$ is not uniformly convergent on $(-1, 1)$. ◊

Remark 5.2.3 We have seen that if $\sum_{n=1}^{\infty} f_n(x)$ is a dominated series on an interval I, then it converges uniformly and absolutely and that an absolutely convergent series need not be a dominated series. Are there series which converge uniformly but not dominated? The answer is in the affirmative. Look at the following series:

$$\sum_{n=1}^{\infty} (-1)^{n+1} \frac{x^n}{n}, \quad x \in [0, 1].$$

Since $\sum_{n=1}^{\infty} \frac{1}{n}$ is divergent, the given series is not absolutely convergent at $x = 1$, and hence, it is not a dominated series. However, the given series converges uniformly on $[0, 1]$.   ◊

5.3   Power Series

Power series is a particular case of series of functions.

Notation 5.3.1 We shall use the notation $\mathbb{N}_0 := \mathbb{N} \cup \{0\}$, and by a sequence $(a_n)_{n \in \mathbb{N}_0}$ we shall also mean a function $\varphi : \mathbb{N}_0 \to \mathbb{R}$ such that $\varphi(n) = a_n$ for $n \in \mathbb{N}_0$.   ◊

Definition 5.3.1 Given a sequence $(a_n)_{n \in \mathbb{N}_0}$ of real numbers and a point $c \in \mathbb{R}$, a series of the form

$$\sum_{n=0}^{\infty} a_n (x - c)^n$$

is called a **power series** around the point c (Fig. 5.5).   ◊

The power series $\sum_{n=0}^{\infty} a_n (x - c)^n$ is also written as

$$a_0 + a_1 (x - c) + a_2 (x - c)^2 + \cdots .$$

We observe that every polynomial is a power series, but the converse is not true. In fact:

> A power series $\sum_{n=0}^{\infty} a_n (x - c)^n$ is a polynomial if and only if there exists $k \in \mathbb{N} \cup \{0\}$ such that $a_n = 0$ for all $n \geq k$

Note that every power series $\sum_{n=0}^{\infty} a_n (x - c)^n$ converges at one point, namely at $x = c$.

Now, we consider some results on the *domain of convergence* and also on the nature of convergence of power series.

5.3.1   Convergence and Absolute Convergence

Theorem 5.3.1 (Abel's theorem)[1] *If the power series $\sum_{n=0}^{\infty} a_n (x - c)^n$ converges at a point $x_0 \neq c$, then it converges absolutely at every x with $|x - c| < |x_0 - c|$.*

---

[1] Named after the Norwegian mathematician Niels Henrik Abel (1802–1829).

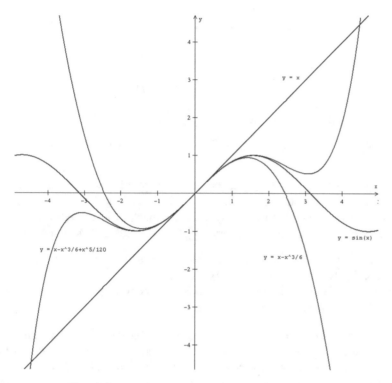

$y = x$

$y = \sin(x)$

$y = x - x^3/6 + x^5/120$

$y = x - x^3/6$

Fig. 5.5 $s_n(x) = \sum_{k=1}^{n} x^{k-1}, \quad 0 < x < 1$

Proof Suppose $\sum_{n=0}^{\infty} a_n(x - c)^n$ converges at a point $x_0 \neq c$. Let x be such that $|x - c| < |x_0 - c|$. Then, we have

$$|a_n(x - c)^n| = |a_n(x_0 - c)^n| \left| \frac{x - c}{x_0 - c} \right|^n \quad \forall n \in \mathbb{N}_0.$$

Since $\sum_{n=0}^{\infty} a_n(x_0 - c)^n$ converges, $|a_n(x_0 - c)^n| \to 0$ as $n \to \infty$. Hence, there exists $M > 0$ such that $|a_n(x_0 - c)^n| \leq M$ for all $n \in \mathbb{N}_0$, so that we have

$$|a_n(x - c)^n| \leq M \left| \frac{x - c}{x_0 - c} \right|^n \quad \forall n \in \mathbb{N}_0.$$

Now, since $|\frac{x-c}{x_0-c}| < 1$, the series $\sum_{n=0}^{\infty} |\frac{x-c}{x_0-c}|^n$ converges. Hence, by comparison test, $\sum_{n=0}^{\infty} |a_n(x - c)^n|$ also converges. ∎

Exercise 5.3.1 If $\sum_{n=0}^{\infty} a_n(x - c)^n$ diverges at a point x_0, then it diverges at every x with $|x - c| > |x_0 - c|$. Why?   ◁

Definition 5.3.2 For a given sequence $(a_n)_{n \in \mathbb{N}_0}$ of real numbers and $c \in \mathbb{R}$, the **domain of convergence** of a power series $\sum_{n=0}^{\infty} a_n(x - c)^n$ is the set

$$D := \{x \in \mathbb{R} : \sum_{n=0}^{\infty} a_n(x - c)^n \text{ converges at } x\}$$

and

$$R := \sup\{|x - c| : x \in D\}$$

is called the **radius of convergence** of $\sum_{n=0}^{\infty} a_n(x - c)^n$.   ◊

By Abel's theorem (Theorem 5.3.1), the domain of convergence of a power series $\sum_{n=0}^{\infty} a_n(x - c)^n$ is either the singleton set $\{c\}$ or an interval.

If the power series $\sum_{n=0}^{\infty} a_n(x - c)^n$ converges only at the point c, then the radius of convergence is 0, and if it converges at all points in $\mathbb{R}$, then $\sup\{|x - c| : x \in D\}$ does not exist, and in that case we say that the radius of convergence is ∞, i.e., we write $R = \infty$. Thus, for the power series $\sum_{n=0}^{\infty} a_n(x - c)^n$, radius of convergence is

- 0 if and only if it does not converge at any $x \neq c$, and
- ∞ if and only if it converges at every $x \in \mathbb{R}$,
- R with $0 < R < \infty$ if and only if $\sum_{n=0}^{\infty} a_n(x - c)^n$ converges for all x with $|x - c| < R$, and diverges at all x with $|x - c| > R$.

Remark 5.3.1 If the power series $\sum_{n=0}^{\infty} a_n(x - c)^n$ converges at some point other than c, and if R is its radius of convergence, then the interval $(c - R, c + R)$ is also called the *interval of convergence* of $\sum_{n=0}^{\infty} a_n(x - c)^n$. If $0 < R < \infty$, then the series can converge or diverge at the points $c + R$ and $c - R$, and hence, the domain of convergence of the series $\sum_{n=0}^{\infty} a_n(x - c)^n$ can be any of the intervals $(c - R, c + R)$, $[c - R, c + R)$, $(c - R, c + R]$, $[c - R, c + R]$.   ◊

Definition 5.3.3 Given a power series $\sum_{n=0}^{\infty} a_n(x - c)^n$ with its domain of convergence D, let $f : D \to \mathbb{R}$ be defined by

$$f(x) = \sum_{n=0}^{\infty} a_n(x - c)^n, \quad x \in D.$$

Then we say that f **represents the power series** on D.   ◊

Example 5.3.1 Consider the power series $\sum_{n=0}^{\infty} x^n$. In this case, we know that the series converges for x with $|x| < 1$, and diverges for all x with $|x| > 1$. Also, the series diverges at $x = 1$ and at $x = -1$. Hence, its radius of convergence is 1, and its domain of convergence is the open interval $(-1, 1)$. Note that in this case the function

$$f(x) = \frac{1}{1 - x}, \quad |x| < 1,$$

represents the given power series.   ◊

Example 5.3.2 Consider the power series $\sum_{n=0}^{\infty} \frac{x^n}{n}$. In this case, we know that the series converges at $x = -1$ and diverges at $x = 1$. Hence, its radius of convergence is 1, and its domain of convergence is $[-1, 1)$. ◇

Example 5.3.3 Consider the power series $\sum_{n=0}^{\infty} \frac{x^n}{n^2}$. We know that this series converges at $x = 1$ and $x = -1$. Since

$$\frac{|x^{n+1}/(n+1)^2|}{|x^n/n^2|} = |x|\frac{n^2}{(n+1)^2} \to |x| \quad \text{as} \quad n \to \infty,$$

by ratio test the series $\sum_{n=0}^{\infty} \left|\frac{x^n}{n^2}\right|$ converges for x with $|x| < 1$ and diverges for x with $|x| > 1$. Therefore, the radius of convergence is 1, and the domain of convergence is $[-1, 1]$. ◇

Example 5.3.4 Consider the power series $\sum_{n=0}^{\infty} \frac{x^n}{n!}$. Since

$$\frac{|x^{n+1}/(n+1)!|}{|x^n/n!|} = \frac{|x|}{n+1} \to 0 \quad \text{as} \quad n \to \infty,$$

by ratio test the series converges at every $x \in \mathbb{R}$. Hence, the radius of convergence is ∞, and the domain of convergence is $\mathbb{R}$. Note that in this case the function

$$f(x) = e^x, \quad x \in \mathbb{R},$$

represents the given power series. ◇

Example 5.3.5 Consider the power series $\sum_{n=0}^{\infty} n!x^n$. Since

$$\frac{|(n+1)!x^{n+1}|}{|n!x^n|} = (n+1)|x| \to \infty \quad \text{as} \quad n \to \infty,$$

by ratio test, the series diverges at every nonzero $x \in \mathbb{R}$. Hence, the radius of convergence is 0, and the domain of convergence is the singleton set $\{0\}$. ◇

For finding the radius of convergence and the domain of convergence, Theorem 5.3.2 below will be useful.

CONVENTION: We shall use the convention, that if $a \in \{0, \infty\}$, then $\frac{1}{a} :=$
$$\begin{cases} \infty & \text{if } a = 0, \\ 0 & \text{if } a = \infty. \end{cases}$$

Theorem 5.3.2 *Consider the power series $\sum_{n=0}^{\infty} a_n(x-c)^n$, and let R be its radius of convergence.*

(i) *If $\lim_{n\to\infty} \left|\frac{a_{n+1}}{a_n}\right| = L \in [0, \infty]$, then $R = 1/L$.*

(ii) *If $\lim_{n\to\infty} |a_n|^{1/n} = \ell \in [0, \infty]$, then $R = 1/\ell$.*

Proof Taking $u_n(x) = a_n(x - c)^n$, $n \in \mathbb{N}_0$, the proofs follow from Abel's theorem (Theorem 5.3.1) and applying ratio test and root test. ∎

Example 5.3.6 Let us find out the radius of convergence and domain of convergence of some power series

$$\sum_{n=1}^{\infty} \frac{x^n}{2n - 1} \quad \text{and} \quad \sum_{n=1}^{\infty} \frac{x^n}{n4^n}.$$

(i) $\sum_{n=1}^{\infty} \frac{x^n}{2n-1}$: In this case $a_n = 1/(2n - 1)$ and $|a_{n+1}/a_n| \to L = 1$. Hence $R = 1/L = 1$. Note that the series diverges at $x = 1$ and, by using Leibnitz's theorem, the series converges at $x = -1$. Hence, the domain of convergence is $[-1, 1)$.

(ii) $\sum_{n=1}^{\infty} \frac{x^n}{n4^n}$: In this case $a_n = 1/(n4^n)$ and $|a_{n+1}/a_n| \to L = 1/4$. Hence, $R = 1/L = 4$. Note that the series diverges at $x = 4$ and, by using Leibnitz's theorem, the series converges at $x = -4$. Hence, the domain of convergence is $[-4, 4)$.   ◊

Suppose $(a_n)_{n\in\mathbb{N}_0}$ and $(b_n)_{n\in\mathbb{N}_0}$ are sequences of numbers such that $|a_n| \le |b_n|$ for all $n \in \mathbb{N}_0$. Then, using the comparison test for the convergence of series of functions, and by Abel's theorem, we can assert the following:

1. If $\sum_{n=0}^{\infty} b_n(x - c)^n$ converges for $|x - c| < r$ for some $r > 0$, then $\sum_{n=0}^{\infty} a_n(x - c)^n$ converges for $|x - c| < r$.
2. If the radii of convergence of $\sum_{n=0}^{\infty} a_n(x - c)^n$ and $\sum_{n=0}^{\infty} b_n(x - c)^n$ are R_1 and R_2, respectively, then $R_1 \ge R_2$.

Exercise 5.3.2 Let $(a_{k_n})_{n\in\mathbb{N}_0}$ be a subsequence of a sequence (a_n) of numbers.

(i) If the power series $\sum_{n=0}^{\infty} a_n(x - c)^n$ has radius of convergence R, then $\sum_{n=0}^{\infty} a_{k_n}(x - c)^{k_n}$ is a power series and its radius of convergence is at least R— Why?

(ii) Using (i), and knowing the radii of convergence of $\sum_{n=1}^{\infty} x^n$ and $\sum_{n=1}^{\infty} \frac{x^n}{n}$, find the radii of convergence of $\sum_{n=1}^{\infty} x^{n^2}$ and $\sum_{n=1}^{\infty} \frac{x^{n!}}{n!}$.   ◁

Remark 5.3.2 Although the domain of convergence of a power series is an interval, the domain of the function to which the power series converges can be larger than the domain of convergence of the series. For instance, the domain of convergence of the power series $\sum_{n=0}^{\infty} x^n$ is $(-1, 1)$, whereas the function $f(x) := 1/(1 - x)$ to which the series converges has domain of definition $\mathbb{R} \setminus \{0\}$. Moreover, the function has a power series representation around any point $c \ne 1$. Indeed, if $c \ne 1$,

$$\frac{1}{1 - x} = \frac{1}{(1 - c) - (x - c)}$$

$$= \frac{1}{(1 - c)} \frac{1}{\left[1 - \frac{x-c}{1-c}\right]}$$

$$= \frac{1}{(1 - c)} \sum_{n=0}^{\infty} \left(\frac{x - c}{1 - c}\right)^n$$

whenever $|x - c| < |1 - c|$. Thus,

$$\frac{1}{1-x} = \sum_{n=0}^{\infty} a_n(x-c)^n \quad \text{for} \quad |x-c| < |1-c|,$$

where $a_n := 1/(1-c)^{n+1}$ for $n \in \mathbb{N}_0$.   ◊

At this point one may ask whether a function can have different power series expansion around the same point or not. We shall see in the next subsection that, that is not possible.

5.3.2 Term by Term Differentiation and Integration

In this subsection, we shall consider, without loss of generality, that a power series is of the form $\sum_{n=0}^{\infty} a_n x^n$ with domain of convergence and radius of convergence are D and R, respectively. Let

$$f(x) = \sum_{n=0}^{\infty} a_n x^n, \quad x \in D.$$

Now, we address the following questions on the power series $\sum_{n=0}^{\infty} a_n x^n$.

1. Is f continuous on D?
2. Is f integrable on every interval $[a, b] \subseteq D$?
3. If f is integrable on $[a, b] \subseteq D$, then do we have the equality

$$\int_a^b f(x)dx = \sum_{n=0}^{\infty} a_n \int_a^b x^n \, dx \; ?$$

4. Is f differentiable on $(-R, R)$?
5. If f is differentiable on $(-R, R)$, and if $g = f'$, then do we have the equality

$$g(x) = \sum_{n=1}^{\infty} n a_n x^{n-1} \text{ on } (-R, R)?$$

We show that the answers to all the above questions are in the affirmative. In this regard the following theorem is very crucial.

Theorem 5.3.3 *Let $R > 0$ be the radius of convergence of a power series $\sum_{n=0}^{\infty} a_n x^n$. Then the series converges uniformly on $[-\rho, \rho]$ for any ρ with $0 < \rho < R$, and the function f defined by*

$$f(x) := \sum_{n=0}^{\infty} a_n x^n, \quad x \in (-R, R)$$

is continuous on the interval $(-R, R)$.

Proof Let $0 < \rho < R$, and r be such that $\rho < r < R$. Then for every x with $|x| \le \rho$, we have

$$|a_n x^n| \le |a_n r^n|\left(\frac{x}{r}\right)^n| \le |a_n r^n|\left(\frac{\rho}{r}\right)^n.$$

Since the series $\sum_{n=0}^{\infty} a_n r^n$ is convergent, $a_n r^n \to 0$ as $n \to \infty$, and hence the sequence $(a_n r^n)$ is bounded, say $|a_n r^n| \le M$ for all $n \in \mathbb{N}$, for some $M > 0$. Thus, $|a_n x^n| \le M (\rho/r)^n$ with $\sum_{n=0}^{\infty} (\rho/r)^n$ converges, since $\frac{\rho}{r} < 1$, so that $\sum_{n=0}^{\infty} a_n x^n$ is a dominated series on $[-\rho, \rho]$. Hence the series $\sum_{n=0}^{\infty} a_n x^n$ is uniformly convergent on $[-\rho, \rho]$, and by Theorem 5.2.4, the function f defined by

$$f(x) := \sum_{n=0}^{\infty} a_n x^n$$

is continuous on $[-\rho, \rho]$. Since, this is true for any $\rho < R$, the function f is continuous at every $x \in (-R, R)$. Indeed, for any $x_0 \in (-R, R)$, we may take ρ such that $|x_0| < \rho < R$. ∎

Definition 5.3.4 Given a power series $\sum_{n=0}^{\infty} a_n x^n$, the power series

$$\sum_{n=1}^{\infty} n a_n x^{n-1}$$

is called the **derived series** of $\sum_{n=0}^{\infty} a_n x^n$.   ◇

Next theorem is about the derived series $\sum_{n=1}^{\infty} n a_n x^{n-1}$.

Theorem 5.3.4 *The power series and its derived series have the same radius of convergence.*

Proof Let $R > 0$ be the radius of convergence of a power series $\sum_{n=0}^{\infty} a_n x^n$. Let $x_0 \in (-R, R)$, and let ρ be such that $|x_0| < \rho < R$. Then, we have

$$|n a_n x_0^{n-1}| = n|a_n \rho^{n-1}|\left(\frac{|x_0|}{\rho}\right)^{n-1} \quad \forall n \in \mathbb{N}.$$

Since $\sum_{n=0}^{\infty} a_n \rho^n$ converges, $|a_n \rho^{n-1}| \to 0$ so that it is bounded, say $|a_n \rho^{n-1}| \le M$ for all $n \in \mathbb{N}$. Thus,

$$|n a_n x_0^{n-1}| \le Mn\left(\frac{|x_0|}{\rho}\right)^{n-1} \quad \forall n \in \mathbb{N}.$$

Since $\sum_{n=1}^{\infty} n(|x_0|/\rho)^{n-1}$ converges, the series $\sum_{n=1}^{\infty} |na_n x_0^{n-1}|$ also converges. Thus, $\sum_{n=1}^{\infty} na_n x^{n-1}$ converges for every $x \in (-R, R)$.

It remains to show that R is the radius of convergence of $\sum_{n=1}^{\infty} na_n x^{n-1}$. Suppose $\sum_{n=1}^{\infty} na_n x^{n-1}$ converges at some point y_0 with $|y_0| > R$. Then taking r with $|y_0| > r > R$, we see that $\sum_{n=1}^{\infty} na_n r^{n-1}$ converges (by Abel's theorem applied to this series). But, $|na_n r^{n-1}| \geq |a_n r^n|/r$ so that by comparison test $\sum_{n=1}^{\infty} a_n r^n$ converges. This is not possible since $r > R$. Thus, the radius of convergence of $\sum_{n=1}^{\infty} na_n x^{n-1}$ is R. ∎

Corollary 5.3.1 *Let $R > 0$ be the radius of convergence of $\sum_{n=0}^{\infty} a_n x^n$. Then for any ρ with $0 < \rho < R$, the derived series $\sum_{n=1}^{\infty} n\, a_n x^{n-1}$ converges uniformly on $[-\rho, \rho]$ and it represents a continuous function on the interval $(-R, R)$.*

Proof By Theorem 5.3.4, the radius of convergence of the derived series $\sum_{n=1}^{\infty} n\, a_n x^{n-1}$ is R. Hence, Theorem 5.3.3 applied to the series $\sum_{n=1}^{\infty} n\, a_n x^{n-1}$ gives the result.

∎

Theorem 5.3.5 *Suppose $R > 0$ is the radius of convergence of a power series $\sum_{n=0}^{\infty} a_n x^n$, and let*

$$f(x) := \sum_{n=0}^{\infty} a_n x^n \ \text{ for } \ |x| < R.$$

Then the following hold.

(i) f is a continuous function on $(-R, R)$ and for $[a, b] \subseteq (-R, R)$,

$$\int_a^b f(x)dx = \sum_{n=0}^{\infty} \frac{a_n}{n+1}[b^{n+1} - a^{n+1}].$$

(ii) f is differentiable on $(-R, R)$ and

$$\frac{d}{dx}f(x) = \sum_{n=1}^{\infty} na_n x^{n-1} \ \ \forall x \in (-R, R).$$

Proof (i) Let $[a, b] \subseteq (-R, R)$. By Theorem 5.3.3, the series $\sum_{n=0}^{\infty} a_n x^n$ converges uniformly on $[a, b]$. Hence, by Theorem 5.2.2, the series can be integrated term by term over $[a, b]$, i.e.,

$$\int_a^b f(x)dx = \sum_{n=0}^{\infty} a_n \int_a^b x^n \, dx = \sum_{n=0}^{\infty} \frac{a_n}{n+1}[b^{n+1} - a^{n+1}].$$

(ii) By Corollary 5.3.1, the derived series $\sum_{n=1}^{\infty} na_n x^{n-1}$ converges uniformly on any interval of the form $[-\rho, \rho]$, where $0 < \rho < R$. Hence, by Theorem 5.2.3, we have

$$f'(x) = \sum_{n=1}^{\infty} n a_n x^{n-1} \qquad (*)$$

for every $x \in (-\rho, \rho)$, where $0 < \rho < R$. Now, if $x_0 \in (-R, R)$, we may take ρ such that $x_0 \in (-\rho, \rho)$, so that the equation $(*)$ holds for $x = x_0$. This is true for any $x_0 \in (-R, R)$. Hence, $(*)$ holds for all $x \in (-R, R)$. ∎

Every power series can be integrated and differentiated term by term in its interval of convergence

---

In view of Theorems 5.3.5 and 5.3.4, we have the following.

Theorem 5.3.6 *Let $R > 0$ be the radius of convergence of the power series $\sum_{n=0}^{\infty} a_n$ $(x - c)^n$ and let*

$$f(x) = \sum_{n=0}^{\infty} a_n (x - c)^n, \quad |x - c| < R.$$

Then f is infinitely differentiable on $(c - R, c + R)$ and

$$a_n = \frac{f^{(n)}(c)}{n!} \quad \forall n \in \mathbb{N}.$$

Proof By Theorems 5.3.5 and 5.3.4, the function f is infinitely differentiable on $(-R, R)$ and for any $k \in \mathbb{N}$,

$$f^{(k)}(x) = \sum_{n=k}^{\infty} n(n-1) \cdots (n-k+1) a_n (x-c)^{n-k} \quad \forall x \in (-R, R).$$

Hence, $f^{(k)}(c) = k! a_k$ so that $a_k = f^{(k)}(c)/k!$ for all $k \in \mathbb{N}$. ∎

In view of the above theorem, we can assert the following:

If $\sum_{n=0}^{\infty} a_n (x - c)^n = \sum_{n=0}^{\infty} b_n (x - c)^n$ for x in a neighbourhood of the point c, then $a_n = b_n$ for all $n \in \mathbb{N} \cup \{0\}$

Definition 5.3.5 If f is infinitely differentiable in a neighbourhood of a point c, and if the series

$$\sum_{n=0}^{\infty} \frac{f^{(n)}(c)}{n!} (x - c)^n$$

has positive radius of convergence, then this series is called the **Taylor series** of f. If $c = 0$, then the Taylor series of f is called the **Maclaurin series** of f. ◊

Example 5.3.7 It can be easily seen that the Maclaurin series of $\sin x$ and $\cos x$ are given by

$$\sin x = \sum_{n=0}^{\infty} \frac{(-1)^n x^{2n+1}}{(2n+1)!}, \quad \cos x = \sum_{n=0}^{\infty} \frac{(-1)^n x^{2n}}{(2n)!},$$

respectively, for all $x \in \mathbb{R}$. ◊

Remark 5.3.3 It is to be remarked that any infinitely differentiable function in a neighbourhood of a point need not have the Taylor series expansion around that point. To see this consider the function

$$f(x) = \begin{cases} e^{-x^2}, & x \neq 0, \\ 0, & x = 0. \end{cases}$$

It can be easily seen that f is infinitely differentiable on $\mathbb{R}$ and $f^{(n)}(0) = 0$ for all $n \in \mathbb{N}$. Hence,

$$f(x) \neq \sum_{n=0}^{\infty} \frac{f^{(n)}(0)}{n!} x^n \text{ for } x \neq 0. \qquad ◊$$

Example 5.3.8 Recall that for $|x| < 1$,

$$\frac{1}{1+x} = \sum_{n=0}^{\infty} (-1)^n x^n. \tag{1}$$

Integrating term by term, we get

$$\log(1+x) = \sum_{n=0}^{\infty} (-1)^n \frac{x^{n+1}}{n+1}. \tag{2}$$

Since the above series converges at $x = 1$, we obtain

$$\log 2 = \sum_{n=0}^{\infty} \frac{(-1)^n}{n+1} = 1 - \frac{1}{2} + \frac{1}{3} - \frac{1}{4} + \cdots + \frac{(-1)^n}{n+1} + \cdots,$$

which we have observed in Sect. 1.2.4. Observe that the series in (1) and (2) have the same radii of convergence 1, but different domains of convergence, namely $(-1, 1)$ and $(-1, 1]$, respectively. ◊

Example 5.3.9 Recall that for $|x| < 1$,

$$\frac{1}{1+x^2} = \sum_{n=0}^{\infty} (-1)^n x^{2n}. \tag{1}$$

Integrating term by term, we get

$$\tan^{-1} x = \int_0^x \frac{dt}{1+t^2} = \sum_{n=0}^{\infty} (-1)^n \frac{x^{2n+1}}{2n+1}, \quad |x| < 1. \tag{2}$$

Note that the above series converges at $x = 1$ and also at $x = -1$. Hence, taking $x = 1$, we obtain

$$\frac{\pi}{4} = \sum_{n=0}^{\infty} \frac{(-1)^n}{2n+1}.$$

which is the *Madhava–Nilakantha* series for $\pi/4$ (cf. Sect. 1.2.4). Observe that the series in (1) and (2) have the same radii of convergence 1, but different domains of convergence, namely $(-1, 1)$ and $[-1, 1]$, respectively.   ◊

Example 5.3.10 Let $|x| < 1$. Then we have

$$\frac{1}{1+x} = \sum_{n=0}^{\infty} (-1)^n x^n.$$

Observe that

$$\frac{1}{(1+x)^2} = -\frac{d}{dx} \frac{1}{(1+x)}.$$

Hence, term by term differentiation of the series for $1/(1+x)$ gives

$$\frac{1}{(1+x)^2} = -\sum_{n=1}^{\infty} (-1)^n n x^{n-1} = \sum_{n=1}^{\infty} (-1)^{n+1} n x^{n-1}. \qquad ◊$$

Remark 5.3.4 We know that a series of functions need not be a power series of the form $\sum_{n=0}^{\infty} a_n x^n$. But, in certain cases, it may be possible to convert them into this form after some change of variable, and obtain their domain of convergence. For example, consider the series

(i) $\displaystyle\sum_{n=0}^{\infty} a_n (x - x_0)^n$ (ii) $\displaystyle\sum_{n=0}^{\infty} a_n x^{3n}$

(iii) $\displaystyle\sum_{n=0}^{\infty} \frac{a_n}{x^n}$ (iv) $\displaystyle\sum_{n=0}^{\infty} a_n \sin^n x$

In each of these cases, we may take a new variable as follows: In (i) $y = x - x_0$, in (ii) $y = x^3$, in (iii) $y = \frac{1}{x}$, and in (iv) $y = \sin x$. Suppose the radius of convergence of $\sum_{n=0}^{\infty} a_n y^n$ is R. Then

(i) $|y| < R \iff |x - x_0| < R,$
(ii) $|y| < R \iff |x|^3 < R \iff |x| < R^{1/3}$
(iii) $|y| < R \iff 1/|x| < R \iff |x| > 1/R.$
(iv) $|y| < R \iff |\sin x| < R.$

Of course one can apply some of the tests for convergence for the series of numbers and obtain the values of x for which the series in (1)–(iv) converge or diverge.   ◊

5.4 Additional Exercises

In the following I denotes an interval.

1. Prove that, if (f_n) converges pointwise on I, then there exists a unique function $f : I \to \mathbb{R}$ such that $f_n \to f$ pointwise.
2. Prove that, for every $x \geq 0$,
$$\lim_{n \to \infty} x e^{-nx} = 0, \quad \lim_{n \to \infty} n^2 x^2 e^{-nx} = 0.$$

3. Let $f_n(x) = (1 - x)x^n$ for $x \in [0, 1]$. Show that (f_n) converges uniformly. [*Hint*: Check $\sup\{|f_n(x) - f(x)| : x \in [0, 1]\} \to 0$.]
4. Suppose $f_n \to f$ uniformly on $[a, b]$. Show that, if f is continuous on $[a, b]$, then for every (x_n) in $[a, b]$, $x_n \to x$ implies $f_n(x_n) \to f(x)$.
5. Let f_n for every $n \in \mathbb{N}$ and f be continuous functions on $[a, b]$. Justify the statement: $f_n \to f$ uniformly if and only if for every sequence (x_n) in $[a, b]$, $|f_n(x_n) - f(x_n)| \to 0$.
6. Justify the statement: If (f_n) converges uniformly to f on I, then (f_n^2) need not converge uniformly to f^2. [*Hint*: Consider $f_n(x) = x + 1/n$ on $[0, \infty)$.]
7. Justify the statement: If (f_n) is a sequence of continuous functions on an interval I which converges pointwise to a continuous function f on I, then the convergence need not be uniform. [*Hint*: Consider $f_n(x) = x/n$ on $[0, \infty)$.]
8. For $n \in \mathbb{N}$, let $f_n : [-1, 1] \to \mathbb{R}$ be defined by

$$f_n(x) = \begin{cases} n(1 - n|x|), & 0 < |x| < 1/n, \\ 0, & x = 0 \ \& \ |x| \geq 1/n \end{cases}$$

for $x \in [-1, 1]$. Show that each f_n is integrable and (f_n) converges pointwise to an integrable function f, but $\left(\int_{-1}^{1} f_n(x) dx \right)$ does not converge to $\int_{-1}^{1} f_n(x) dx$.
9. For $n \in \mathbb{N}$, let
$$f_n(x) = n^2 x(1 - x^2)^n, \quad x \in [0, 1].$$

Show that $f_n \to 0$ pointwise , but $\left(\int_{-1}^{1} f_n(x) dx \right)$ does not converge to 0.

10. For $n \in \mathbb{N}$, let

$$f_n(x) = \frac{\log(1 + n^3 x^2)}{n^2}, \quad g_n(x) = \frac{2nx}{1 + n^3 x^2}, \quad x \in [0, 1].$$

Show that $g_n \to 0$ uniformly on $[0, 1]$. Using this fact, show also that $f_n \to 0$ uniformly on $[0, 1]$.

11. For $n \in \mathbb{N}$, let

$$f_n(x) = \begin{cases} n^2 x, & 0 \le x \le 1/n, \\ 2n - n^2 x, & 1/n \le x \le 2/n, \\ 0, & 2/n \le x \le 1. \end{cases}$$

Show that (f_n) does not converge uniformly of $[0, 1]$. [*Hint*: Use the relation between uniform convergence and integrals.]

12. For $n \in \mathbb{N}$, let $f_n(x) = \sqrt{x^2 + \frac{1}{n}}$, $n \in \mathbb{N}$. Show the following.

(a) (f_n) converges uniformly on $(-1, 1)$.
(b) Each f_n is continuously differentiable on $(-1, 1)$.
(c) (f_n') converge point wise, but not uniformly.

13. Let I be an open interval and for $n \in \mathbb{N}$, let

$$f_n(x) = \frac{2x}{1 + n^3 x^2}, \quad x \in I.$$

Find all possible intervals I such that $f_n \to f$ and $f_n' \to f'$ uniformly for some differentiable function f on I.
[Hint: Refer example 5.1.16.]

14. Suppose (a_n) is such that $\sum_{n=1}^{\infty} a_n$ is absolutely convergent. Show that $\sum_{n=1}^{\infty} \frac{a_n x^{2n}}{1 + x^{2n}}$ is a dominated series on $\mathbb{R}$.

15. Show that for each $p > 1$, the series $\sum_{n=1}^{\infty} \frac{x^n}{n^p}$ is convergent on $[-1, 1]$ and the limit function is continuous.

16. Show that the series $\sum_{n=1}^{\infty} \{(n + 1)^2 x^{n+1} - n^2 x^n\}(1 - x)$ converges to a continuous function on $[0, 1]$, but it is not dominated.

17. Show that the series $\sum_{n=1}^{\infty} \left[\frac{1}{1 + (k+1)x} - \frac{1}{1 + kx} \right]$ is convergent on $[0, 1]$, but it is not a dominated series. Show also that the series can be integrated term by term over the interval $[0, 1]$.

18. Show that the series $\sum_{n=1}^{\infty} nx^{n-1}$ converges pointwise, but not uniformly, on $(-1, 1)$.

19. Find the interval of convergence of the following power series.

(i) $\displaystyle\sum_{n=1}^{\infty} \frac{(-1)^n}{nx^n}$ (ii) $\displaystyle\sum_{n=1}^{\infty} \frac{(-1)^n 3^n}{(4n - 1)x^n}$

(iii) $\displaystyle\sum_{n=1}^{\infty} \frac{n(x + 5)^n}{(2n + 1)^3}$ (iv) $\displaystyle\sum_{n=1}^{\infty} \frac{2^n \sin^n x}{n^2}$

[Answers: (i): $(-\infty, -1) \cup [1, \infty)$; (ii): $(-\infty, -3) \cup [3, \infty)$;
(iii): $[-6, -4]$; (iv): $[-\frac{\pi}{6} + k\pi, \frac{\pi}{6} + k\pi]$, $k \in \mathbb{Z}$.]

20. Find the radius of convergence of $\sum_{n=0}^{\infty} \frac{(n-1)!}{n^n} x^n$.

21. Find the radius of convergence and interval of convergence of the following series:

(i) $\displaystyle\sum_{n=0}^{\infty} \alpha^n x^n$ for $\alpha > 0$ (ii) $\displaystyle\sum_{n=0}^{\infty} \alpha^{n^2} x^n$ for $\alpha > 0$

(iii) $\displaystyle\sum_{n=0}^{\infty} x^{n!}$ (iv) $\displaystyle\sum_{n=0}^{\infty} \frac{(-1)^n}{n} x^{n(n+1)}$

(v) $\displaystyle\sum_{n=0}^{\infty} \frac{\sin(n\pi/6)}{2^n} (x-1)^n$ (vi) $\displaystyle\sum_{n=0}^{\infty} \frac{(-i)^n}{4^n n^\alpha} x^{2n}$

22. Using the limit of known series find the Maclaurin series for the functions

(i) $\dfrac{x^2}{1 + x^3}$ (ii) $\dfrac{x^2}{(1 + x^3)^2}$.

23. The functions $|x| x^k$ has no power series expansion around 0 for any value $k \in \mathbb{N}$. Why?

Chapter 6
Fourier Series

In the final section of the last chapter we have considered a special case of series of functions, namely, the *power series*. In this chapter we consider another special case of series of functions, namely, the *Fourier series* which is in many respects similar to power series. In fact, while studying the heat conduction problem in the year 1804, Fourier found it necessary to use a special type of function series associated with certain functions f, later known as *Fourier series* of f. As it is a special case, all the results known for general series of functions are true for these series as well. In addition, we obtain many more interesting and important results.

6.1   Fourier Series of 2π-Periodic Functions

Definition 6.1.1 Let (a_n) and (b_n) be sequences of real numbers. Then a series of functions of the form

$$c_0 + \sum_{n=1}^{\infty}(a_n \cos nx + b_n \sin nx), \quad x \in \mathbb{R}, \qquad \Diamond$$

is called a **trigonometric series**.

The trigonometric series in Definition 6.1.1 is also written as

$$c_0 + a_1 \cos x + b_1 \sin x + a_2 \cos 2x + b_2 \sin 2x + \cdots.$$

If $a_n = 0$ and $b_n = 0$ for all $n > k$ for some $k \in \mathbb{N} \cup \{0\}$, then the resulting trigonometric series takes the form

© The Author(s), under exclusive license to Springer Nature Switzerland AG 2021
M. T. Nair, *Calculus of One Variable*,
https://doi.org/10.1007/978-3-030-88637-0_6

$$c_0 + \sum_{n=1}^{k}(a_n \cos nx + b_n \sin nx).$$

Definition 6.1.2 Given $c_0, a_n, b_n \in \mathbb{R}$ for $n = 1, \ldots, k$, a function of the form

$$c_0 + \sum_{n=1}^{k}(a_n \cos nx + b_n \sin nx), \quad x \in \mathbb{R}, \qquad\qquad \Diamond$$

is called a **trigonometric polynomial** of degree at most k.

We observe that trigonometric polynomials are 2π-*periodic* on $\mathbb{R}$, i.e., if $f(x)$ is a trigonometric polynomial, then

$$f(x + 2\pi) = f(x) \quad \forall x \in \mathbb{R}.$$

This, in particular, implies that for any $x \in \mathbb{R}$,

$$f(x + 2n\pi) = f(x) \quad \forall n \in \mathbb{Z}.$$

From this, we can infer that, if the trigonometric series

$$c_0 + \sum_{n=1}^{\infty}(a_n \cos nx + b_n \sin nx)$$

converges at a point $x \in \mathbb{R}$, then it has to converge at $x + 2\pi$ as well; and hence at $x + 2n\pi$ for all integers n. Indeed, if we write

$$f_k(x) = c_0 + \sum_{n=1}^{k}(a_n \cos nx + b_n \sin nx)$$

for $k \in \mathbb{N}$, then each $f_k(x)$ is a trigonometric polynomial, and hence

$$\lim_{k \to \infty} f_k(x) = \lim_{k \to \infty} f_k(x + 2\pi).$$

This shows that we can restrict the discussion of convergence of a trigonometric series to an interval of length 2π.

Definition 6.1.3 A function $f : \mathbb{R} \to \mathbb{R}$ is said to be **periodic** with **period** $T > 0$ if $f(x + T) = f(x)$ for all $x \in \mathbb{R}$, and in that case we also say that f is a T-**periodic** function. $\qquad\qquad \Diamond$

Remark 6.1.1 Suppose f is a function defined on an interval $[a, b]$ with $f(a) = f(b)$. Then f can be extended to all of $\mathbb{R}$ as a T-periodic function $\tilde{f}$ with $T = b - a$, by defining

$$\tilde{f}(x + nT) = f(x), \quad x \in I, \quad n \in \mathbb{Z}.$$

Indeed, corresponding to each $y \in \mathbb{R}$, there exists a unique $x \in I$ and $n \in \mathbb{Z}$ such that $y = x + nT$, so that $\tilde{f}$ is defined at all points in $\mathbb{R}$ and its restriction to $[a, b]$ is f.

In case, the value of f is not specified at one or both of the end-points, then we may assign the value(s) such that $f(a) = f(b)$. In the due course, we shall denote the extended function $\tilde{f}$ by the same notation f. Thus, if we say that f is a T-periodic function with its values specified on an interval I, then we mean the T-periodic extension of $f : I \rightarrow \mathbb{R}$.   ◊

We know that, if a trigonometric series converges, then the limit function is 2π-periodic. What about the converse?

Does every 2π-periodic function have a trigonometric series representation?

Suppose, for a moment, that f is a 2π-periodic function and that we can write

$$f(x) = c_0 + \sum_{n=1}^{\infty} (a_n \cos nx + b_n \sin nx)$$

for all $x \in \mathbb{R}$. Then what should be the coefficients c_0, a_n, b_n? Before, discussing this question, let us consider a simpler problem:

Suppose f is a trigonometric polynomial, say

$$f(x) = c_0 + \sum_{n=1}^{k} (a_n \cos nx + b_n \sin nx) \qquad (*)$$

for some $k \in \mathbb{N}$. Then, we have

$$\int_{-\pi}^{\pi} f(x)\, dx = 2c_0\pi + \sum_{n=1}^{k} \left(a_n \int_{-\pi}^{\pi} \cos nx\, dx + b_n \int_{-\pi}^{\pi} \sin nx\, dx \right)$$

$$= 2c_0\pi,$$

since $\int_{-\pi}^{\pi} \cos nx\, dx = 0 = \int_{-\pi}^{\pi} \sin nx\, dx$. Thus,

$$c_0 = \frac{1}{2\pi} \int_{-\pi}^{\pi} f(x)\, dx.$$

Also, for $n, m \in \mathbb{N}$, we have $\int_{-\pi}^{\pi} \cos nx \sin mx dx = 0$ and

$$\int\limits_{-\pi}^{\pi} \cos nx \cos mx dx = \int\limits_{-\pi}^{\pi} \sin nx \sin mx dx = \begin{cases} 0, & \text{if } n \neq m \\ \pi, & \text{if } n = m. \end{cases}$$

Therefore, from (∗) we obtain

$$\int\limits_{-\pi}^{\pi} f(x) \cos mx \, dx = a_m \pi, \quad \int\limits_{-\pi}^{\pi} f(x) \sin mx \, dx = b_m \pi$$

for $m = 1, 2 \ldots, k$. Thus, for $n = 1, \ldots, k$, we have

$$a_n = \frac{1}{\pi} \int\limits_{-\pi}^{\pi} f(x) \cos nx \, dx, \quad b_n = \frac{1}{\pi} \int\limits_{-\pi}^{\pi} f(x) \cos nx \, dx.$$

If a trigonometric series $\frac{a_0}{2} + \sum_{n=1}^{\infty} (a_n \cos nx + b_n \sin nx)$ converges pointwise to a function f, which is integrable over $[-\pi, \pi]$, and if the series can be integrated term by term over $[-\pi, \pi]$, then the above procedure can be adopted to obtain the coefficients $c_0, a_1, b_1, a_2, b_2, \ldots$.

Motivated by the above consideration, we have the following definition.

Definition 6.1.4 If f is a 2π-periodic function integrable over $[-\pi, \pi]$, then the **Fourier series** of f is the trigonometric series

$$\frac{a_0}{2} + \sum_{n=1}^{\infty} (a_n \cos nx + b_n \sin nx),$$

where

$$a_n = \frac{1}{\pi} \int\limits_{-\pi}^{\pi} f(x) \cos nx dx, \quad b_n = \frac{1}{\pi} \int\limits_{-\pi}^{\pi} f(x) \sin nx dx,$$

and this fact is written as

$$f(x) \sim \frac{a_0}{2} + \sum_{n=1}^{\infty} (a_n \cos nx + b_n \sin nx).$$

The numbers a_n and b_n are called the **Fourier coefficients** of f.   ◊

Remark 6.1.2 In the above, and in what follows, by integrability of a function f over an interval $[a, b]$ we mean that it is Riemann integrable over $[a, b]$. The notion of Fourier series also can be considered in a more general context of Lebesgue integration (see, e.g., Nair [9]), so as to be applicable to functions that are not necessarily Riemann integrable. In this book, we shall deal only Riemann integrable functions. ◊

We have already observed that if a trigonometric series converges pointwise to a function f, which is integrable over $[-\pi, \pi]$, and if the series can be integrated term by term over $[-\pi, \pi]$, then it is the Fourier series of f. For instance, if a trigonometric series is uniformly convergent to a function f in $[-\pi, \pi]$, then f is continuous on $[-\pi, \pi]$ and the series can be integrated term by term. Here is a sufficient condition for the uniform convergence of the trigonometric series, which follows from Theorem 5.2.4 (Weierstrass test):

If the series $\sum_{n=0}^{\infty} a_n$ and $\sum_{n=0}^{\infty} b_n$ are absolutely convergent, then the trigonometric series $c_0 + \sum_{n=1}^{\infty} (a_n \cos nx + b_n \sin nx)$ is uniformly convergent.

> If f is a trigonometric polynomial, then its Fourier series is itself

The following two theorems show that there is a large class of functions which can be represented by their Fourier series. Interested readers may look for their proofs in books on Fourier series; for example Bhatia [2].

Theorem 6.1.1 *Suppose f is a 2π-periodic function which is bounded and monotonic on $[-\pi, \pi]$. Then the Fourier series of f converges, and the limit function $\tilde{f}(x)$ is given by*

$$\tilde{f}(x) = \begin{cases} f(x) & \text{if } f \text{ is continuous at } x, \\ \frac{1}{2}[f(x-) + f(x+)] & \text{if } f \text{ is not continuous at } x. \end{cases}$$

Theorem 6.1.2 *(Dirichlet's theorem) Let $f : \mathbb{R} \to \mathbb{R}$ be a 2π-periodic function which is piecewise differentiable on $(-\pi, \pi)$. Then the Fourier series of f converges, and the limit function $\tilde{f}(x)$ is given by*

$$\tilde{f}(x) = \begin{cases} f(x) & \text{if } f \text{ is continuous at } x, \\ \frac{1}{2}[f(x-) + f(x+)] & \text{if } f \text{ is not continuous at } x. \end{cases}$$

In Theorem 6.1.2 we used the terminology *piecewise differentiable* as per the following definition.

Definition 6.1.5 A function $f : [a, b] \to \mathbb{R}$ is said to be **piecewise differentiable** if f' exists and is piecewise continuous on $[a, b]$ except possibly at a finite number of points. ◊

Remark 6.1.3 It is known that there are continuous functions f defined on $[-\pi, \pi]$ whose Fourier series does not converge pointwise to f. Its proof relies on concepts from advanced mathematics (cf. [2, 8]) ◊

Recall that, in the case of a power series, the partial sums are polynomials and the limit function is infinitely differentiable. In the case of a Fourier series, the partial sums are infinitely differentiable, but, the limit function, if exists, need not be even continuous at certain points. This fact is best illustrated by the following example.

Example 6.1.1 Let f be a 2π-periodic function defined on $(-\pi, \pi]$ by

$$f(x) = \begin{cases} 0, & -\pi < x \le 0, \\ 1, & 0 < x \le \pi. \end{cases}$$

Note that this function satisfies the conditions in Dirichlet's theorem (Theorem 6.1.2). Hence, its Fourier series converges to $f(x)$ for every $x \ne 0$, and at the point 0, the series converges to $1/2$. Note that, for $n \in \mathbb{N} \cup \{0\}$,

$$a_n = \frac{1}{\pi} \int_0^\pi \cos nx dx = \begin{cases} 1, & n = 0, \\ 0, & n \ne 0, \end{cases}$$

and for $n \in \mathbb{N}$,

$$b_n = \frac{1}{\pi} \int_0^\pi \sin nx dx = \frac{1}{\pi} \left[\frac{1 - \cos n\pi}{n} \right] = \frac{1}{\pi} \left[\frac{1 - (-1)^n}{n} \right].$$

Thus,

$$b_n = \begin{cases} \dfrac{2}{\pi n}, & n \text{ odd}, \\ 0, & n \text{ even}. \end{cases}$$

Thus, the Fourier series of f is

$$\frac{1}{2} + \frac{2}{\pi} \sum_{n=0}^\infty \frac{\sin(2n+1)x}{(2n+1)}.$$

In particular, for $x = \pi/2$,

$$1 = \frac{1}{2} + \frac{2}{\pi} \sum_{n=0}^\infty \frac{\sin[(2n+1)\pi/2]}{(2n+1)} = \frac{1}{2} + \frac{2}{\pi} \sum_{n=0}^\infty \frac{(-1)^n}{(2n+1)}$$

which leads to the *Madhava–Nilakantha* series

$$\frac{\pi}{4} = \sum_{n=0}^\infty \frac{(-1)^n}{(2n+1)}. \qquad \diamond$$

6.2 Best Approximation Property

Let f be a 2π-periodic function which is integrable on $[-\pi, \pi]$. We know that the Fourier series of f need not converge to f pointwise. However, we have the following theorem on the *best approximation property* for the truncations of the Fourier series.

Theorem 6.2.1 *Let f be a 2π-periodic function which is integrable on $[-\pi, \pi]$, and for $k \in \mathbb{N}$, let*

$$f_k(x) = \frac{a_0}{2} + \sum_{n=1}^{k}(a_n \cos nx + b_n \sin nx),$$

where $a_0, a_1, a_2, \ldots$ and $b_1, b_2, \ldots$ are the Fourier coefficients of f. Then

$$\int_{-\pi}^{\pi} |f(x) - f_k(x)|^2 dx \leq \int_{-\pi}^{\pi} |f(x) - g(x)|^2 dx$$

for any trigonometric polynomial g of the form

$$g(x) = c_0 + \sum_{n=1}^{k}(c_n \cos nx + d_n \sin nx)$$

with $c_0, c_1, \ldots, c_k, d_1, \ldots, d_k$ are in $\mathbb{R}$.

Proof Let $g(x) = c_0 + \sum_{n=1}^{k}(c_n \cos nx + d_n \sin nx)$ for some numbers $c_0, c_1, \ldots, c_k, d_1, \ldots, d_k$ in $\mathbb{R}$. Then

$$\int_{-\pi}^{\pi} |f(x) - g(x)|^2 dx = \int_{-\pi}^{\pi} |(f(x) - f_k(x)) + (f_k(x) - g(x))|^2 dx$$

$$= \int_{-\pi}^{\pi} |f(x) - f_k(x)|^2 + \int_{-\pi}^{\pi} |f_k(x) - g(x)|^2 dx$$

$$+ 2\int_{-\pi}^{\pi}(f(x) - f_k(x))(f_k(x) - g(x))dx.$$

Note that

$$f_k(x) - g(x) = \frac{a_0}{2} - c_0 + \sum_{n=1}^{k}[(a_n - c_n)\cos nx + (b_n - d_n)\sin nx)].$$

Multiplying by $f(x) - f_k(x)$ and integrating, and observing the facts that $\int\limits_{-\pi}^{\pi} (f(x) - f_k(x))d = 0$ and

$$\int\limits_{-\pi}^{\pi} (f(x) - f_k(x)) \cos nx \, dx = 0, \quad \int\limits_{-\pi}^{\pi} (f(x) - f_k(x)) \sin nx \, dx = 0$$

for $n = 1, \ldots, k$, we obtain

$$\int\limits_{-\pi}^{\pi} (f(x) - f_k(x))(f_k(x) - g(x))dx = 0.$$

Hence, we have

$$\int\limits_{-\pi}^{\pi} |f(x) - g(x)|^2 dx = \int\limits_{-\pi}^{\pi} |f(x) - f_k(x)|^2 + \int\limits_{-\pi}^{\pi} |f_k(x) - g(x)|^2 dx$$

$$\geq \int\limits_{-\pi}^{\pi} |f(x) - f_k(x)|^2.$$

Thus, the proof is complete.   ∎

The conclusion in the above theorem can also be written as

$$\int\limits_{-\pi}^{\pi} |f(x) - f_k(x)|^2 dx = \inf_{g \in \mathcal{T}_k} \int\limits_{-\pi}^{\pi} |f(x) - g(x)|^2 dx,$$

where $\mathcal{T}_k$ is the set of all trigonometric polynomials of degree at most k. In fact, in advanced courses (see, e.g., Rudin [13] or Nair [8], one learns that

$$\int\limits_{-\pi}^{\pi} |f(x) - f_k(x)|^2 dx \to 0 \quad \text{as} \quad k \to \infty.$$

6.3 Fourier Series for Even and Odd Functions

Definition 6.3.1 Suppose f is a function defined on an interval containing $(-\ell, \ell)$ for some $\ell > 0$. Then on the interval $(-\ell, \ell)$, f is said to be an

(i) **even function**, if $f(-x) = f(x)$ for all $x \in (-\ell, \ell)$,

(ii) **odd function**, if $f(-x) = -f(x)$ for all $x \in (-\ell, \ell)$.   ◊

An even or odd function f defined on $(-\ell, \ell)$ can be extended to all of $\mathbb{R}$ as a 2ℓ-periodic function by assigning the same value for f at $-\ell$ and ℓ, and defining

$$\tilde{f}(x + 2n\ell) = f(x), \quad -\ell \leq x \leq \ell.$$

Note that if f is an odd function on $(-\ell, \ell)$, then for the extended 2ℓ-periodic function $\tilde{f}$ to be odd on $\mathbb{R}$, we require

$$\tilde{f}(\ell) = 0 = \tilde{f}(-\ell),$$

since $\tilde{f}(\ell) = \tilde{f}(-\ell) = -\tilde{f}(\ell)$. This shows that, for an odd function on $(-\ell, \ell)$ to have an odd 2ℓ-periodic extension, the values at $-\ell$ and ℓ cannot be assigned arbitrarily. However, if f is an even function on $(-\ell, \ell)$, then we can assign any value at ℓ and then assign the same value at $-\ell$ so that the 2ℓ-periodic extension remains even.

Now, let us consider the cases of even and odd on $(-\pi, \pi)$ separately:

Case(i): f is even on $(-\pi, \pi)$ and integrable function on $[-\pi, \pi]$.

In this case, $f(x) \cos nx$ is an even function and $f(x) \sin nx$ is an odd function. Hence $b_n = 0$ for all $n \in \mathbb{N}$, so that in this case

$$f(x) \sim \frac{a_0}{2} + \sum_{n=1}^{\infty} a_n \cos nx \quad \text{with} \quad a_n := \frac{2}{\pi} \int_0^{\pi} f(x) \cos nx dx.$$

Note that at the points $x = 0$ and $x = \pi$, the above series takes the forms

$$\frac{a_0}{2} + \sum_{n=0}^{\infty} a_n \quad \text{and} \quad \frac{a_0}{2} + \sum_{n=1}^{\infty} (-1)^n a_n,$$

respectively.

Case(ii): f is odd on $(-\pi, \pi)$ and integrable on $[-\pi, \pi]$.

In this case, $f(x) \sin nx$ is an even function and $f(x) \cos nx$ is an odd function. Hence $a_n = 0$ for all $n \in \mathbb{N} \cup \{0\}$, so that

$$f(x) \sim \sum_{n=1}^{\infty} b_n \sin nx \quad \text{with} \quad b_n := \frac{2}{\pi} \int_0^{\pi} f(x) \sin nx dx,$$

At the point $x = \pi/2$, the series takes the form

$$\sum_{n=0}^{\infty}(-1)^n b_{2n+1}.$$

Let us illustrate the above observations by some examples.

Example 6.3.1 Consider the 2π-periodic function f defined on $[-\pi, \pi]$ by

$$f(x) = |x|, \quad x \in [-\pi, \pi].$$

Note that f is an even function. Hence, $b_n = 0$ for $n = 1, 2, \ldots$, and

$$f(x) \sim \frac{a_0}{2} + \sum_{n=1}^{\infty} a_n \cos nx, \quad x \in [-\pi, \pi]$$

with

$$a_0 = \frac{2}{\pi} \int_0^{\pi} x \, dx = \pi$$

and for $n = 1, 2, \ldots$,

$$a_n = \frac{2}{\pi} \int_0^{\pi} x \cos nx dx = \frac{2}{\pi} \left\{ \left[x \frac{\sin nx}{n} \right]_0^{\pi} - \int_0^{\pi} \frac{\sin nx}{n} dx \right\}$$

$$= \frac{2}{\pi} \left[\frac{\cos nx}{n^2} \right]_0^{\pi} = \frac{2}{\pi} \left[\frac{(-1)^n - 1}{n^2} \right]$$

Thus,

$$a_{2n} = 0, \quad a_{2n+1} = \frac{-4}{\pi(2n+1)^2}, \quad n = 0, 1, 2, \ldots$$

so that

$$|x| \sim \frac{\pi}{2} - \frac{4}{\pi} \sum_{n=0}^{\infty} \frac{\cos(2n+1)x}{(2n+1)^2}, \quad x \in [-\pi, \pi].$$

Taking $x = 0$ (using Dirichlet's theorem), we obtain

$$\frac{\pi^2}{8} = \sum_{n=0}^{\infty} \frac{1}{(2n+1)^2}. \qquad\qquad \Diamond$$

Example 6.3.2 Consider the 2π-periodic function f defined on $(-\pi, \pi]$ by

$$f(x) = x, \quad x \in (-\pi, \pi].$$

Note that f is an odd function on $(-\pi, \pi)$. Hence, $a_n = 0$ for $n \in \mathbb{N} \cup \{0\}$ and the Fourier series is

$$\sum_{n=1}^{\infty} b_n \sin nx, \quad x \in (-\pi, \pi]$$

with

$$b_n = \frac{2}{\pi} \int_0^{\pi} x \sin nx \, dx = \frac{2}{\pi} \left\{ \left[-x \frac{\cos nx}{n} \right]_0^{\pi} + \int_0^{\pi} \frac{\cos nx}{n} \, dx \right\}$$

$$= \frac{2}{\pi} \left\{ -\pi \frac{\cos n\pi}{n} \right\} = \frac{(-1)^{n+1} 2}{n}.$$

Thus the Fourier series is

$$2 \sum_{n=1}^{\infty} \frac{(-1)^{n+1}}{n} \sin nx.$$

In particular (using Dirichlet's theorem), with $x = \pi/2$ we have

$$\frac{\pi}{4} = \sum_{n=1}^{\infty} \frac{(-1)^{n+1}}{n} \sin \frac{n\pi}{2} = \sum_{n=0}^{\infty} \frac{(-1)^n}{2n + 1},$$

the *Madhava–Nilakantha* series.   ◊

Example 6.3.3 Consider the 2π-periodic function f defined on $(-\pi, \pi]$ by

$$f(x) = \begin{cases} -1, & -\pi < x < 0, \\ 1, & 0 \le x \le \pi. \end{cases}$$

Note that f is an odd function on $(-\pi, \pi)$. Hence, $a_n = 0$ for $n \in \mathbb{N} \cup \{0\}$ and the Fourier series is

$$\sum_{n=1}^{\infty} b_n \sin nx,$$

with

$$b_n = \frac{2}{\pi} \int_0^{\pi} \sin nx \, dx = \frac{2}{n\pi} (1 - \cos n\pi) = \frac{2}{n\pi} [1 - (-1)^n].$$

Thus

$$f(x) \sim \frac{4}{\pi} \sum_{n=0}^{\infty} \frac{\sin(2n + 1)x}{2n + 1}.$$

Taking $x = \pi/2$, we have again the *Madhava–Nilakantha* series

$$\frac{\pi}{4} = \sum_{n=0}^{\infty} \frac{(-1)^n}{2n+1}.$$

◇

Example 6.3.4 Consider the 2π-periodic function f defined on $[-\pi, \pi]$ by

$$f(x) = x^2, \quad x \in [-\pi, \pi].$$

Note that f is an even function. Hence, $b_n = 0$ for $n = 1, 2, \ldots$, and the Fourier series is

$$\frac{a_0}{2} + \sum_{n=1}^{\infty} a_n \cos nx, \quad x \in [-\pi, \pi), \quad a_n = \frac{2}{\pi} \int_0^{\pi} x^2 \cos nx \, dx.$$

It can be see that $a_0 = 2\pi^2/3$, and $a_n = (-1)^n 4/n^2$. Thus

$$x^2 \sim \frac{\pi^2}{3} + 4 \sum_{n=1}^{\infty} \frac{(-1)^n \cos nx}{n^2}, \quad x \in [-\pi, \pi].$$

Taking $x = 0$ and $x = \pi$ (using Dirichlet's theorem), we have

$$\frac{\pi^2}{12} = \sum_{n=1}^{\infty} \frac{(-1)^{n+1}}{n^2}, \quad \frac{\pi^2}{6} = \sum_{n=1}^{\infty} \frac{1}{n^2}$$

respectively (Fig. 6.1).

◇

6.4   Sine and Cosine Series Expansions

We have seen that the Fourier series of an odd function involves only sine functions, whereas Fourier series of an even function involves only cosine functions. The above observation points to the following:

Suppose a function f is defined on $[0, \pi]$. In order to get sine series expansion for f, we may extend it to $[-\pi, \pi]$ so that the extended function is an odd 2π-periodic function. Similarly, to get cosine series expansion for f, we may extend it to $[-\pi, \pi]$ so that the extended function is an even 2π-periodic function.

Fig. 6.1 Fourier
approximation of
$f(x) = x/|x|$, $0 < |x| \le 3$

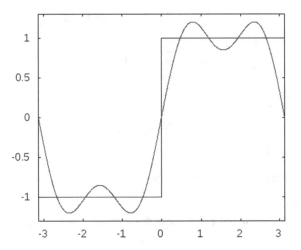

For obtaining such 2π-periodic functions based on a given integrable function
$f : [0, \pi] \to \mathbb{R}$, we first define an odd function f_{od} and even function f_{ev} on $(-\pi, \pi)$
as follows:

$$f_{od}(x) = \begin{cases} f(x) & \text{if } 0 \le x < \pi, \\ -f(-x) & \text{if } -\pi < x < 0, \end{cases}$$

$$f_{ev}(x) = \begin{cases} f(x) & \text{if } 0 \le x < \pi, \\ f(-x) & \text{if } -\pi < x < 0. \end{cases}$$

Clearly,

$$f_{od}(-x) = -f_{od}(x), \quad f_{ev}(-x) = f_{ev}(x)$$

for all $x \in (-\pi, \pi)$, so that f_{od} is an odd function and f_{ev} is an even function on
$(-\pi, \pi)$. These functions f_{od} and f_{ev} can be extended to 2π-periodic functions with
their values at $-\pi$ and π coincide with $f(\pi)$, and thus we obtain the corresponding
get corresponding Fourier series on $[0, \pi]$ as

$$f(x) \sim \sum_{n=1}^{\infty} b_n \sin nx \quad \text{and} \quad f(x) \sim \frac{a_0}{2} + \sum_{n=1}^{\infty} a_n \cos nx$$

with

$$a_n = \frac{2}{\pi} \int_0^{\pi} f(x) \cos nx \, dx, \quad b_n = \frac{2}{\pi} \int_0^{\pi} f(x) \sin nx \, dx.$$

Definition 6.4.1 Let $f : [0, \pi] \to \mathbb{R}$ be an integrable function. Then the series

$$f(x) \sim \sum_{n=1}^{\infty} b_n \sin nx \quad \text{and} \quad f(x) \sim \frac{a_0}{2} + \sum_{n=1}^{\infty} a_n \cos nx$$

with

$$a_n = \frac{2}{\pi} \int_0^{\pi} f(x) \cos nx \, dx, \quad b_n = \frac{2}{\pi} \int_0^{\pi} f(x) \sin nx \, dx$$

are called, respectively, the **sine series expansion** and the **cosine series expansion** of f. $\diamondsuit$

Example 6.4.1 Consider the function $f(x) = x^2$, $x \in [0, \pi]$. Then

$$f_{od}(x) = \begin{cases} x^2, & \text{if } 0 \le x < \pi, \\ -x^2, & \text{if } -\pi < x < 0, \end{cases}$$
$$f_{ev}(x) = x^2, \quad x \in (-\pi, \pi).$$

Hence, the cosine series expansion of f on $[0, \pi]$ is (cf. Example 6.3.4)

$$x^2 \sim \frac{\pi^2}{3} + 4 \sum_{n=1}^{\infty} \frac{(-1)^n \cos nx}{n^2}, \quad x \in [0, \pi].$$

Its sine series expansion is

$$f(x) \sim \sum_{n=1}^{\infty} b_n \sin nx, \quad x \in [0, \pi],$$

where

$$b_n = \frac{2}{\pi} \int_0^{\pi} x^2 \sin nx \, dx = \frac{2}{\pi} \left\{ \left[-x^2 \frac{\cos nx}{n} \right]_0^{\pi} + \int_0^{\pi} 2x \frac{\cos nx}{n} \, dx \right\}.$$

Note that

$$\left[-x^2 \frac{\cos nx}{n} \right]_0^{\pi} = -\pi^2 \frac{\cos n\pi}{n} = \pi^2 \frac{(-1)^{n+1}}{n},$$

$$\int_0^{\pi} 2x \frac{\cos nx}{n} \, d = \frac{2}{n} \left[x \frac{\sin nx}{n} \right]_0^{\pi} - \frac{2}{n} \int_0^{\pi} \frac{\sin nx}{n} \, dx$$

$$= \frac{2}{n^2} \left[\frac{\cos nx}{n} \right]_0^{\pi} = \frac{2}{n^2} \left[\frac{(-1)^n - 1}{n} \right].$$

Thus,

$$b_n = 2\pi \frac{(-1)^{n+1}}{n} + \frac{4}{\pi} \left[\frac{(-1)^n - 1}{n^3} \right]. \qquad \diamond$$

Example 6.4.2 Consider the function $f(x) = x$, $x \in [0, \pi]$. Note that

$$f_{od}(x) = x, \quad x \in (-\pi, \pi)$$

and

$$f_{ev}(x) = \begin{cases} x, & \text{if } 0 \le x < \pi, \\ -x, & \text{if } -\pi < x < 0. \end{cases}$$

Thus,

$$f_{ev}(x) = |x|, \quad x \in (-\pi, \pi).$$

From Examples 6.3.2 and 6.3.1, we obtain

$$x \sim 2 \sum_{n=1}^{\infty} \frac{(-1)^{n+1}}{n} \sin nx, \quad x \in [0, \pi]$$

and

$$|x| \sim \frac{\pi}{2} - \frac{4}{\pi} \sum_{n=0}^{\infty} \frac{\cos(2n + 1)x}{(2n + 1)^2}, \quad x \in [0, \pi]. \qquad \diamond$$

Example 6.4.3 Let us consider sine and cosine expansions of the function

$$f(x) = \begin{cases} 0, & \text{if } 0 \le x < \pi/2, \\ 1, & \text{if } \pi/2 \le x < \pi. \end{cases}$$

Then the sine series of f is given by

$$f(x) \sim \sum_{n=1}^{\infty} b_n \sin nx, \quad x \in [0, \pi],$$

where

$$b_n = \frac{2}{\pi} \int_{\pi/2}^{\pi} \sin nx \, dx = -\frac{2}{\pi} \left[\frac{\cos nx}{n} \right]_{\pi/2}^{\pi} = \frac{2}{\pi} \left[\frac{\cos n\pi/2 - \cos n\pi}{n} \right].$$

Note that $b_{2n-1} = \dfrac{2}{(2n-1)\pi}$ and

$$b_{2n} = \frac{2}{2n\pi}[(-1)^n - 1] = \begin{cases} -\frac{2}{n\pi} & \text{if } n \text{ odd,} \\ 0 & \text{if } n \text{ even.} \end{cases}$$

Thus, for $x \in [0, \pi]$, we have

$$\frac{\pi}{2} f(x) \sim \frac{\sin x}{1} - \frac{\sin 2x}{1} + \frac{\sin 3x}{3} + \frac{\sin 5x}{5} + \cdots + \frac{\sin(4n-3)x}{4n-3}$$
$$- \frac{\sin(4n-2)x}{4n-2} + \frac{\sin(4n-1)x}{4n-1} + \frac{\sin(4n+1)x}{4n+1} + \cdots . \qquad \diamond$$

6.5   Fourier Series of 2ℓ-Periodic Functions

Suppose f is a T-periodic function. We may write $T = 2\ell$. Then we may consider the change of variable $t = \pi x/\ell$ so that the function

$$f(x) = f(\ell t/\pi),$$

as a function of t, it is 2π-periodic. Hence, its Fourier series is

$$\frac{a_0}{2} + \sum_{n=1}^{\infty} (a_n \cos nt + b_n \sin nt)$$

where

$$a_n = \frac{1}{\pi} \int_{-\pi}^{\pi} f(\ell t/\pi) \cos nt \, dt = \frac{1}{\ell} \int_{-\ell}^{\ell} f(x) \cos \frac{n\pi x}{\ell} dx,$$

$$b_n = \frac{1}{\pi} \int_{-\pi}^{\pi} f(\ell t/\pi) \sin nt \, dt = \frac{1}{\ell} \int_{-\ell}^{\ell} f(x) \sin \frac{n\pi x}{\ell} dx.$$

In particular, on the interval $(-\ell, \ell)$,

- f even implies $b_n = 0$ and $a_n = \dfrac{2}{\ell} \int_0^{\ell} f(x) \cos \dfrac{n\pi x}{\ell} dx,$
- f odd implies $a_n = 0$ and $b_n = \frac{2}{\ell} \int_0^{\ell} f(x) \sin \frac{n\pi x}{\ell} dx.$

Example 6.5.1 Consider the function

$$f(x) = 1 - |x|, \quad -1 \le x \le 1.$$

Here, $\ell = 1$, so that

$$a_n = \int_{-1}^{1} (1 - |x|) \cos n\pi x \, dx = 2 \int_{0}^{1} (1 - x) \cos n\pi x \, dx,$$

$$b_n = \int_{-1}^{1} (1 - |x|) \sin n\pi x \, dx = 0.$$

Note that

$$\int_{0}^{1} \cos n\pi x \, dx = \left[\frac{\sin n\pi x}{n\pi} \right]_{0}^{1} = 0,$$

$$\int_{0}^{1} x \cos n\pi x \, dx = \left[x \frac{\sin n\pi x}{n\pi} \right]_{0}^{1} - \int_{0}^{1} \frac{\sin n\pi x}{n\pi} \, dx = \left[\frac{\cos n\pi x}{n^2\pi^2} \right]_{0}^{1}$$

$$= \frac{(-1)^n - 1}{n^2\pi^2}.$$

Hence,

$$a_n = 2 \int_{0}^{1} (1 - x) \cos n\pi x \, dx = \frac{2}{n^2\pi^2} [1 - (-1)^n] = \begin{cases} 0, & n \text{ even}, \\ 4/n^2\pi^2, & n \text{ odd}. \end{cases}$$

Thus,

$$f(x) \sim \frac{4}{\pi^2} \sum_{n=0}^{\infty} \frac{\cos(2n+1)\pi x}{(2n+1)^2}. \qquad \Diamond$$

6.6 Fourier Series on Arbitrary Intervals

Suppose a function f is defined in an interval $[a, b]$ such that $f(a) = f(b)$. We can obtain Fourier expansion of it as follows:

Method 1: Let us consider a change of variable as $y = x - \frac{a+b}{2}$. Let

$$\varphi(y) := f(x) = f\left(y + \frac{a+b}{2}\right), \quad \text{where} \quad -\ell \le y \le \ell$$

with $\ell = (b-a)/2$. We can extend φ as a 2ℓ-periodic function and obtain its Fourier series as

$$\varphi(y) \sim \frac{a_0}{2} + \sum_{n=1}^{\infty} (a_n \cos \frac{n\pi}{\ell} y + b_n \sin \frac{n\pi}{\ell} y)$$

where

$$a_n = \frac{1}{\ell} \int_{-\ell}^{\ell} \varphi(y) \cos \frac{n\pi y}{\ell} dy, \quad b_n = \frac{1}{\ell} \int_{-\ell}^{\ell} \varphi(y) \sin \frac{n\pi y}{\ell} dy.$$

Method 2: Considering the change of variable as $y = x - a$ and $\ell := b - a$, we define $\varphi(y) := f(x) = f(y + a)$ where $0 \le y < \ell$. We can extend φ as a 2ℓ-periodic function in any manner and obtain its Fourier series. Here are two specific cases:

(a) Extending φ to $(-\ell, \ell)$ as an even function, we obtain

$$\varphi(y) \sim \frac{a_0}{2} + \sum_{n=1}^{\infty} a_n \cos \frac{n\pi}{\ell} y$$

where $\ell = (b-a)/2$ and

$$a_n = \frac{1}{\ell} \int_{-\ell}^{\ell} \varphi(y) \cos \frac{n\pi y}{\ell} dy.$$

(b) Extending φ to $(-\ell, \ell)$ as an odd function, we obtain

$$\varphi(y) \sim \sum_{n=1}^{\infty} b_n \sin \frac{n\pi}{\ell} y$$

where

$$b_n = \frac{1}{\ell} \int_{-\ell}^{\ell} \varphi(y) \sin \frac{n\pi y}{\ell} dy.$$

From the series of φ we can recover the corresponding series of f on $[a, b]$ by writing $y = x - a$.

6.7 Additional Exercises

1. Find the Fourier series of the 2π- period function f such that:

   (a) $f(x) = \begin{cases} 1, & -\frac{\pi}{2} \le x < \frac{\pi}{2} \\ 0, & \frac{\pi}{2} < x < \frac{3\pi}{2}. \end{cases}$

   (b) $f(x) = \begin{cases} x, & -\frac{\pi}{2} \le x < \frac{\pi}{2} \\ \pi - x, & \frac{\pi}{2} < x < \frac{3\pi}{2}. \end{cases}$

   (c) $f(x) = \begin{cases} 1 + \frac{2x}{\pi}, & -\pi \le x \le 0 \\ 1 - \frac{2x}{\pi}, & 0 \le x \le \pi. \end{cases}$

   (d) $f(x) = \frac{x^2}{4}, -\pi \le x \le \pi.$

2. Using the Fourier series in Exercise 1, find the sum of the following series:

   (a) $1 - \frac{1}{3} + \frac{1}{5} - \frac{1}{7} + \dots,$ (b) $1 + \frac{1}{4} + \frac{1}{9} + \frac{1}{16} + \dots$

   (c) $1 - \frac{1}{4} + \frac{1}{9} - \frac{1}{16} + \dots,$ (d) $1 + \frac{1}{3^2} + \frac{1}{5^2} + \frac{1}{7^2} + \dots$

3. If $f(x) = \begin{cases} \sin x, & 0 \le x \le \frac{\pi}{4} \\ \cos x, & \frac{\pi}{4} \le x < \frac{\pi}{2} \end{cases}$, then show that

$$f(x) \sim \frac{8}{\pi} \cos \frac{\pi}{4} \left[\frac{\sin x}{1.3} + \frac{\sin 3x}{5.7} + \frac{\sin 10x}{9.11} + \dots \right].$$

4. Show that for $0 < x < 1$,

$$x - x^2 = \frac{8}{\pi^2} \left[\frac{\sin x\pi}{1^3} + \frac{\sin 3\pi x}{3^3} + \frac{\sin 5\pi x}{5^3} + \dots \right].$$

5. Show that for $0 < x < \pi$,

$$\sin x + \frac{\sin 3x}{3} + \frac{\sin 5x}{5} + \dots = \frac{\pi}{4}.$$

6. Show that for $-\pi < x < \pi$,

$$x \sin x = 1 - \frac{1}{2} \cos x - \frac{2}{1.3} \cos 2x + \frac{2}{2.4} \cos 3x - \frac{2}{3.5} \cos 4x + \dots,$$

and find the sum of the series

$$\frac{1}{1.3} - \frac{1}{3.5} + \frac{1}{5.7} - \frac{1}{7.9} + \dots.$$

7. Show that for $0 \leq x \leq \pi$,

$$x(\pi - x) = \frac{\pi^2}{6} - \left[\frac{\cos 2x}{1^2} + \frac{\cos 4x}{2^2} + \frac{\cos 6x}{3^2} + \cdots \right],$$

$$x(\pi - x) = \frac{8}{\pi} \left[\frac{\sin x}{1^3} + \frac{\sin 3x}{3^3} + \frac{\sin 5x}{5^3} + \cdots \right].$$

8. Find the Fourier series of the 2-periodic functions f given below defined on $[-1, 1)$ by

   (a) $f(x) = \begin{cases} -1, & -1 \leq x < 0, \\ 1, & 0 \leq x < 1, \end{cases}$

   (b) $f(x) = \begin{cases} 0, & x^2 \leq x < 0, \\ 1, & 0 \leq x < 1, \end{cases}$

9. Find the cosine series and sine series for the functions f given below:

   (a) $f(x) = \begin{cases} 1, & 0 \leq x < 1, \\ 0, & 1 \leq x < 2. \end{cases}$

   (b) $f(x) = x,\ 0 \leq x < 1$.

10. Justify the statement: If f is an odd 2ℓ-periodic function, then $f(-\ell) = f(0) = f(\ell) = 0$.

11. If $f : [0, \ell] \to \mathbb{R}$, then show that

    (a) $g : [-\ell, \ell] \to \mathbb{R}$ defined by $g(x) = \begin{cases} f(x), & 0 \leq x \leq \ell, \\ f(-x), & -\ell \leq x < 0, \end{cases}$ is an even function on $[-\ell, \ell]$, and

    (b) $h : [-\ell, \ell] \to \mathbb{R}$ defined by $g(x) = \begin{cases} f(x), & 0 < x < \ell, \\ -f(-x), & -\ell < x < 0, \\ 0, & x \in \{0, \ell, -\ell\} \end{cases}$ is an odd function on $[-\ell, \ell]$.

12. If f is (Riemann) integrable on $[-\pi, \pi]$, then show that

$$\frac{a_0^2}{2} + \sum_{n=1}^{\infty} (a_n^2 + b_n^2) \leq \frac{1}{\pi} \int_{-\pi}^{\pi} [f(x)]^2 dx.$$

13. If f is (Riemann) integrable on $[-\pi, \pi]$, then justify the following:

$$\int_{-\pi}^{\pi} f(x) \cos nx dx \to 0 \quad \text{and} \quad \int_{-\pi}^{\pi} f(x) \sin nx dx \to 0$$

as $n \to \infty$.

14. Assuming that the Fourier series of f converges uniformly on $[-\pi, \pi]$, show that

$$\frac{1}{\pi} \int_{-\pi}^{\pi} [f(x)]^2 dx = \frac{a_0^2}{2} + \sum_{n=1}^{\infty} (a_n^2 + b_n^2).$$

15. Using Exercises 7 and 14 show that

(a) $\displaystyle\sum_{n=1}^{\infty} \frac{1}{n^4} = \frac{\pi^4}{90}$,

(b) $\displaystyle\sum_{n=1}^{\infty} \frac{(-1)^{n-1}}{n^2} = \frac{\pi^2}{12}$

(c) $\displaystyle\sum_{n=1}^{\infty} \frac{1}{n^6} = \frac{\pi^6}{945}$

(d) $\displaystyle\sum_{n=1}^{\infty} \frac{(-1)^{n-1}}{(2n-1)^3} = \frac{\pi^3}{32}$

References

1. R.G. Bartle, D.R. Sherbert, *Introduction to Real Analysis*, 3rd edn (Wiley & Sons, Inc., 2000)
2. R. Bhatia, *Fourier Series*, TRIM, 2nd edn, (Hindustan Book Agnency, New Delhi 1993) (2003)
3. K.G. Binmore, *Mathematical Analysis: A Straight Forward Approach* (Cambridge University Press, 1991)
4. C.G. Delninger, *Elements of Real Analysis* (Jones & Martlett Learning, 2011)
5. P.M. Fitzpatric, *Advanced Calculus* (American Mathematical Society, 2006)
6. S. Ghorpade, B.V. Limaye, *A Course in Calculus and Analysis*, 2nd edn (Springer, Berlin, 2018)
7. G.G. Joseph, *Indian Mathematics: Engaging with the World from Ancient to Modern Times* (World Scientific, 2016)
8. M.T. Nair, *Functional Analysis: A First Course.* Prentice-Hall of India, New Delhi, 2002 (Fourth Print, 2014)
9. M.T. Nair, *Measure and Integration: A First Course* (Taylor & Francis, CRC Press, 2019)
10. D. Perkins, φ, π, e, I (MAAPRESS, The Mathematical Association of America, 2017)
11. N. Piskunov, *Differential Integral Calculus*, vol. I & vol. II (Mir Publishers, Moscow, 1974)
12. C.H. Pugh, *Real Mathematical Analysis* (Springer, Berlin, 2004)
13. W. Rudin, *Principles of Mathematical Analysis*, 3rd edn (McGraw-Hill, International Student Edition, 1964)

© The Editor(s) (if applicable) and The Author(s), under exclusive license to Springer
Nature Switzerland AG 2021
M. T. Nair, *Calculus of One Variable*,
https://doi.org/10.1007/978-3-030-88637-0

Index

© The Editor(s) (if applicable) and The Author(s), under exclusive license to Springer
Nature Switzerland AG 2021
M. T. Nair, *Calculus of One Variable*,
https://doi.org/10.1007/978-3-030-88637-0

Printed in the United States
by Baker & Taylor Publisher Services